Ocean and Hydropower

Guest Editors

Yuquan Zhang
Jianhua Zhang
Ling Zhou
Bin Huang

MDPI

Basel • Beijing • Wuhan • Barcelona • Belgrade • Novi Sad • Cluj • Manchester

Guest Editors
Yuquan Zhang
School of Electrical and Power Engineering
Hohai University
Nanjing
China

Jianhua Zhang
College of Aerospace and Civil Engineering
Harbin Engineering University
Harbin
China

Ling Zhou
National Research Center of Pumps
Jiangsu University
Zhenjiang
China

Bin Huang
Ocean College
Zhejiang University
Zhoushan
China

Editorial Office
MDPI AG
Grosspeteranlage 5
4052 Basel, Switzerland

This is a reprint of the Special Issue, published open access by the journal *Sustainability* (ISSN 2071-1050), freely accessible at: https://www.mdpi.com/journal/sustainability/special_issues/Ocean_hydropower.

For citation purposes, cite each article independently as indicated on the article page online and as indicated below:

Lastname, A.A.; Lastname, B.B. Article Title. *Journal Name* **Year**, *Volume Number*, Page Range.

ISBN 978-3-7258-4004-5 (Hbk)
ISBN 978-3-7258-4003-8 (PDF)
https://doi.org/10.3390/books978-3-7258-4003-8

Contents

About the Editors

Yuquan Zhang

Zhang Yuquan, male, born in 1989 in Huai'an, Jiangsu Province, is a doctor, professor, doctoral supervisor, and vice president of the School of Electrical and Power Engineering. He studied at both Hohai University and the University of Manchester, UK, and was selected for the National Youth Talent Program, Outstanding Young Energy Science and Technology Worker of China Energy Research Association, Outstanding Youth Fund winner of Jiangsu Provincial Natural Science Foundation, Outstanding Young Backbone Teacher of Jiangsu University's "Green and Blue Project", Youth Position Model of Hohai University, and Outstanding Graduate Supervisor Team of Hohai University. He currently serves as an editorial board member for SCI journal *Sustainability* and Chinese core journal *Journal of Irrigation and Drainage Machinery Engineering*.

Jianhua Zhang

Jianhua Zhang is a professor, doctoral supervisor, member of the Talent Program, Key R&D Program, National Natural Science Foundation of China of the Ministry of Science and Technology, member of the Natural Science Foundation of Beijing, Shandong Province and Zhejiang Province, and recipient of the Science and Technology Award of the Ministry of Natural Resources. Prof. Zhang is the current executive director of the Ocean Engineering Council, an Editorial Board Member of *Maritime Technology and Research*, executive editor of *Sustainable Ocean Energies*, guest editor of *Journal of Marine Science and Engineering*, member of the Ocean Engineering Terminology Review Subcommittee of the Third National Ocean Science and Technology Terminology Review Committee, member of the Wind Resistance and Disaster Reduction and Wind Energy Utilization Professional Committee of the China Engineering Construction Standardization Association, member of the Wind Power Engineering Materials Special Committee of the China Composite Materials Society, etc. Prof. Zhang is engaged in long-term research on the mechanical performance analysis, design, and safe operation and maintenance technology of offshore wind power and tidal energy structures, and has achieved innovative results in coupling response mechanisms, disaster failure modes, and vibration control and protection technologies. Prof. Zhang has hosted/completed over 20 projects, including the National Natural Science Foundation Enterprise Innovation Joint Fund, National Key R&D Program, and Heilongjiang Provincial Natural Science Foundation; has participated in multiple scientific research projects such as the National 863 Program, National Science and Technology Support Program, and horizontal projects; and has been involved in the design and analysis of multiple practical engineering projects.

Ling Zhou

Ling Zhou is a professor, doctoral supervisor, and national-level young talent, as well as the Deputy Director of Discipline Construction and Standardized Development Department at Jiangsu University (on-the-job training), Deputy Director of Jiangsu Special Pump and System Engineering Research Center, and Deputy Director of China Petroleum and Chemical Industry Pump and System Energy Saving Technology Key Laboratory. Jiangsu University Washington University in St. Louis of the United States jointly cultivated doctors; Prof. Zhou was selected as the leader of industrial innovation in Mount Taishan, Shandong Province, alongside 333 high-level talents in Jiangsu Province; presided over the programs of Jiangsu Outstanding Youth Fund and Jiangsu Outstanding Youth Fund; and won the outstanding doctoral and master's degree theses in Jiangsu Province. For four consecutive years, he was selected as one of the world's top 2% scientists in Elsevier. Prof. Zhou is mainly engaged in pump optimization design and multiphase flow research in fluid machinery, and has published more than 60 academic papers and won scientific and technological awards such as the National Science and Technology Progress Second Prize (7th), the China Machinery Industry Science and Technology First Prize (2nd), and the Ministry of Education Science and Technology Progress Second Prize (1st). He was selected for the Newton Foundation Sino British Innovation Award (2017), the United Nations Sustainable Development Goals Young Engineer Award (2018), the Jiangsu Engineering Thermophysics Society Outstanding Young Science and Technology Worker Award (2019), the Jiangsu Youth Job Expert Award (2020), the Digital Simulation Youth Science and Technology Award (2021), the Jiangsu Youth May Fourth Medal (2022), the I&EC Research Influential Researcher Award (2023), and the Jiangsu Youth Innovation Talent Award (2024).

Bin Huang

Professor Bin Huang mainly engages in research on fluid mechanical and hydrodynamic technologies such as ocean current energy generation equipment and low-noise ship pumps. Prof. Huang has undertaken more than 30 projects, including the National Natural Science Foundation of China Youth Project, General Project, Key Project Sub Project, Equipment Pre research Joint Fund of the Ministry of Education Youth Project, Military Commission Science and Technology Commission Basic Strengthening Plan Technology Field Fund, National Key R&D Plan Sub Task, Zhejiang Province Key R&D Project Sub Topic, Zhejiang Province "Vanguard" Plan Sub Project, and various military civilian enterprise horizontal projects. Prof. Huang has published 15 SCI papers, authorized 8 national invention patents, and won the second prize of Zhejiang Provincial Science and Technology Progress Award in 2012 and the third prize of Mechanical Industry Science and Technology Award in 2019. Prof. Huang is serving as a young editorial board member of "Big Motor Technology", a member of the organizing committee of the 5th National Marine Technology Conference, and a Special Issue editorial board member of the international SCI journals *Renewable Energy*, *Sustainability*, and *Processes*.

Preface

At a critical time when humanity is seeking sustainable energy solutions, a Special Issue focusing on ocean and hydropower technology has emerged. Ocean and water bodies contain enormous potential for renewable resources, providing strong support for the decarbonization goals of the global power system. From rough waves and regular tides to strong and stable offshore wind power and hydropower technology driven by long-standing water currents, these forms of energy are gradually becoming the core force of global energy transformation.

Marine energy technology, including wave, tidal, and offshore wind power converters, although still in the early stages of vigorous development, shows unlimited prospects. This technology expected to provide stable and predictable power output, injecting new vitality into the global energy structure. As a "veteran" in the field of renewable energy, hydropower still occupies an important position in today's energy landscape with its mature utilization methods and technological accumulation. Especially in the EU region, it accounts for up to 36% of renewable electricity consumption and continues to provide solid power for the operation of human society.

With the rapid development of related technologies, designing and optimizing innovative energy conversion devices has become a key task in achieving the commercialization of ocean and hydropower. At the same time, control strategies, smart grid technology, and fluid/aerodynamic characteristics play a decisive role in the development of these sustainable energy sources, affecting the performance and robustness of energy systems.

In this context, this Special Issue focuses on this field of great scientific value and practical significance, and promoting in-depth exchanges and cooperation between academia and industry. By gathering global scientific research forces and deeply exploring the enormous potential of the ocean and hydropower, this Special Issue aims to open up a new path for the sustainable development of global energy and help humanity move towards a cleaner and more efficient energy future. Let us witness together the glorious chapter of ocean and hydropower technology moving from scientific breakthroughs to widespread applications.

Yuquan Zhang, Jianhua Zhang, Ling Zhou, and Bin Huang
Guest Editors

sustainability

MDPI

Article

A Parameter Sensitivity Analysis of Hydropower Units under Full Operating Conditions Considering Turbine Nonlinearity

Dong Liu *, Xinxu Wei, Jingjing Zhang, Xiao Hu and Lihong Zhang

College of Energy and Power Engineering, North China University of Water Resources and Electric Power, Zhengzhou 450045, China; weixinxu@ncwu.edu.cn (X.W.); zhangjingjing@ncwu.edu.cn (J.Z.); huxiao@ncwu.edu.cn (X.H.); zhanglihong@ncwu.edu.cn (L.Z.)

* Correspondence: liudong@ncwu.edu.cn

Abstract: A parameter sensitivity analysis is an important part of the stability study of hydro turbine regulation systems, which helps operators to deepen their understanding of the characteristics and connections among the various parts of these systems. Considering that large hydropower stations undertake an essential regulation task in the power grid, the safety and stability of their operation cannot be ignored. To this end, taking a unit in a giant hydropower station in China as an example, a hydraulic–mechanical–electrical coupling model of the hydraulic turbine regulation system is established. A comprehensive parameter sensitivity indicator and parameter sensitivity analysis framework are proposed. On this basis, the sensitivity of the main system variables to parameter changes under full operating conditions is investigated by considering two different control modes of the unit (i.e., corresponding to different grid types). The results show that the sensitivity of the system state to the mechanical parameters of the generator is the highest in the power control mode, while the sensitivity to the electrical parameters of the generator and excitation system is higher in the frequency control mode. The sensitivity of the system with these key parameters also shows different patterns of change with a change in the unit operating conditions. The relevant findings can provide some theoretical guidance for the operation of hydropower stations and help to reduce the risk of system instability.

Keywords: hydropower unit; parameter sensitivity analysis; operating condition; nonlinear modeling; power system

Citation: Liu, D.; Wei, X.; Zhang, J.; Hu, X.; Zhang, L. A Parameter Sensitivity Analysis of Hydropower Units under Full Operating Conditions Considering Turbine Nonlinearity. *Sustainability* **2023**, *15*, 11691. https://doi.org/10.3390/su151511691

Academic Editors: Bin Huang, Ling Zhou, Yuquan Zhang and Jianhua Zhang

Received: 1 July 2023
Revised: 26 July 2023
Accepted: 27 July 2023
Published: 28 July 2023

1. Introduction

In recent years, the rapid development of variable renewable energy sources (VRES) has effectively reduced the increasing load burden on power systems. However, the variability and unpredictability of energy sources such as wind and solar have made power systems with hydropower generation as their main energy supply face great challenges in terms of power balance and frequency stability [1]. On the one hand, although hydropower units have a fast load regulation capability, their traditional operation finds it difficult to cope with the negative impact of VRES fluctuations, which is not conducive to high VRES penetration [2,3]. On the other hand, as the proportion of VRES in power grids increases, it is necessary to conduct research on the stability of hydro-turbine regulation systems (HTRS) to optimize the operation of their units, ensure the stability of their power systems, and improve their frequency quality [4].

Generally speaking, the research focusing on the stability problems of HTRSs mainly includes a stability analysis, nonlinear dynamics analysis, and parameter sensitivity analysis. From the perspective of research methods, a time-frequency domain analysis [5], root trajectory analysis [6], the Hopf bifurcation theory [7,8], and parameter sensitivity indicators [9] are commonly used methods. From the perspective of the research object, the influence of individual systems (e.g., hydraulic [10–13], mechanical [7,14], and electromagnetic subsystems [15]) or coupled systems (hydraulic–mechanical–electrical coupling [8,16,17]) on the stability and dynamic characteristics of a whole HTRS can be analyzed. Existing studies have mainly considered the influence of servo-systems, water diversion systems, and power grids on the system stability, while less attention is paid to the characteristics of prime movers. The nonlinear characteristics of prime movers have the most significant impact on the overall transition process and are, to some extent, an important reflection of the system's dynamics [18]. Meanwhile, the models of non-major subsystems are over-simplified for the convenience of analysis. For example, the grid model is simplified to a constant load when considering the characteristics and effects of the hydraulic system [19–22], while the hydro-turbine is simplified to an ideal model when considering the grid characteristics. On the other hand, the operating conditions studied are relatively single, accounting for only a small portion of all operating conditions [23], and the conclusions obtained are not comprehensive enough. Hydropower units often undertake important regulation tasks, leading to a wide range of operating conditions [24]. In the case of hydro-turbines, changes in their operating conditions cause changes in the discharge and torque characteristics of the turbine, which, in turn, affect the dynamic process of the whole system [25]. In addition, related studies have generally been based on system models in the frequency control mode (FCM) [26–28]. When the unit is connected to a large grid, the governor may also operate in the opening control mode (OCM) or power control mode (PCM) [15,20,29,30]. Therefore, studying the stability and dynamic characteristics of a HTRS under full operating conditions (FOC) and different control modes can provide a more comprehensive understanding of the effects of turbine nonlinearity.

A parameter sensitivity analysis (PSA) is a method for evaluating the influence of the uncertainty of system parameters. Its purpose is to identify the sensitive parameters that have an important influence on the system state from multiple uncertain parameters, analyze and measure the degree of their influence on the system variables, and then judge the ability of the system to withstand the risk of stability reduction due to parameter changes. In the field of engineering, parameter sensitivity analyses have been practically applied in river ecosystems, geological systems, power systems, and mechanical systems, and meaningful results have been achieved [31]. In the parameter sensitivity analyses of HTRSs, current research focuses on the effects of certain parameters on certain systems. For instance, the effects of the shaft system parameters (such as shaft misalignment and guide bearing stiffness) on the unit speed, discharge, and water head stability [32] or the effects of the runner parameters (such as the relative height of the guide blades, diameter ratio, blade angle, and displacement coefficient, etc.) on the unit efficiency and vibration [33]. These works have purposefully investigated the mechanism of certain key parameters on specific variables. However, there is a lack of systematic research on the primary and secondary effects as well as the action patterns of these parameters on the system variables. The operating conditions and control modes studied are also limited.

Based on the above analysis, this paper takes the Xiluodu hydropower station as an example to study the parameter sensitivity of a HTRS under FOC and different control modes. The influence of turbine nonlinearity on the parameter sensitivity is the focus. To this end, the neural network model of the hydro-turbine is transformed into a segmental linearized model through the neural network derivation (NND) method, and the hydro-mechanical-electrical coupling model of the HTRS is established. Parameter sensitivity indicators and the corresponding analysis process are proposed considering FOC and two control modes. Based on the sensitivity indicators, the sensitivity of the main system variables to the system parameters is studied under short-circuit fault conditions, and the primary and secondary effects of the parameters on the system variables are analyzed. The relevant conclusions are made to deepen the understanding of the influence law of the system parameters on the variables, which will have a certain theoretical guidance significance for system modeling and controller design.

2. Model

This section describes the basic models of each subsystem of the HTRS for the construction of the overall hydro-mechanical-electrical coupled nonlinear model, with a consideration of the grid-side characteristics and a detailed model of the hydropower unit on the MATLAB/Simulink platform.

2.1. Hydro-Turbine

The dynamic characteristics of the hydro-turbine can be approximated by the discharge and torque characteristics when it is in the steady state, which are nonlinear functions of the guide vane opening (GVO), unit speed, and water head for Francis turbines, as shown in Equation (1).

$$\begin{cases} Q = Q(Y, X, H) \\ M_t = M_t(Y, X, H) \end{cases} \quad (1)$$

where Q is the turbine discharge; Y is the GVO; X is the rotor speed; H is the water head; and M_t is the turbine torque.

Under the small fluctuation operation condition, the turbine discharge and torque characteristics are expanded by Taylor at a certain operating condition, respectively, and the higher-order differential above the second order is neglected, in order to obtain the algebraic equation of the turbine, as shown in Equation (2).

$$\begin{cases} q = e_{qy}y + e_{qx}x + e_{qh}h \\ m_t = e_y y + e_x x + e_h h \end{cases} \quad (2)$$

where q, m_t, y, x, and h are the deviation relative values of the discharge, torque, GVO, rotor speed, and water head, respectively; e_{qy}, e_{qx}, and e_{qh} are the transfer coefficients of the discharge to the GVO, rotor speed, and water head, respectively; and e_y, e_x, and e_h are the transfer coefficients of the torque to the GVO, rotor speed, and water head, respectively.

Neural networks are an important fitting method with a powerful ability to map nonlinear relations and approximate any nonlinear function with an arbitrary accuracy. The hydro turbine characteristics can be represented by the model comprehensive characteristic curve and runaway characteristic curve. Therefore, neural networks can be used to build a non-linear model of the hydro-turbine by combining the data on these characteristic curves. For the Francis turbine, the network inputs are the unit speed and GVO. The number of neurons in the hidden layer depends on the complexity of the model. The network output is the unit discharge or torque. The process of building a nonlinear model of the hydro-turbine can be found in the related literature [34]. Based on the turbine neural network model, the transfer coefficients under different operating conditions can be obtained via the NND method [35].

2.2. Water Diversion System

Commonly used models for water diversion systems generally include equivalent simplified models (i.e., rigid and elastic water hammer models) and characteristic line models. These models are derived through the basic equations that describe the non-constant flow in pipelines. Usually, rigid water hammer models are used for power stations with short diversion pipelines, while elastic water hammer models or characteristic line models are used for the opposite. Due to the short length of the diversion pipeline of the hydropower station studied in this paper, the accuracy requirement can be met by using a rigid water hammer model, as shown in Equation (3).

$$G_h(s) = -T_w s \quad (3)$$

where T_w is the flow inertia time constant.

2.3. *Governor*

The hydro-turbine governor includes two parts: a controller and servo-system. At present, controllers with PID regulation laws are commonly used in domestic and foreign hydropower stations, including parallel PID and series PID. The HTRS studied in this paper has a parallel-PID-type controller containing an output amplitude limitation, including FCM and PCM, and its logic block diagram is shown in Figure 1, where b_p is the permanent droop coefficient; e_p is the regulation rate; the subscript c of the symbols indicates the given value, i.e., ω_c for a given speed or frequency, y_c for a given opening, and P_c for a given power; T_{1v} is the derivative time constant; e is the total control error; and u is the controller output.

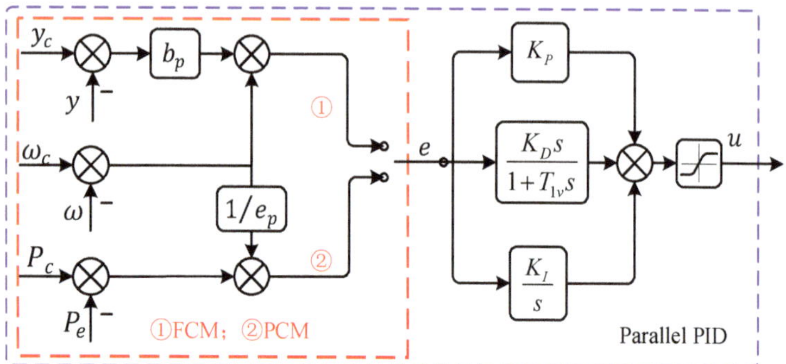

Figure 1. Parallel-PID-type controller.

The servo-system consists of components such as the main servomotor, pressure distribution valve, and electro-hydraulic converter, where the nonlinear links mainly include dead zone, saturation, speed limit, and delay. For the Xiluodu power station unit, the nonlinear simulation model of the servo-system adopts the structure shown in Figure 2, where T_{y1} and T_y denote the response time constant of the pilot servomotor and servomotor, respectively. The expression block related to T_{y1} represents the mathematical model of the pilot servomotor. Since the value of T_{y1} is very small (here is 0.041), it can be approximated that the transfer function of the pilot servomotor is 1.

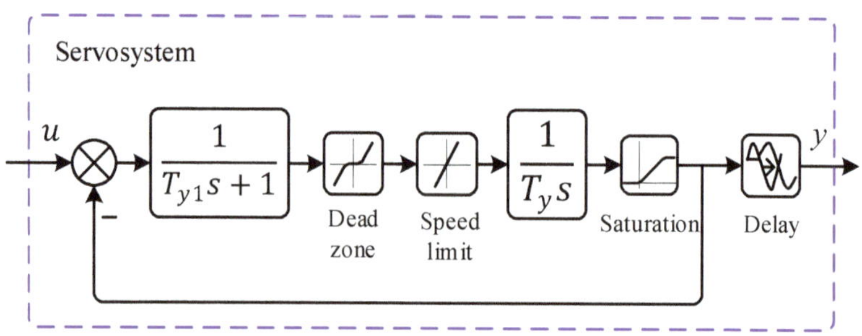

Figure 2. Nonlinear model of servo-systems.

2.4. *Generator and Network*

The fifth-order model of generators is used in this paper, which does not consider the stator transients, considers the effect of the rotor damping winding, but neglects the

G-winding of the q-axis. It can be represented by a set of differential equations, as shown in Equation (4) (neglecting the stator resistance) [36].

$$\begin{cases} \dot{\delta} = \omega_0 \omega \\ T_a \dot{\omega} = M_t - M_e \\ T'_{d0} \dot{E}'_q = E_f - E'_q - I_d (X_d - X'_d) \\ T''_{d0} \dot{E}''_q = E'_q - E''_q - I_d (X'_d - X''_d) \\ T''_{q0} \dot{E}''_d = -E''_d + I_q (X_q - X''_q) \end{cases} \tag{4}$$

where δ is the rotor angle; ω_0 is the synchronous angular velocity; T_a is the unit inertia time constant; E_f is the excitation emf proportional to the excitation voltage; E'_q is the q-axis component of the transient internal emf proportional to the field winding flux linkages; E''_q is the q-axis component of the sub-transient internal emf proportional to the total flux linkage in the d-axis damper winding and field winding; E''_d is the d-axis component of the sub-transient internal emf proportional to the total flux linkage in the q-axis damper winding and q-axis solid steel rotor body; T'_{d0} and T''_{d0} are open-circuit d-axis transient and sub-transient time constants, respectively; T''_{q0} is an open-circuit q-axis sub-transient time constant; X_d, X'_d, and X''_d are the d-axis synchronous, transient, and sub-transient generator reactance, respectively; X_q and X''_q are the q-axis synchronous and sub-transient reactance of the generator, respectively; and I_d and I_q are d- and q-axis components of the armature current, respectively. The calculation of these three variables M_e (or P_e), V_d, and V_q can be found in the relevant literature [9].

For the convenience of analysis, a simplified model of the first-order excitation system, without considering the effect of a power system stabilizer (PSS), is used, as shown in Equation (5) [15].

$$\frac{E_f(s)}{V_{ref}(s) - V_g(s)} = \frac{K_a}{T_r s + 1} \tag{5}$$

where K_a and T_r are the gain and time constant of the excitation system, respectively; V_g is the voltage at the generator terminal; and V_{ref} is the reference value of V_g.

In the FCM, the grid side can be approximated as a transformer, transmission line, and isolated load in the form of impedance, and the network equation is [9]:

$$(R_{sl} + jX_{sl})(I_x + jI_y) = U_x + jU_y \tag{6}$$

where R_{sl} and X_{sl} are the total resistance and reactance of the transmission line and load, respectively; I_x and I_y are the real and imaginary parts of the network current; U_x and U_y are the real and imaginary parts of the network voltage; and j represents the imaginary unit, i.e., the square root of -1.

In the PCM, the grid side can be approximated as an infinite bus power system, and the network equation is [9]:

$$V_s + (R_l + jX_l)(I_x + jI_y) = U_x + jU_y \tag{7}$$

where R_l is the resistance of the transmission line (between the generator and infinite bus); X_l is the reactance of the transmission line; and $X_{d\Sigma} = X''_d + X_l$; $X_{q\Sigma} = X''_q + X_l$; V_s is the infinite bus voltage.

It should be noted that the models described in this section are known basic or refined models and will be integrated to establish a hydraulic–mechanical–electrical coupling model of a HTRS in Section 4.

3. Method

This section introduces the parameter sensitivity indicators that measure the degree of influence of the system parameters on the variables, and on this basis, proposes a general framework for a parameter sensitivity analysis of a HTRS under FOC and different control modes.

3.1. Parameter Sensitivity Indicator

Parameter sensitivity is the degree of change in a system state variable caused by a small change in a parameter of the system. According to different coordinate systems, parameter sensitivity can be divided into time domain sensitivity and frequency domain sensitivity. The most commonly used indicator for time domain sensitivity is the trajectory sensitivity, while the sensitivity indicator calculated in the frequency domain is mainly the amplitude phase sensitivity. For the convenience of a solution, this paper adopts the trajectory sensitivity indicator for the PSA of a HTRS.

The dynamic behavior of a HTRS containing grid characteristics can generally be expressed in the form of an algebraic differential equation, as follows:

$$\dot{z} = f(z, a, r, p) \tag{8}$$

$$g(z, a, r, p) = 0 \tag{9}$$

where z is the state variable of the system, such as the rotor speed; a is the algebraic variable or output of the system, such as GVO; r is the control input or disturbance of the system, such as a given frequency or load change; and p is the parameter of the system, such as the flow inertia time constant.

The solution trajectory of the state or algebraic variables is simultaneously Taylor expanded at the current value of the parameter p to obtain Equations (10) and (11):

$$z(p + \Delta p, t) = z(p, t) + \frac{\partial z(p, t)}{\partial p}\Delta p + o\left[(\Delta p)^n\right] \tag{10}$$

$$a(p + \Delta p, t) = a(p, t) + \frac{\partial a(p, t)}{\partial p}\Delta p + o\left[(\Delta p)^n\right] \tag{11}$$

where Δp is the tiny change in parameter p near the current value and $o\left[(\Delta p)^n\right]$ is the second- and higher-order differential of Δp.

Neglecting the higher-order infinitesimal terms, the trajectory sensitivity of any variable can be calculated by Equations (12) and (13):

$$\frac{\partial z(p, t)}{\partial p} = \frac{z(p + \Delta p, t) - z(p - \Delta p, t)}{2\Delta p} \tag{12}$$

$$\frac{\partial a(p, t)}{\partial p} = \frac{a(p + \Delta p, t) - a(p - \Delta p, t)}{2\Delta p} \tag{13}$$

To facilitate comparison and analysis, the relative trajectory sensitivity is used and its absolute mean value is taken as the indicator, as shown in Equation (14):

$$MAS = \frac{1}{N}\sum_{i=1}^{N}\left|S_p\right| = \frac{1}{N}\sum_{t=1}^{N}\left|\frac{\partial z(p, t)/z_0}{\partial p/p_0}\right| \tag{14}$$

where N is the data length of the simulation, S_p is the relative trajectory sensitivity, and MAS can be interpreted as the slope of the function at the value of a parameter when the absolute mean of the relative change of a system variable during the transition is taken as the dependent variable and the value of the parameter is taken as the independent variable. The subscript 0 indicates the initial value of the variable or parameter.

The relative trajectory sensitivity of all the parameters is normalized to obtain the relative degree of influence of a parameter change on the dynamic of a specified system state variable near a specific operating condition, namely:

$$\overline{MAS_j^i} = \frac{MAS_j^i - min\left\{MAS_j^1, MAS_j^2, \cdots, MAS_j^n\right\}}{max\left\{MAS_j^1, MAS_j^2, \cdots, MAS_j^n\right\} - min\left\{MAS_j^1, MAS_j^2, \cdots, MAS_j^n\right\}} \quad (15)$$

where i is the parameter label of the system, j is the state variable label of the system, n is the number of parameters, and MAS_j^i indicates the influence degree of the i-th parameter of the system on the j-th variable of the system.

The influence degree of a parameter on the system state variables should also vary under different operating conditions. When studying the variation in parameter sensitivity indicators with operating conditions, the following comprehensive parameter sensitivity indicator, which can be used for any operating condition, is introduced to facilitate the analysis:

$$CMAS^i = \frac{1}{m}\sum_{j=1}^{m} \overline{MAS_j^i} \quad (16)$$

where m is the total number of system variables considered in the PSA.

3.2. Parameter Sensitivity Analysis Process under FOC

A HTRS considering grid characteristics is affected by various factors such as hydraulic, mechanical, and electromagnetic factors, etc. The study of the parameter sensitivity of each subsystem helps to comprehensively understand the influence degree of various physical factors on the dynamic of the system. At the same time, the sensitivity indicator of the same parameter under different operating conditions may vary. Therefore, for the nonlinearity of the hydro-turbine, this paper analyzes the parameter sensitivity of the system under two control modes and FOC based on the hydraulic–mechanical–electrical coupling model of the HTRS, which considers the grid characteristics, as shown in Figure 3. The detailed procedure is as follows.

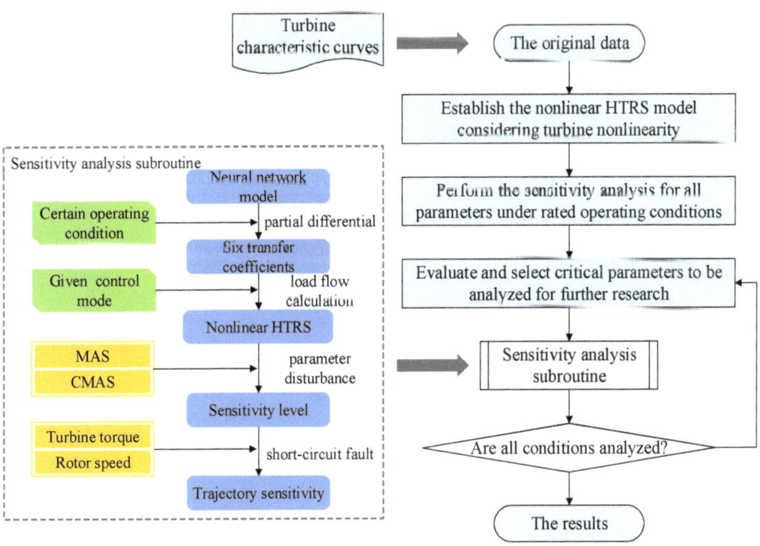

Figure 3. PSA process of the HTRS under FOC and different control modes.

(1) The nonlinear model of the hydro-turbine is established using model characteristic curves, actual measurement data, and model correction methods [34].

(2) The hydraulic–mechanical–electrical coupling nonlinear model of the HTRS is established considering the grid characteristics.

(3) The highly sensitive parameters can be determined by analyzing the influence degree of each parameter on the state variables of each subsystem under the rated operating condition.

(4) The operating conditions are classified according to the water head and GVO of the hydropower unit.

(5) According to the subroutine described in Figure 3, the parameter sensitivity level and trajectory sensitivity under FOC are investigated for a given control mode.

(5.1) The transfer coefficients under a specified operating condition are calculated, and the values are substituted into the coupled nonlinear model of the HTRS.

(5.2) The value of a specified parameter is disturbed under short-circuit fault conditions, and the comprehensive parameter sensitivity indicator for this parameter under a specified operating condition is determined using Equation (16).

(5.3) The comprehensive indicator under FOC can be obtained by repeating the above two steps. The result is then utilized to analyze the variation in the parameter sensitivity of the HTRS with the operating conditions.

(5.4) The trajectory sensitivity of the system variables to the parameters under short-circuit fault conditions is further investigated. The trajectory sensitivity of the turbine torque and rotor speed to specified parameters (e.g., highly sensitive parameters) is observed and analyzed considering the change in operating conditions.

4. Results

In this section, the FCM (i.e., isolated operation) and PCM (i.e., grid-connected operation) of the Xiluodu hydropower station are taken as examples to analyze the parameter sensitivity under FOC, that is, to study the influence degree of the parameters on each subsystem in the process of a short-circuit fault of the HTRS under FOC.

4.1. Hydraulic–Mechanical–Electrical Coupling Model of HTRS

Since the PSA does not require solving higher-order differential algebraic equations, a hydraulic–mechanical–electrical coupling model of a HTRS is adopted in this section. This model can simulate the load characteristics on the grid side and includes a refined hydro-turbine model and a high-order generator model, as well as a servo-system model with nonlinear links. The basic structure and interrelation of the whole system are shown in Figure 4.

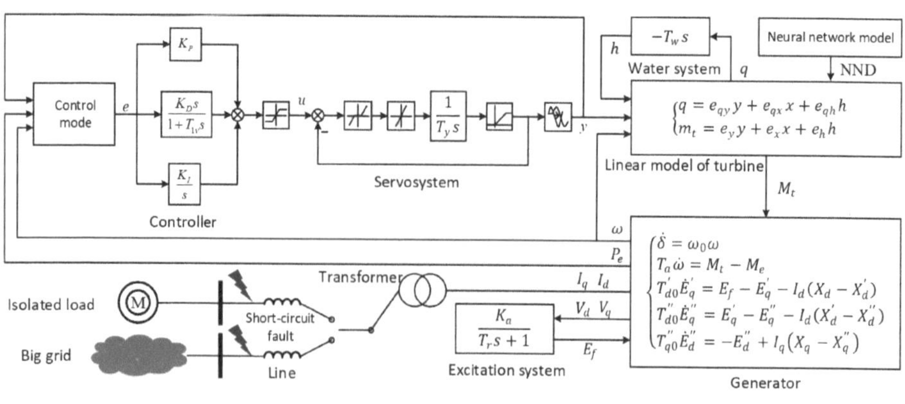

Figure 4. Schematic diagram of hydraulic–mechanical–electrical coupling model of HTRS.

It can be seen from the figure that the hydraulic, mechanical, and electrical factors in the system interact with and influence each other. The mechanical factors and hydraulic factors are connected through hydro-turbines, while the mechanical factors and electrical factors are connected through generators. Although the hydraulic factors and electrical factors are not directly connected, they can form indirect interactions through mechanical transmission. The hydropower unit can operate in the isolated grid mode or connect into a large power grid according to its needs, and adopt the corresponding control mode. The models of each subsystem in the system were described in detail in Section 2. Except for the turbine parameters, all the parameters of the system are constant, i.e., they do not change with the operating conditions, as shown in Table 1. It is worth noting that these parameters are not static and can change to varying degrees due to usage time, fault maintenance, or other reasons [37]. When the parameters change significantly, it is necessary to redetermine the parameter values and make corrections to the results of the previous parameter sensitivity analysis.

Table 1. Main parameters of HTRS.

Subsystem	Parameter
Controller	$e_p = 0.04$; $b_p = 0.01$; $T_{1v} = 0.118$
Servo-system	$T_y = 0.459$s; Delay $T_d = 0.1$s
Water system	$T_w = 0.873$s
Generator and network	$T_a = 12.239$s; $X_d = 0.985$; $X'_d = 0.326$; $X''_d = 0.241$; $X_q = 0.665$; $X''_q = 0.260$; $T'_{d0} = 11.72$; $T''_{d0} = 0.12$; $T''_{q0} = 0.25$; $V_s = 1.0$; $K_a = 100$; $T_r = 0.12$s
Hydro-turbine	The transfer coefficients under FOC are can be found in our previous work [35].

To calculate the trajectory sensitivity and the related sensitivity indicators, a three-phase short-circuit fault occurs at the end of the transmission line, and the fault recovers after 0.1 s [36]. Different from step disturbance, the equilibrium point of the system does not change in this process, and the test conditions can fully activate the dynamic of the system so that all the state variables can undergo a rich change process.

4.2. Parameter Sensitivity Analysis under FOC and PCM

According to the calculation method and process in Section 3, the PSA of the Xiluodu power station under PCM and a rated operating condition is first conducted. Then, the operating condition range is extended to FOC, and the variation in the sensitivity of the system states to the key parameters with operating conditions is investigated. The parameters to be analyzed are determined according to the model structure of each subsystem. The controller parameters mainly include the control parameters K_P and K_I. The servo-system parameters mainly include T_y and T_d. The water diversion system parameter is T_w. The hydro-turbine parameters include e_{qy}, e_{qh}, e_{qx}, e_y, e_h, and e_x, while the generator parameters include T_a, X_d, X'_d, X_q, and T'_{d0}. The excitation system parameters mainly include K_a and T_r.

4.2.1. Parameter Sensitivity Analysis under Rated Operating Condition

The rated operating condition of the hydropower unit has a water head of 195 m, GVO of 80%, and rated power of 700 MW. Under this condition, the transfer coefficient of the hydro-turbine and the other parameters of the unit can be found in Table 1, and the control parameters of the governor take the actual values of the power station (i.e., $K_P = 0.28$, $K_I = 0.2$, and $K_D = 0$).

Each parameter to be analyzed is disturbed by ±5% at the current set value. The short-circuit start time is 1 s and the simulation time is 25 s. The simulation waveform of the model output variables before parameter disturbance is shown in Figure 5.

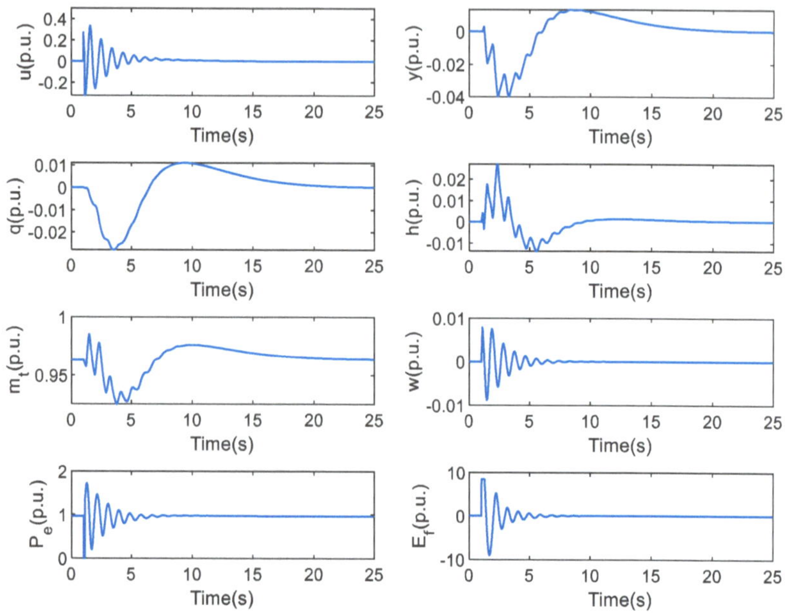

Figure 5. Simulation waveforms of model output variables before parameter perturbation in PCM.

The simulation waveform after parameter disturbance is consistent with Figure 5, but there are local differences, which show the influence of a parameter change on the system variables. According to the simulation waveforms after upper and lower disturbances of the parameters, the sensitivity indicator *MAS* of each state variable of the system to each parameter is calculated, and the results are shown in Table 2. As can be seen from Table 2, among all the parameters analyzed, almost all the state variables are the most sensitive to a change in T_a. According to the contrast of the color shades, the following conclusions can be drawn: (1) the controller output is mainly sensitive to a change in the generator parameter T_a; (2) the servo-system output is not only sensitive to the generator parameters T_a and T'_{d0}, but is also influenced by the controller parameter K_P; (3) the sensitive parameters of the turbine discharge are same as those of the servo-system output and are also more sensitive to the self-parameter e_{qy}, while the turbine torque is only sensitive to the generator parameter T_a; (4) the water head in water diversion system is sensitive to the generator parameter T_a and turbine parameter e_y; (5) the rotor speed and active power of the generator are most sensitive to its own parameter T_a, and the sub-sensitive parameters are the generator parameters X'_q, X'_d, and T'_{do}, and the excitation system parameters K_a and T_r; and (6) the excitation system output is also most sensitive to T_a, and the sub-sensitive parameters are mainly the self-parameters K_a and T_r, as well as the generator parameters X'_q X'_d, and T'_{do}. Considering the hydraulic–mechanical–electrical interaction, it can be concluded that: (1) the state variables of the electromagnetic system are insensitive to the parameter changes in the hydraulic–mechanical system in the PCM, which also shows that the electromagnetic state of the unit is mainly maintained by the regulation of the power grid; and (2) in the PCM, the state variables of the hydraulic–mechanical system are sensitive to the parameters of itself and the electromagnetic system, which indicates that the hydraulic–mechanical state variables are jointly regulated by the unit and the power grid.

Table 2. Quantitative assessment of the effect of parameter changes on subsystem output variation (*MAS* values in PCM).

		Controller	Servo-System	Hydro-Turbine		Water System	Generator		Excitation System
		u	y	q	m_t	h	ω	P_e	E_f
Controller	K_P	2.23×10^{-2}	7.57×10^{-3}	6.31×10^{-3}	2.38×10^{-3}	7.40×10^{-3}	2.42×10^{-5}	7.75×10^{-3}	1.46×10^{-2}
	K_I	6.85×10^{-3}	4.72×10^{-3}	3.94×10^{-3}	1.12×10^{-3}	4.45×10^{-3}	9.23×10^{-6}	4.83×10^{-3}	6.04×10^{-3}
Servo-system	T_y	2.76×10^{-3}	4.45×10^{-3}	3.64×10^{-3}	2.07×10^{-3}	4.50×10^{-3}	1.79×10^{-5}	4.71×10^{-3}	1.07×10^{-2}
	T_d	2.04×10^{-3}	6.00×10^{-4}	2.56×10^{-4}	7.08×10^{-4}	7.71×10^{-4}	4.98×10^{-5}	4.31×10^{-3}	2.51×10^{-2}
Water system	T_w	2.40×10^{-3}	1.46×10^{-3}	2.16×10^{-3}	2.93×10^{-3}	4.86×10^{-3}	1.89×10^{-5}	5.05×10^{-3}	1.03×10^{-2}
Hydro-turbine	e_{qy}	5.81×10^{-3}	1.92×10^{-3}	5.87×10^{-3}	2.78×10^{-3}	3.52×10^{-3}	1.25×10^{-4}	1.15×10^{-2}	6.32×10^{-2}
	e_{qh}	4.13×10^{-3}	1.80×10^{-3}	2.83×10^{-3}	1.82×10^{-3}	3.12×10^{-3}	9.03×10^{-5}	8.59×10^{-3}	4.67×10^{-2}
	e_{qx}	1.55×10^{-3}	6.82×10^{-4}	5.91×10^{-4}	4.00×10^{-4}	9.80×10^{-4}	3.32×10^{-5}	3.12×10^{-3}	1.71×10^{-2}
	e_y	5.97×10^{-3}	3.79×10^{-3}	3.19×10^{-3}	8.73×10^{-4}	8.99×10^{-3}	5.39×10^{-5}	9.93×10^{-3}	2.47×10^{-2}
	e_h	4.45×10^{-3}	1.47×10^{-3}	1.22×10^{-3}	4.37×10^{-4}	3.91×10^{-3}	9.29×10^{-5}	8.89×10^{-3}	4.68×10^{-2}
	e_x	2.30×10^{-3}	1.06×10^{-3}	8.81×10^{-4}	3.51×10^{-4}	1.48×10^{-3}	4.85×10^{-5}	4.59×10^{-3}	2.51×10^{-2}
Generator	T_a	1.10×10^{-1}	9.31×10^{-3}	5.69×10^{-3}	6.86×10^{-3}	1.16×10^{-2}	2.79×10^{-3}	2.38×10^{-1}	1.38×10^{0}
	X_d	1.34×10^{-3}	5.20×10^{-4}	4.30×10^{-4}	1.72×10^{-4}	5.23×10^{-4}	3.11×10^{-5}	2.84×10^{-3}	5.59×10^{-2}
	X'_d	1.78×10^{-2}	3.86×10^{-3}	3.08×10^{-3}	1.54×10^{-3}	4.12×10^{-3}	4.61×10^{-4}	4.05×10^{-2}	4.72×10^{-1}
	X_q	2.42×10^{-2}	6.74×10^{-3}	5.60×10^{-3}	2.45×10^{-3}	7.00×10^{-3}	6.04×10^{-4}	5.49×10^{-2}	3.43×10^{-1}
	T'_{d0}	1.77×10^{-2}	7.52×10^{-3}	6.30×10^{-3}	2.54×10^{-3}	7.56×10^{-3}	4.12×10^{-4}	3.63×10^{-2}	5.66×10^{-1}
Excitation system	K_a	1.46×10^{-2}	5.65×10^{-3}	4.74×10^{-3}	1.95×10^{-3}	5.76×10^{-3}	3.54×10^{-4}	3.13×10^{-2}	7.07×10^{-1}
	T_r	9.46×10^{-3}	4.44×10^{-3}	3.69×10^{-3}	1.54×10^{-3}	4.50×10^{-3}	2.15×10^{-4}	1.97×10^{-2}	4.92×10^{-1}

Different colours in the table are used to distinguish different state variables. The darkness of the same colour depends on the sensitivity of the state variable to the parameter. The larger the value of the parameter sensitivity, the darker the corresponding colour.

Several key system states (including the GVO, water head, and rotor speed) are further analyzed, and the trajectory sensitivity of each state under the action of the three most sensitive parameters is drawn. The results are shown in Figure 6a–c. Here, the GVO is taken as an example to illustrate: the trajectory sensitivity of the GVO to the controller parameter K_P is negative in the range of 1–6s, which means that, when the controller parameter K_P increases, the value of the GVO in this time period is less than that of the parameter not being changed. That is, in the initial stage after the short-circuit fault occurs, the guide vane closes quickly under the regulation of the controller, and the increase in K_P accelerates the closing speed of the guide vane, which makes the change curve of the GVO move down compared to the original, and the corresponding trajectory sensitivity appears negative. By the same token, an increase in T'_{d0} slows down the closing speed of the guide vane, which is opposite to the effect of K_P. The GVO is sensitive to a change in the parameters T'_{d0} and K_P, that is, the parameter sensitivity is only related to the influence degree of the parameters on the system state, but has nothing to do with the action effect of the parameters. The comprehensive parameter sensitivity indicator of the system to each parameter is shown in Figure 6d. It is obvious that the system is most sensitive to changes in T_a. The sub-sensitive parameters are mainly T'_{d0}, X_q, and K_a, and the insensitive parameters are mainly e_{qx}, X_d, and T_d.

4.2.2. Parameter Sensitivity Analysis under FOC

In the PCM, a change in the comprehensive parameter sensitivity with the operating conditions is further considered. Since there are many parameters to be analyzed, only the first four parameters with a high sensitivity under the rated operating condition (according to the previous analysis, they are T_a, T'_{d0}, X_q, and K_a) are selected. The comprehensive parameter sensitivity for these parameters under FOC is shown in Figure 7. It can be seen from the figure that: (1) with an increase in the water head, the sensitivity of the system to parameter T_a gradually increases, while the sensitivity to T'_{d0} gradually weakens; (2) with an increase in the GVO, the sensitivity of the system to T'_{d0} gradually weakens; (3) the sensitivity of the system to X_q under a small GVO (less than 50%), is higher than that under the condition of a large GVO; and (4) the system is more sensitive to a change in K_a under a low water head and small GVO.

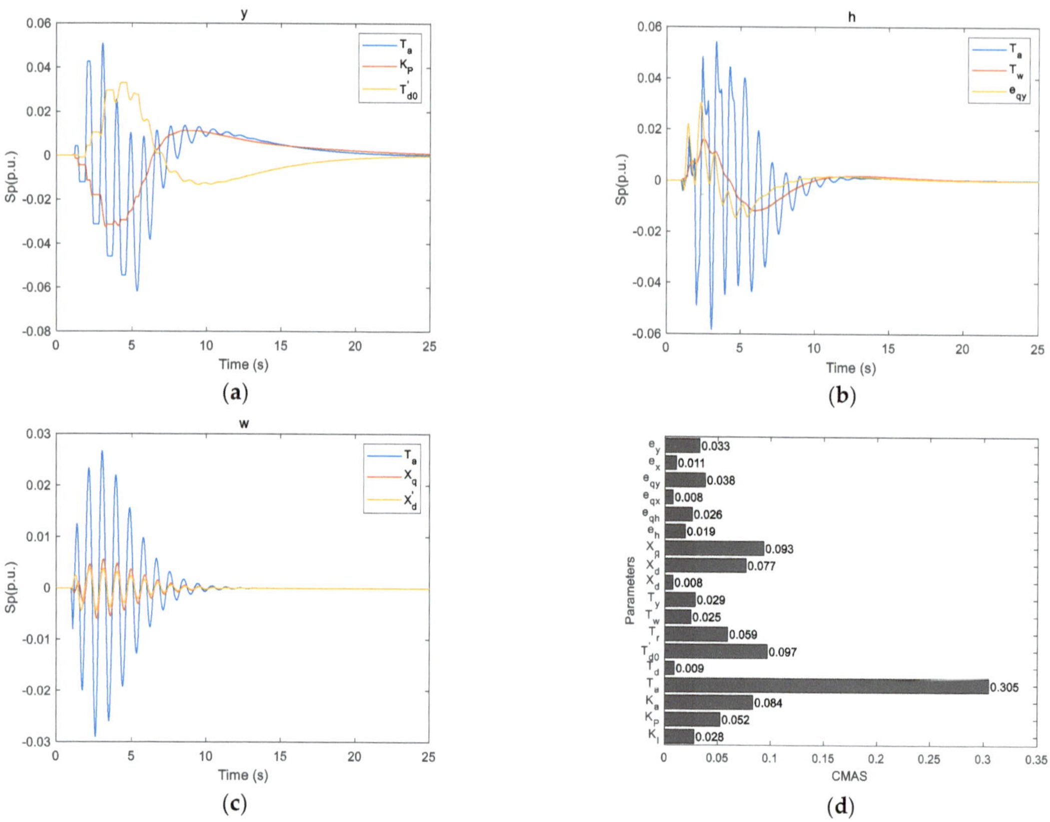

Figure 6. Trajectory sensitivity and comprehensive parameter sensitivity under rated operating condition and PCM. (**a**) Trajectory sensitivity of GVO to parameters. (**b**) Trajectory sensitivity of water head to parameters. (**c**) Trajectory sensitivity of rotor speed to parameters. (**d**) Comprehensive parameter sensitivity.

Finally, the parameter with the highest sensitivity (i.e., T_a) is taken as an example to analyze the trajectory sensitivity of the turbine torque and rotor speed under different operating conditions. Under various typical operating conditions, the trajectory sensitivities of the turbine torque and rotor speed to parameter T_a are shown in Figures 8 and 9, respectively. It can be seen from Figure 8 that, under the same water head, with an increase in the GVO, the vibration amplitude of the trajectory sensitivity of the turbine torque and unit speed to T_a increases gradually, indicating that the sensitivity of the two states to a variation in T_a increases with an increase in the GVO. This trend is particularly obvious for the rotor speed, which shows that the sensitivity of the rotor speed to T_a greatly depends on the current load situation. For the same GVO, the variation in the trajectory sensitivity with the water head in Figure 9 is similar to that of Figure 8. Therefore, the sensitivity of the turbine torque and rotor speed to T_a increases with an increase in the water head.

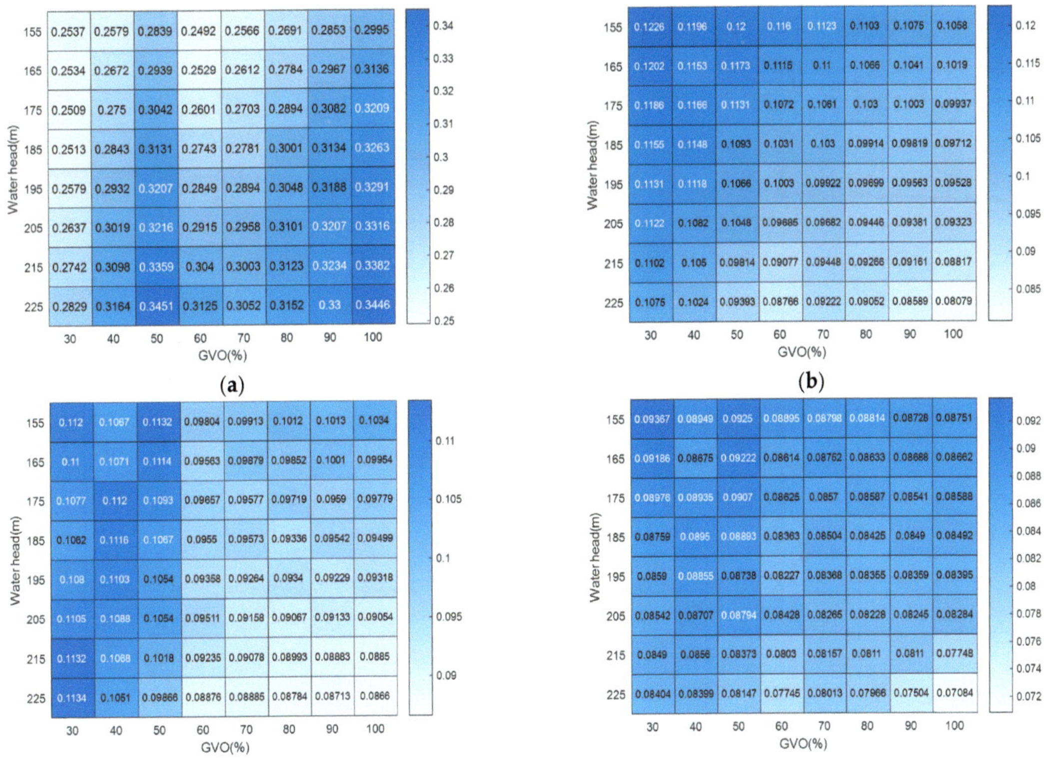

(a) (b) (c) (d)

Figure 7. Comprehensive parameter sensitivity under FOC and PCM. (**a**) T_a, (**b**) T'_{d0}, (**c**) X_q, and (**d**) K_a. The white font indicates operating conditions with relatively high parameter sensitivity.

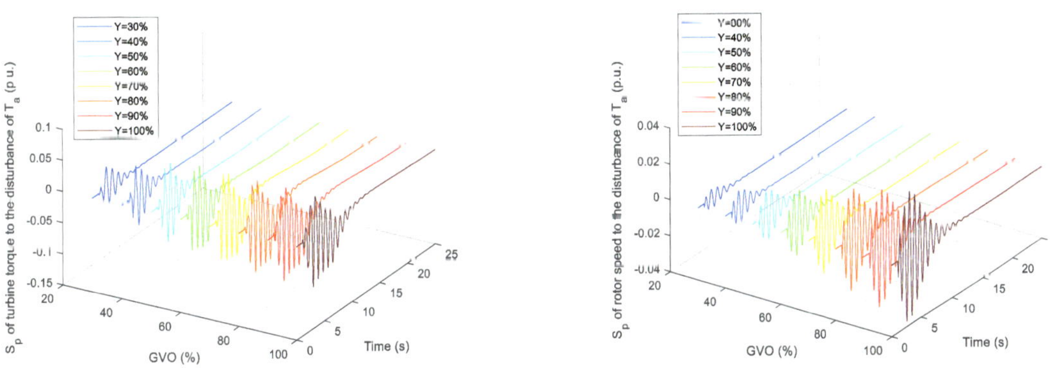

(a) (b)

Figure 8. Trajectory sensitivity of system states to changes in T_a at the water head of 195m (in PCM). (**a**) Turbine torque. (**b**) Rotor speed.

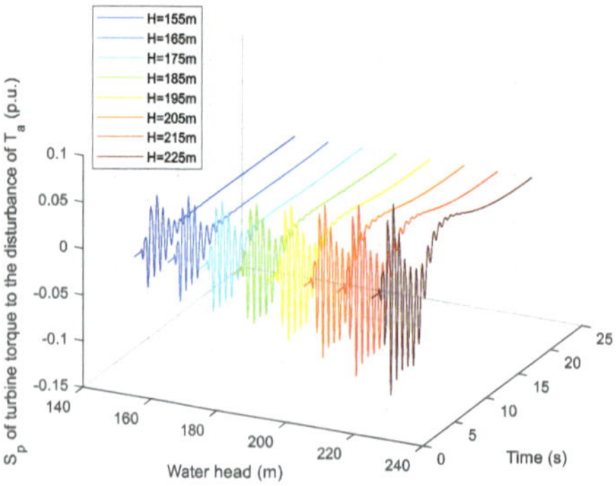

(a)

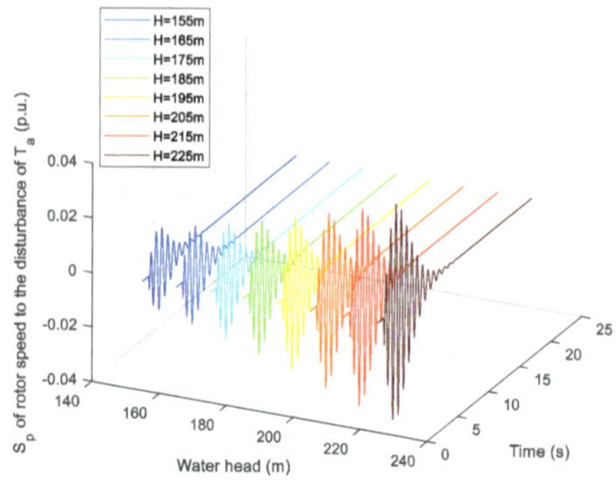

(b)

Figure 9. Trajectory sensitivity of system states to changes in T_a at the GVO of 70% (in PCM). (**a**) Turbine torque. (**b**) Rotor speed.

4.3. *Parameter Sensitivity Analysis under FOC and FCM*

Similar to Section 4.2, this section is focused on performing calculations and analyses for the system parameter sensitivities under the FCM and FOC. Among the parameters to be analyzed, the parameters of the controller mainly include the control parameters K_P, K_I, and K_D, and the other parameters to be analyzed are the same as those in the PCM.

4.3.1. Parameter Sensitivity Analysis under Rated Operating Condition

The load characteristics on the grid side under isolated grid conditions (i.e., FCM) are different from those in the PCM, and the corresponding initial power flow calculation results are also different. Therefore, the sensitivity of each system variable to each parameter in the FCM and the change in the parameter sensitivity with the operating conditions also need to be studied pertinently. Under the rated operating condition, the values of the turbine transfer coefficients and the other parameters of the unit are the same as those in Section 4.2.1, and the control parameters of the governor are taken from the actual values of the power station (i.e., K_P = 2.0, K_I= 0.2, and K_D = 0.2).

Under the same short-circuit fault, the simulation waveforms of the model output variables before parameter disturbance are shown in Figure 10. Similarly, according to the simulation waveform after the parameter disturbance, the sensitivity indicator MAS of each variable to each parameter is calculated, and the results are shown in Table 3.

Table 3. Quantitative assessment of the effect of parameter changes on subsystem output variation (MAS values in FCM).

		Controller	Servo-System	Hydro-Turbine		Water System	Generator		Excitation System
		u	y	q	m_t	h	ω	P_e	E_f
Controller	K_P	2.20×10^{-3}	2.17×10^{-3}	1.86×10^{-3}	9.16×10^{-4}	2.45×10^{-3}	6.02×10^{-4}	5.10×10^{-12}	3.19×10^{-10}
	K_I	5.31×10^{-4}	5.24×10^{-4}	4.47×10^{-4}	1.16×10^{-4}	5.26×10^{-4}	2.25×10^{-4}	5.26×10^{-12}	3.13×10^{-10}
	K_D	1.64×10^{-4}	1.01×10^{-4}	7.46×10^{-5}	8.48×10^{-5}	1.34×10^{-4}	1.42×10^{-5}	4.84×10^{-12}	3.03×10^{-10}
Servo-system	T_y	2.28×10^{-4}	5.49×10^{-4}	4.50×10^{-4}	4.20×10^{-4}	7.28×10^{-4}	8.51×10^{-5}	4.62×10^{-12}	3.26×10^{-10}
	T_d	5.26×10^{-5}	1.58×10^{-4}	1.03×10^{-4}	1.49×10^{-4}	2.13×10^{-4}	1.95×10^{-5}	4.85×10^{-12}	3.15×10^{-10}
Water system	T_w	6.61×10^{-4}	6.52×10^{-4}	8.07×10^{-4}	8.91×10^{-4}	1.61×10^{-3}	2.44×10^{-4}	5.14×10^{-12}	3.36×10^{-10}

Table 3. *Cont.*

		Controller	Servo-System	Hydro-Turbine		Water System	Generator		Excitation System
		u	y	q	m_t	h	ω	P_e	E_f
Hydro-turbine	e_{qy}	5.97×10^{-4}	5.89×10^{-4}	1.87×10^{-3}	8.24×10^{-4}	1.52×10^{-3}	2.19×10^{-4}	4.99×10^{-12}	3.35×10^{-10}
	e_{qh}	1.86×10^{-4}	1.78×10^{-4}	5.37×10^{-4}	4.47×10^{-4}	8.24×10^{-4}	6.92×10^{-5}	5.21×10^{-12}	3.52×10^{-10}
	e_{qx}	8.92×10^{-5}	8.77×10^{-5}	3.40×10^{-4}	1.42×10^{-4}	2.60×10^{-4}	3.40×10^{-5}	5.14×10^{-12}	3.38×10^{-10}
	e_y	1.83×10^{-3}	1.83×10^{-3}	1.63×10^{-3}	3.90×10^{-4}	2.16×10^{-3}	5.83×10^{-4}	4.73×10^{-12}	3.08×10^{-10}
	e_h	6.76×10^{-4}	6.67×10^{-4}	6.31×10^{-4}	2.88×10^{-4}	1.74×10^{-3}	2.51×10^{-4}	4.99×10^{-12}	3.11×10^{-10}
	e_x	7.97×10^{-4}	7.87×10^{-4}	7.58×10^{-4}	1.98×10^{-4}	1.14×10^{-3}	3.06×10^{-4}	5.05×10^{-12}	3.25×10^{-10}
	T_a	2.32×10^{-3}	2.28×10^{-3}	2.17×10^{-3}	1.07×10^{-3}	3.17×10^{-3}	8.90×10^{-4}	5.33×10^{-12}	3.52×10^{-10}
Generator	X_d	2.78×10^{-4}	2.77×10^{-4}	2.46×10^{-4}	9.83×10^{-5}	3.45×10^{-4}	1.11×10^{-4}	7.58×10^{-4}	4.41×10^{-2}
	X'_d	3.01×10^{-4}	2.63×10^{-4}	2.33×10^{-4}	1.75×10^{-4}	3.53×10^{-4}	1.20×10^{-4}	4.51×10^{-3}	2.02×10^{-1}
	X_q	6.01×10^{-5}	2.98×10^{-5}	2.66×10^{-5}	3.33×10^{-5}	4.54×10^{-5}	2.25×10^{-5}	1.25×10^{-3}	5.73×10^{-2}
	T'_{d0}	7.56×10^{-4}	3.95×10^{-4}	2.82×10^{-4}	3.88×10^{-4}	4.59×10^{-4}	2.56×10^{-4}	1.45×10^{-2}	7.13×10^{-1}
Excitation system	K_a	1.03×10^{-3}	9.61×10^{-4}	8.45×10^{-4}	4.53×10^{-4}	1.22×10^{-3}	3.98×10^{-4}	1.11×10^{-2}	6.96×10^{-1}
	T_r	1.12×10^{-3}	1.11×10^{-3}	9.86×10^{-4}	4.43×10^{-4}	1.39×10^{-3}	4.45×10^{-4}	8.96×10^{-3}	5.49×10^{-1}

Different colours in the table are used to distinguish different state variables. The darkness of the same colour depends on the sensitivity of the state variable to the parameter. The larger the value of the parameter sensitivity, the darker the corresponding colour.

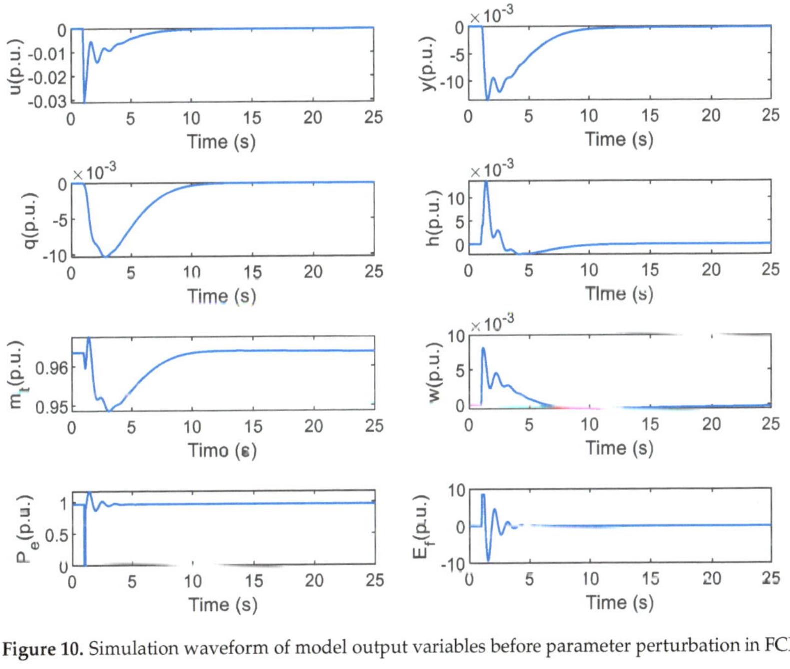

Figure 10. Simulation waveform of model output variables before parameter perturbation in FCM.

It can be seen from Table 3 that the parameters that have the most obvious influence on the system state are not unique and are mainly the controller parameter K_P, generator parameter T_a, and excitation system parameters K_a and T_r. According to the color contrast, the following conclusions can be drawn: (1) the output of the controller and servo-system is sensitive to a change in its own parameter K_P and the generator parameter T_a; (2) in addition to K_P and T_a, the turbine discharge is also sensitive to its own parameter e_{qy}, and the turbine torque is also sensitive to the water system parameter T_w; (3) the water head is sensitive to K_P, T_a, and the turbine parameter e_y to a similar extent; and (4) the rotor speed is most affected by a change in its self-parameter T_a, and it is also highly sensitive to changes in K_P and e_y, while the active power is mainly affected by the excitation system parameters K_a and T_r. Considering the hydraulic–mechanical–electrical interaction, the following conclusions can be drawn: (1) in the FCM, the rotor speed of the unit is sensitive to parameter changes in the hydraulic–mechanical system (i.e., the governor and mechanical part of the hydropower unit). This indicates that the frequency stability of the system in the isolated grid mode is mainly maintained by the mechanical power adjustment, which is also the fundamental difference between the isolated grid operation and the operation connected into a large power grid; (2) in the FCM, the hydraulic–mechanical system is sensitive to its self-parameters, which shows that the state variables related to hydraulic–mechanical system are kept stable mainly by the self-regulation of the unit.

Furthermore, several key system states are analyzed, and the trajectory sensitivity of each state under the action of the three most sensitive parameters is drawn. The results are shown in Figure 11a–c.

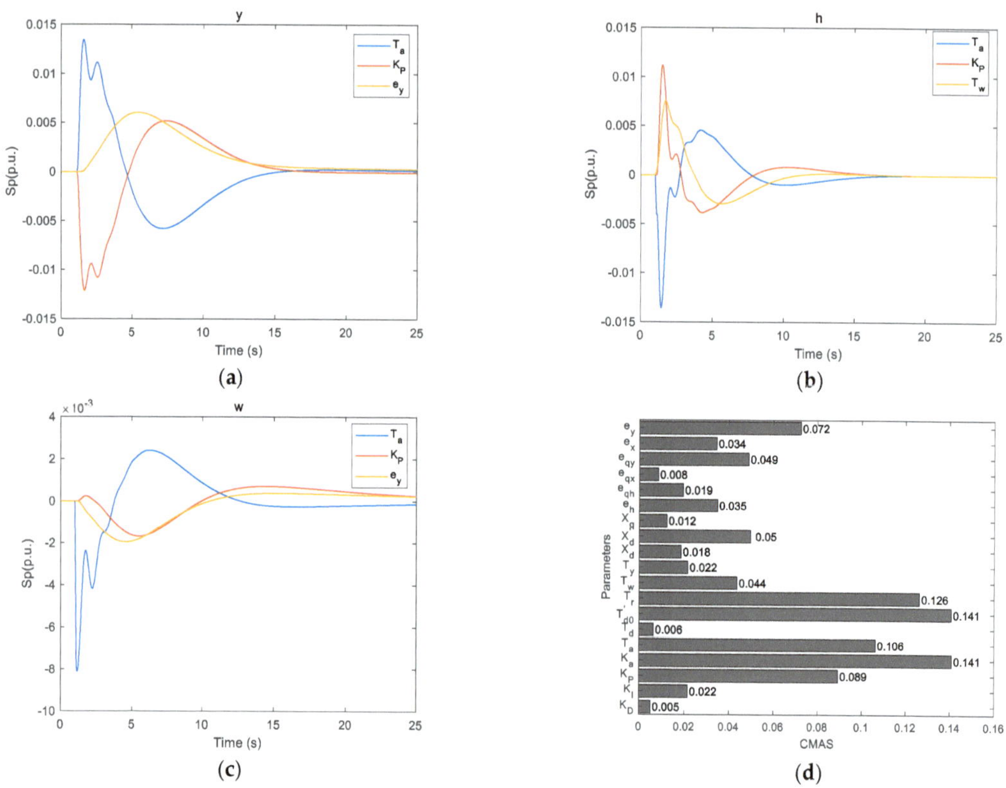

Figure 11. Trajectory sensitivity and comprehensive parameter sensitivity under rated operating condition and FCM. (**a**) Trajectory sensitivity of GVO to parameters. (**b**) Trajectory sensitivity of water head to parameters. (**c**) Trajectory sensitivity of rotor speed to parameters. (**d**) Comprehensive parameter sensitivity.

The GVO is also taken as an example to illustrate. In the FCM, due to the trajectory sensitivity of the GVO to the controller parameters being negative within 1–5 s, an increase in K_P will accelerate the opening and closing speed of the guide vane, which shows that a change in the GVO is positively related to a change in K_P under the two control modes. On the contrary, since the trajectory sensitivity of the GVO to the generator parameter T_a and turbine parameter e_y is positive within 1–5 s, an increase in T_a and e_y will slow down the opening and closing speed of the guide vane. That is, a change in the GVO is negatively correlated with changes in these two parameters. The comprehensive parameter sensitivity indicator of the system to each parameter is shown in Figure 11d. It can be seen that the sensitivity of the system to K_a and T'_{d0} is the highest. The sub-sensitive parameters are mainly T_r, T_a, and K_P. The insensitive parameters are mainly K_D, T_d, and e_{qx}.

4.3.2. Parameter Sensitivity Analysis under FOC

In the FCM, the comprehensive sensitivity of the system to some key parameters under FOC is further considered to analyze its variation with the operating conditions. Similarly, the first four highly sensitive parameters (according to the previous analysis, they are K_a, T'_{d0}, T_r, and T_a) under the rated operating condition are selected for analysis. The parameter with the highest comprehensive sensitivity (i.e., K_a) is used as an example to analyze the variation in the trajectory sensitivity of the turbine torque and rotor speed with the operating conditions. The comprehensive sensitivity for parameters K_a, T'_{d0}, T_r, and T_a under FOC is shown in Figure 12.

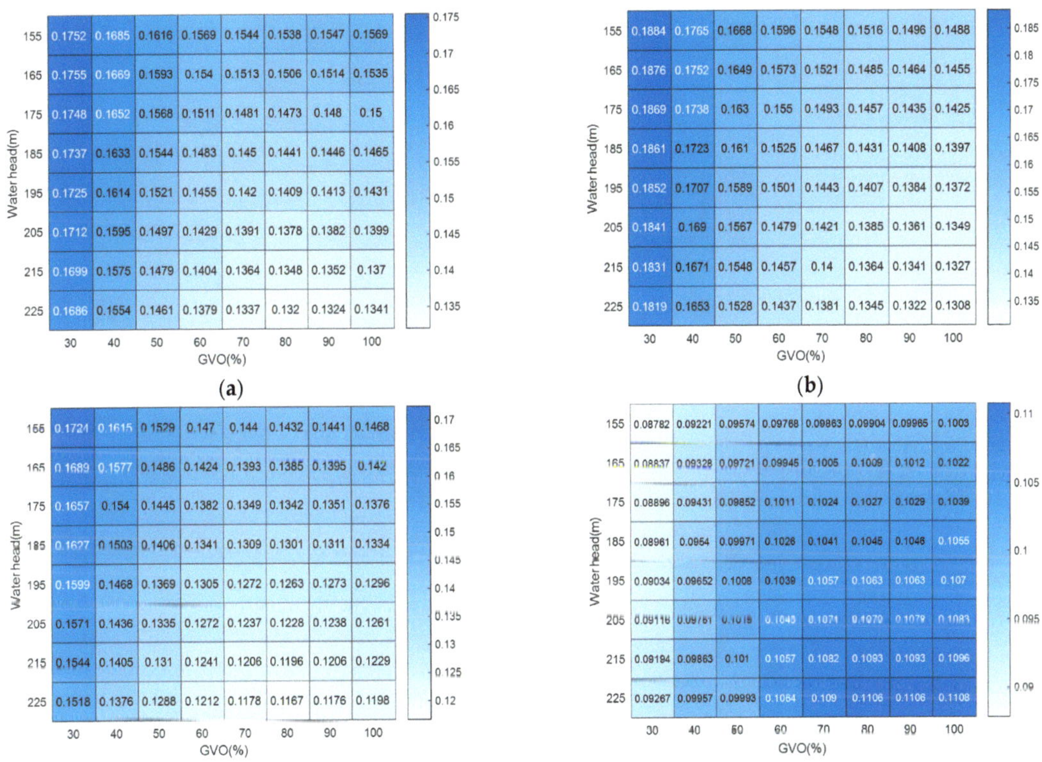

Figure 12. Comprehensive parameter sensitivity under FOC and FCM. (**a**) K_a. (**b**) T'_{d0}. (**c**) T_r. (**d**) T_a. The white font indicates operating conditions with relatively high parameter sensitivity.

It can be seen from Figure 12 that: (1) with an increase in the water head, the sensitivity of the system to parameters K_a, T'_{d0}, and T_r gradually weakens, while its sensitivity to parameter T_a gradually increases; (2) with an increase in the GVO, the sensitivity of the system to parameters K_a, T'_{d0}, and T_r basically shows a decreasing trend, while its sensitivity to parameter T_a gradually increases; and (3) when the GVO is increased from 80% to 100%, the sensitivity of the system to parameters K_a and T_r is enhanced.

Under various typical operating conditions (same as Section 4.2.2), the trajectory sensitivity of the turbine torque and rotor speed to parameter K_a is shown in Figures 13 and 14. It can be seen from Figure 13 that, under the same water head, with an increase in the GVO, the vibration amplitude of the trajectory sensitivity of the turbine torque increases first and then decreases, and the vibration amplitude of the trajectory sensitivity of the rotor speed increases gradually. This phenomenon shows that the sensitivity of the turbine torque to K_a is the strongest when the GVO is about 60%, while the sensitivity of the rotor speed to K_a is the strongest when the GVO is around 100%. As can be seen from Figure 14, with an increase in the water head, the vibration amplitudes of the turbine torque and rotor speed show a gradually increasing trend, indicating that both their sensitivities to K_a increase with an increase in the water head, and the former trend is more obvious.

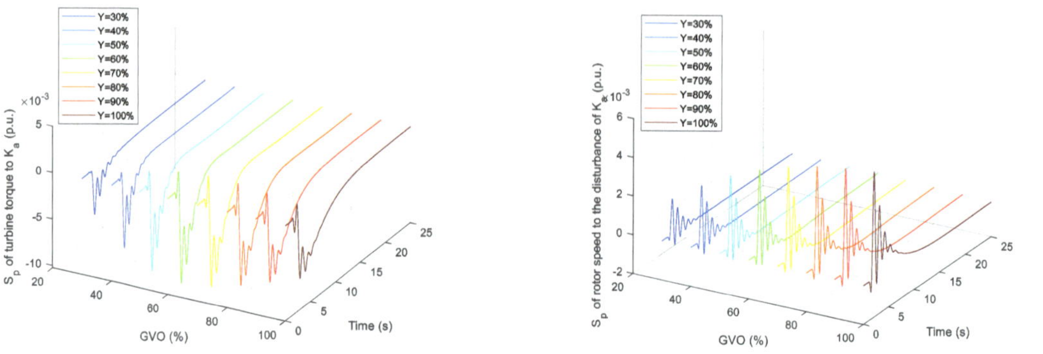

(a) (b)

Figure 13. Trajectory sensitivity of system states to changes in K_a at the water head of 195m (in FCM). (**a**) Turbine torque. (**b**) Rotor speed.

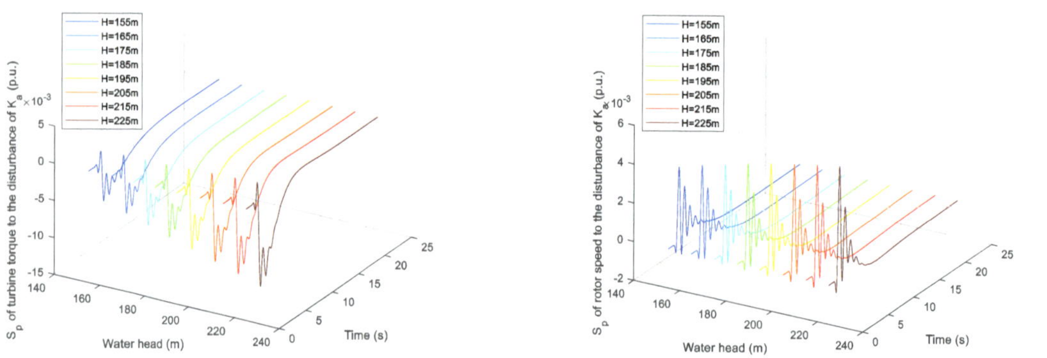

(a) (b)

Figure 14. Trajectory sensitivity of system states to changes in K_a at the GVO of 70% (in FCM). (**a**) Turbine torque. (**b**) Rotor speed.

5. Conclusions

In this paper, the influence of hydro-turbine nonlinearity on the parameter sensitivity of a HTRS was analyzed and discussed in detail under FOC. Taking a unit of the Xiluodu power station as the research object, the parameter sensitivity indicator and analysis framework were proposed, the sensitivity of the system state to parameter changes under the rated operating condition was calculated, the parameters with important influence were identified, and the influence of the operating conditions on the comprehensive parameter sensitivity and trajectory sensitivity of the system state was analyzed. The main conclusions are drawn below:

(1) In the PCM, the unit inertia time constant (i.e., T_a) is the most sensitive parameter of the system, and a small change will significantly affect the dynamic response of the system. At this time, the sensitivity of the system state to this parameter increases significantly under the conditions of a medium–high water head and medium GVO (near the vibration zone) or close to the maximum GVO.
(2) In the FCM, the electrical parameters of the generator and excitation system (i.e., K_a and T'_{d0}) are the most sensitive parameters of the system. The sensitivity of the system state to the two parameters is higher at a small GVO and low water head, which is especially obvious at a small GVO.

Research on the parameter sensitivity of different operating conditions in hydropower stations can help operators to optimize the unit control or operation mode in time, according to changes in the system parameters, and to improve the stability and operating efficiency of these units. In the optimization of the unit control strategy, a parameter sensitivity analysis can be used first to understand to which control parameters the unit transition process is sensitive. When optimizing these control parameters, for control parameters with a high sensitivity, the adjustment speed is as slow as possible and the adjustment amplitude is as small as possible, so as to avoid the unit from entering into a prolonged oscillation or unstable state. Similarly, in the optimization of the unit operation strategy, the operating conditions (i.e., water head or opening) of the unit can be adjusted to make the unit run in conditions that are not sensitive to parameter changes, in order to improve the stability of the system under the premise of meeting the power generation requirements. Although the conclusions drawn in this paper may not be applicable to all power stations, the proposed method of analyzing the sensitivity of parameters for FOC is universal, and the relevant conclusions can be applied to the same type of power stations, which is of great significance for the optimal control and safe and stable operation of actual units.

It is worth noting that the parameters of the system, except for the hydro-turbine, were assumed to be constants in this work. In fact, the parameters of such systems are not static and will change to different degrees with an increase in the usage time or the addition of external foreign matter. For example, the wear and tear of mechanical components in the servo-system, changes in the composition of the water body in the water system, and so on. All these factors may lead to system deterioration and performance degradation. Therefore, for future research work, the focus can be given to the modelling of the relationship between contamination and the system parameters. On this basis, the parameter sensitivity of the system under different contamination conditions can be investigated, so that the results can be closer to the actual situation and a more efficient operation of the unit can be achieved.

Author Contributions: D.L.: Conceptualization, methodology, formal analysis, writing—original draft, supervision, project administration, funding acquisition, resources, software. X.W.: methodology, visualization, formal analysis, writing—original draft. J.Z.: writing—review and editing, investigation, validation, software. X.H.: writing—review and editing, investigation, validation, data curation. L.Z.: writing—review and editing, investigation, validation. All authors have read and agreed to the published version of the manuscript.

Funding: China Postdoctoral Science Foundation (Grant No. 2020M682416).

Data Availability Statement: The data in this paper are available by contacting the corresponding author.

Acknowledgments: The authors acknowledge financial support from the China Postdoctoral Science Foundation (Grant No. 2020M682416) for the research, authorship, and/or publication of this article.

Conflicts of Interest: The authors declare no conflict of interest.

Abbreviations

VRES	Variable Renewable Energy Source
HTRS	Hydro-turbine Regulation System
FCM	Frequency Control Mode
OCM	Opening Control Mode
PCM	Power Control Mode
FOC	Full Operating Conditions
PSA	Parameter sensitivity analysis
NND	Neural Network Derivation
GVO	Guide Vane Opening
PSS	Power System Stabilizer

References

1. Pérez-Díaz, J.I.; Chazarra, M.; García-González, J.; Cavazzini, G.; Stoppato, A. Trends and challenges in the operation of pumped-storage hydro-power plants. *Renew. Sustain. Energy Rev.* **2015**, *44*, 767–784. [CrossRef]
2. Brouwer, A.S.; van den Broek, M.; Seebregts, A.; Faaij, A. Operational flexibility and economics of power plants in future low-carbon power systems. *Appl. Energy* **2015**, *156*, 107–128. [CrossRef]
3. Heard, B.P.; Brook, B.W.; Wigley, T.M.; Bradshaw, C.J. Burden of proof: A comprehensive review of the feasibility of 100% renewa-ble-electricity systems. *Renew. Sustain. Energy Rev.* **2017**, *76*, 1122–1133. [CrossRef]
4. Yang, W.; Norrlund, P.; Saarinen, L.; Witt, A.; Smith, B.; Yang, J.; Lundin, U. Burden on hydropower units for short-term balancing of renewable power systems. *Nat. Commun.* **2018**, *9*, 2633. [CrossRef]
5. Yu, X.; Zhang, J.; Fan, C.; Chen, S. Stability analysis of governor-turbine-hydraulic system by state space method and graph theory. *Energy* **2016**, *114*, 613–622. [CrossRef]
6. Hagihara, S.; Yokota, H.; Goda, K.; Isobe, K. Stability of a Hydraulic Turbine Generating Unit Controlled by P.I.D. Governor. *IEEE Trans. Power Appar. Syst.* **1979**, *98*, 2294–2298. [CrossRef]
7. Ling, D.J.; Tao, Y. An analysis of the Hopf bifurcation in a hydroturbine governing system with saturation. *IEEE Trans. Energy Convers.* **2006**, *21*, 512–515. [CrossRef]
8. Lai, X.; Li, C.; Guo, W.; Xu, Y.; Li, Y. Stability and dynamic characteristics of the nonlinear coupling system of hydropower station and power grid. *Commun. Nonlinear Sci. Numer. Simul.* **2019**, *79*, 104919. [CrossRef]
9. Liu, D.; Li, C.; Malik, O. Operational characteristics and parameter sensitivity analysis of hydropower unit damping under ultra-low frequency oscillations. *Int. J. Electr. Power Energy Syst.* **2022**, *136*, 107689. [CrossRef]
10. Liang, J.; Yuan, X.; Yuan, Y.; Chen, Z.; Li, Y. Nonlinear dynamic analysis and robust controller design for Francis hydraulic turbine regulating system with a straight-tube surge tank. *Mech. Syst. Signal Process.* **2017**, *85*, 927–946. [CrossRef]
11. Yu, X.D.; Yang, X.W.; Zhang, J. Stability analysis of hydro-turbine governing system including surge tanks under interconnected operation during small load disturbance. *Renew. Energy* **2019**, *133*, 1426–1435. [CrossRef]
12. Lu, X.; Li, C.; Liu, D.; Zhu, Z.; Tan, X. Influence of water diversion system topologies and operation scenarios on the damping charac-teristics of hydropower units under ultra-low frequency oscillations. *Energy* **2022**, *239*, 122679. [CrossRef]
13. Lu, X.; Li, C.; Liu, D.; Zhu, Z.; Tan, X.; Xu, R. Comprehensive stability analysis of complex hydropower system under flexible operating conditions based on a fast stability domain solving method. *Energy* **2023**, *274*, 127368. [CrossRef]
14. Wang, F.; Chen, D.; Xu, B.; Zhang, H. Nonlinear dynamics of a novel fractional-order Francis hydro-turbine governing system with time delay. *Chaos Solitons Fractals* **2016**, *91*, 329–338. [CrossRef]
15. Yang, W.; Norrlund, P.; Chung, C.Y.; Yang, J.; Lundin, U. Eigen-analysis of hydraulic-mechanical-electrical coupling mechanism for small signal stability of hydropower plant. *Renew. Energy* **2018**, *115*, 1014–1025. [CrossRef]
16. Guo, W.C.; Peng, Z.Y. Hydropower system operation stability considering the coupling effect of water potential energy in surge tank and power grid. *Renew. Energy* **2019**, *134*, 846–861. [CrossRef]
17. Liu, D.; Li, C.; Malik, O. Nonlinear modeling and multi-scale damping characteristics of hydro-turbine regulation systems under complex variable hydraulic and electrical network structures. *Appl. Energy* **2021**, *293*, 116949. [CrossRef]
18. Wang, Z.; Qin, L.; Zeng, J.; Lin, J.; Yang, J.; Chen, W.; Luo, Y.; Yan, Z. Hydroturbine operating region partitioning based on analyses of unsteady flow field and dynamic response. *Sci. China-Technol. Sci.* **2010**, *53*, 519–528. [CrossRef]
19. Aggidis, G.A.; Židonis, A. Hydro turbine prototype testing and generation of performance curves: Fully automated approach. *Renew. Energy* **2014**, *71*, 433–441. [CrossRef]
20. Yang, W.; Yang, J.; Guo, W.; Zeng, W.; Wang, C.; Saarinen, L.; Norrlund, P. A Mathematical Model and Its Application for Hydro Power Units under Different Operating Conditions. *Energies* **2015**, *8*, 10260–10275. [CrossRef]
21. Giosio, D.R.; Henderson, A.D.; Walker, J.M.; Brandner, P.A. Physics Based Hydraulic Turbine Model for System Dynamics Studies. *IEEE Trans. Power Syst.* **2017**, *32*, 1161–1168. [CrossRef]
22. Li, H.; Chen, D.; Zhang, X.; Wu, Y. Dynamic analysis and modelling of a Francis hydro-energy generation system in the load rejection transient. *IET Renew. Power Gener.* **2016**, *10*, 1140–1148. [CrossRef]

23. Su, W.-T.; Li, X.-B.; Li, F.-C.; Wei, X.-Z.; Han, W.-F.; Liu, S.-H. Experimental Investigation on the Characteristics of Hydrodynamic Stabilities in Francis Hy-droturbine Models. *Adv. Mech. Eng.* **2014**, *6*, 486821. [CrossRef]
24. Zhang, S.F.; Andrews-Speed, P.; Li, S.T. To what extent will China's ongoing electricity market reforms assist the integration of renewable energy? *Energy Policy* **2018**, *114*, 165–172. [CrossRef]
25. Zhang, G.T.; Cheng, Y.C.; Lu, N. Research on Francis Turbine Modeling for Large Disturbance Hydropower Station Transient Process Simulation. *Math. Probl. Eng.* **2015**, *2015*, 971678. [CrossRef]
26. Yuan, X.; Chen, Z.; Yuan, Y.; Huang, Y. Design of fuzzy sliding mode controller for hydraulic turbine regulating system via input state feedback linearization method. *Energy* **2015**, *93*, 173–187. [CrossRef]
27. Zhang, H.; Chen, D.; Wu, C.; Wang, X. Dynamics analysis of the fast-slow hydro-turbine governing system with different time-scale coupling. *Commun. Nonlinear Sci. Numer. Simul.* **2018**, *54*, 136–147. [CrossRef]
28. Chen, Z.; Yuan, X.; Tian, H.; Ji, B. Improved gravitational search algorithm for parameter identification of water turbine regulation system. *Energy Convers. Manag.* **2014**, *78*, 306–315. [CrossRef]
29. Zhang, C.B.; Yang, M.J.; Li, J.Y. Detailed modelling and parameters optimisation analysis on governing system of hydro-turbine generator unit. *IET Gener. Transm. Distrib.* **2018**, *12*, 1045–1051. [CrossRef]
30. Yang, W.; Yang, J.; Guo, W.; Norrlund, P. Frequency Stability of Isolated Hydropower Plant with Surge Tank Under Different Tur-bine Control Modes. *Electr. Power Compon. Syst.* **2015**, *43*, 1707–1716. [CrossRef]
31. Zamora-Cárdenas, E.A.; Fuerte-Esquivel, C.R. Computation of multi-parameter sensitivities of equilibrium points in electric power systems. *Electr. Power Syst. Res.* **2013**, *96*, 246–254. [CrossRef]
32. Xu, B.; Yan, D.; Chen, D.; Gao, X.; Wu, C. Sensitivity analysis of a Pelton hydropower station based on a novel approach of turbine torque. *Energy Convers. Manag.* **2017**, *148*, 785–800. [CrossRef]
33. Xu, B.; Chen, D.; Patelli, E.; Shen, H.; Park, J.H. Mathematical model and parametric uncertainty analysis of a hydraulic generating system. *Renew. Energy* **2019**, *136*, 1217–1230. [CrossRef]
34. Liu, D.; Li, C.; Tan, X.; Lu, X.; Malik, O. Damping characteristics analysis of hydropower units under full operating conditions and control parameters: Accurate quantitative evaluation based on refined models. *Appl. Energy* **2021**, *292*, 116881. [CrossRef]
35. Liu, D.; Wang, X.; Peng, Y.; Zhang, H.; Xiao, Z.; Han, X.; Malik, O. Stability analysis of hydropower units under full operating conditions considering turbine nonlinearity. *Renew. Energy* **2020**, *154*, 723–742. [CrossRef]
36. Yang, W.; Norrlund, P.; Bladh, J.; Yang, J.; Lundin, U. Hydraulic damping mechanism of low frequency oscillations in power systems: Quan-titative analysis using a nonlinear model of hydropower plants. *Appl. Energy* **2018**, *212*, 1138–1152. [CrossRef]
37. Orošnjak, M.; Brkljač, N.; Šević, D.; Čavić, M.; Oros, D.; Penčić, M. From predictive to energy-based maintenance paradigm: Achieving cleaner production through functional-productiveness. *J. Clean. Prod.* **2023**, *408*, 137177. [CrossRef]

sustainability

MDPI

Article

Numerical Study of the Internal Fluid Dynamics of Draft Tube in Seawater Pumped Storage Hydropower Plant

Jianyong Hu [1,2], Qingbo Wang [1,3], Zhenzhu Meng [2,4,*], Hongge Song [1,3], Bowen Chen [5] and Hui Shen [1,6]

1 School of Geomatics and Municipal Engineering, Zhejiang University of Water Resources and Electric Power, Hangzhou 310018, China; hujy@zjweu.edu.cn (J.H.); wqb5113@163.com (Q.W.); shg1069876245@163.com (H.S.); huishen0807@163.com (H.S.)
2 Engineering Research Center of Digital Twin Basin of Zhejiang Province, Hangzhou 310018, China
3 College of Energy and Power Engineering, North China University of Water Resources and Hydropower, Zhengzhou 450045, China
4 School of Water Conservancy & Environment Engineering, Zhejiang University of Water Resources and Electric Power, Hangzhou 310018, China
5 School of Civil and Hydraulic Engineering, Huazhong University of Science and Technology, Wuhan 430000, China; 13083603612@163.com
6 School of Water Resources and Hydropower, Hebei University of Engineering, Handan 056038, China
* Correspondence: mengzhzh@zjweu.edu.cn

Abstract: Pumped storage hydropower plants are renewable energy systems that are effective in saving energy and solving electricity peak-on shortage. Seawater pumped storage hydropower plants are a novel type of pumped storage hydropower plant specifically supplying electric power for ocean islands with the support of solar energy and wind energy. Compared with traditional pumped storage hydropower plants that are constructed on the mainland, seawater pumped storage hydropower plants should take the influence of the complex marine environment, such as extreme waves and winds, into consideration. Taking the characteristics of waves near islands in the East China Sea as an example, we explored the transient hydraulic characteristics in the draft tube of a pump turbine under wave disturbance using a sliding grid interface and the detached eddy simulation (DES) turbulence model. By analyzing the characteristics of unsteady flow in the draft tube, the vortex characteristics under the Q criterion, the frequency characteristics of the pressure pulsation, the evolution law of the internal fluid, and the propagation law of the pressure pulsation were explored. For the situation without wave disturbance, an obvious eccentric vortex in the straight cone section of the draft tube was observed in the case where the opening of the guide vane was small. With the increase in the opening of the guide vane, the eccentric vortex gradually dissipated. For the situation with wave disturbance, the main frequency of the draft tube equaled the frequency of the wave disturbance, the maximum pressure pulsation at the selected monitoring points increased 5 to 15 times, and the superposition of the wave pressure pulsation signals and the draft tube pressure pulsation produced more low-frequency, high-amplitude pressure pulsation signals. Even though the pressure pulsation frequency spectrum varied a lot, the frequency domain of the pressure pulsation without wave disturbance still existed. In addition, the wave disturbance merely varied with the pressure of the draft tube. The influence of wave disturbance on the pressure distribution in the draft tube was relatively small. The results can provide a reference for the operation of seawater pumped storage hydropower plants.

Keywords: wave disturbance; pump turbine; guide vane opening; draft tube; pressure pulsation

Citation: Hu, J.; Wang, Q.; Meng, Z.; Song, H.; Chen, B.; Shen, H. Numerical Study of the Internal Fluid Dynamics of Draft Tube in Seawater Pumped Storage Hydropower Plant. *Sustainability* **2023**, *15*, 8327. https://doi.org/10.3390/su15108327

Academic Editors: Bin Huang, Ling Zhou, Yuquan Zhang and Jianhua Zhang

Received: 22 April 2023
Revised: 15 May 2023
Accepted: 18 May 2023
Published: 19 May 2023

1. Introduction

Pumped storage hydropower plants are renewable energy systems that aim to solve electricity peak-on shortage problems. Seawater pumped storage hydropower plants are a novel type of pumped storage hydropower plant that are constructed with the support of solar energy and wind energy. Seawater pumped storage hydropower plants are used to supply electric power for ocean islands, which are difficult to connect with mainland electricity networks. Under the background of China's efforts to achieve the goal of "peak carbon dioxide emissions and carbon neutrality", the development of the pumped storage industry has ushered in a new stage. Seawater pumped storage hydropower plants are a novel type of pumped storage hydropower plant, which can not only form a stable power supply system with wind energy and solar energy but also solve the fresh water supply problem of ocean islands. According to the latest census reports of seawater pumped storage hydropower plant data released by the Chinese National Energy Administration, 238 points that offer the possibility of constructing seawater pumped storage hydropower plants have been discovered. The potential total capacity could reach 42.083 million kilowatts [1,2]. Similar to conventional pumped storage units, seawater pumped storage units undertake the task of load and frequency control. As the hydraulic unit output power varies frequently, engineers often change the guide vane opening to adjust the volume of flow to adjust the output power [3,4]. However, the difference between seawater pumped storage hydropower plants and conventional pumped storage hydropower plants is that the lower reservoir of seawater pumped storage is located in the marine environment. Many complicated influencing factors including typhoons, tides, and ocean currents may induce rapid changes in the water level, which may increase the operating range of the unit head and accordingly increase the possibility of hydraulic instability in the pump turbine [5–7].

When the pump turbine deviates from the optimal working condition, the draft tube pressure pulsation is one of the main factors affecting the stable operation of the pump turbine. The pressure pulsation of the draft tube is mainly induced by two factors: one is the dynamic and static interference of the runner; the other is the secondary flow in the draft tube. The pressure pulsation caused by the dynamic and static interference mainly includes the frequency conversion, blade frequency, and double frequency of the blade frequency. The pressure pulsation resulting from the secondary flow has the characteristics of a low frequency and high amplitude [8–12]. Previous studies mainly provided insights into the internal dynamics of draft tubes considering different types of units and levels of opening of the guide vanes. Taking a typical partial load condition of a Francis turbine as an example, Zhou et al. simulated the velocity field of the vortex rope in a draft tube. They also predicted the frequency of pressure pulsation in the vortex rope and determined the pressure pulsation on several different sections of the draft tube [13]. Ji et al. simulated the unsteady flow in the draft tube of a Francis turbine and found that a vortex rope with a low pressure and large flow rate may lead to a low-pressure pulsation in the turbine [14]. Gao et al. explored the internal flow dynamics of a tidal turbine with an indicated guide vane opening of the tubular unit and proposed that the rotating flow in the draft tube under low pressure is mainly caused by the air mass and hydraulic imbalance at the outlet of the runner; thus, the rotating flow cannot be simply defined as the vortex rope [15]. Ji et al. pointed out that the large and invisible vortex in the conical cross-section at the inlet of the draft tube gradually changes to a tangible vortex rope with an increase in the guide vane opening [16]. Wang and Cavazzini simulated the pressure fluctuation in the S-shaped region of a pump turbine and captured the low-frequency component caused by the rotating stall phenomenon at a low flow rate, which was induced by the rotating stall phenomenon [17,18]. By analyzing the pressure difference between the inner wall of the draft tube center and the inner wall, Yang et al. determined the relation between the pressure pulsation values of the inner wall and that of the center wall of a draft tube [19]. Guo et al. found that the smaller the opening of the guide vane, the more disordered the vortex generated by each part of the turbine, and the more unstable the flow field. The original shape of the vortex rope in the draft tube is disturbed due to the interaction between vortices, which affects the efficient and stable operation of the turbine [20]. Chen et al. analyzed the influence of upstream disturbance on the draft tube flow of a Francis turbine under partial load conditions using different vortex identification methods [21].

The results indicated that when the guide vane opening is small, the upstream disturbance propagates toward the downstream, which has a significant impact on the flow pattern of the downstream draft tube. When the opening approaches its optimal status, the upstream disturbance becomes weak, and thus the associated influence on the downstream flow pattern is fairly limited.

Numerous numerical simulations of the fluid dynamics of steady and unsteady flows of conventional pumped storage units have been conducted [22,23]. However, studies on wave disturbance in the upstream and downstream of these units have mostly focused on low-head tubular turbines whose water head and output are relatively small [24,25]. Different from the low-head and small-capacity tubular turbines that are commonly installed in conventional pumped storage hydropower plants, seawater pumped storage hydropower plants commonly adopt a vertical-shaft Francis turbine with a high head and large capacity, which has a high possibility of becoming hydraulically unstable when facing the disturbance of periodic waves. Therefore, this study aimed to reveal the influence of the disturbance of periodic waves on a pump turbine using numerical simulation, by exploring the evolution law of the flow field and the propagation law of the pressure pulsation in the draft tube. The laws and mechanisms obtained from the results can provide a reference for maintaining the operating stability of seawater pumped storage units.

2. Simulation Configuration

2.1. Model Parameters and Computing Domain

Using Fluent, we conducted a numerical simulation of a three-dimensional full channel of a pump turbine, which mainly includes five areas: spiral case, stay vanes, guide vanes, runner, and draft tube (see Figure 1). In order to accurately capture the characteristics and propagation law of the pressure pulsation in the draft tube, 9 monitoring points were arranged at every 0.1 m from the center of the draft tube to the bend section, which were labeled P1, P2, P3, P4, P5, P6, P7, P8, and P9. The initial settings of the parameters of the modeled pump turbine under the condition of a hydraulic turbine are shown in Table 1. The calculation was carried out using a rack server with 64-bit high-performance processors, a 36-core single-threaded Intel Xeon Gold 6140 CPU, and a memory of 512 GB.

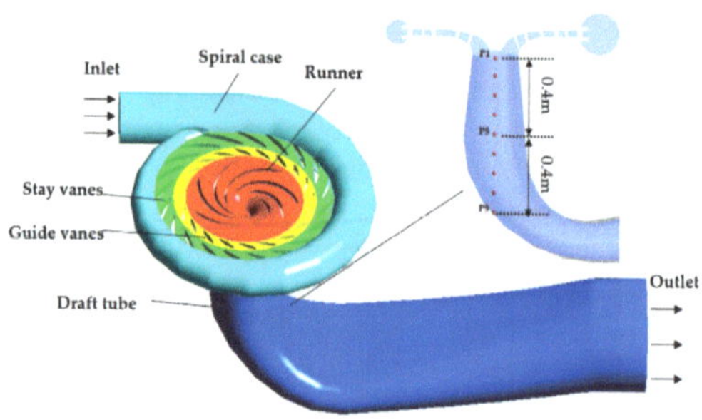

Figure 1. Three-dimensional fluid domain model of the pump turbine.

Table 1. Parameters of the defined pump turbine.

Parameters	Initial Settings
Runner inlet diameter D_1	552.2 mm
Runner outlet diameter D_2	250 mm
Number of runner blades	7
Number of stay vanes	20
Number of guide vanes	20
Rated head	32 m
Rated speed	1000/min

2.2. *Grid Division*

In this study, an unstructured grid was used to divide the flow components. The grid division was conducted using Fluent Meshing. The detached eddy simulation (DES) turbulence model was adopted, in which the grid was divided into a viscous bottom layer, and the boundary of the viscous bottom layer was considered to be y+ less than 10. To improve the accuracy of the calculation, the boundary layer of the straight cone section of the runner and draft tube was locally encrypted to ensure that its average y+ was less than 10. As shown in Figure 2, we selected the steady-state calculation of three-dimensional turbulence under the rated conditions and verified the grid independence based on the parameters of efficiency. In the case of the number of grids larger than 4.5 million, the improvement in efficiency was neglectable with the increase in the grid number. We thus defined 5 million grids in the present case study.

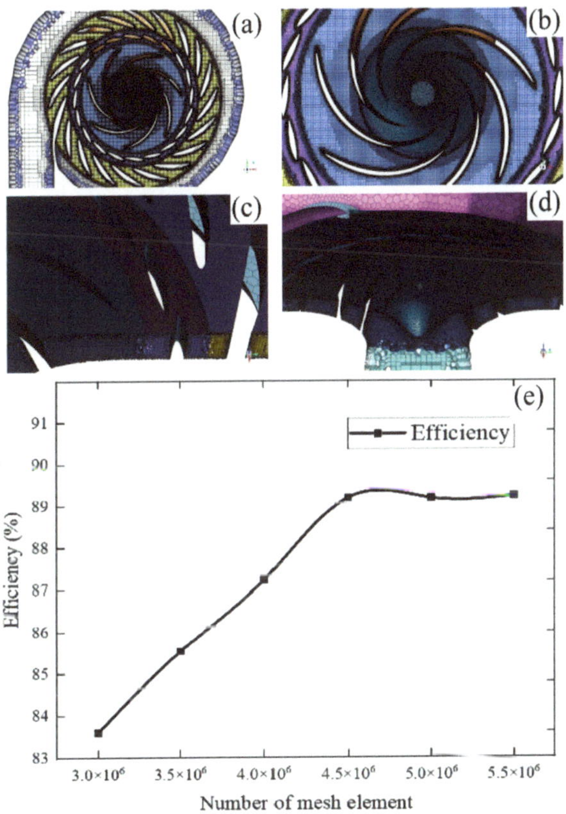

Figure 2. Grid division and grid independence verification: (**a**) the local grid division of spiral case cross section, (**b**) the local grid division of runner cross section, (**c**) the local grid division of stay vanes and guide vanes, (**d**) the local grid division of runner longitudinal section, (**e**) verification of grid number independence.

2.3. *Turbulence Model and Boundary Conditions*

As a type of rotating fluid machinery, pump turbines demonstrate the phenomenon of flow separation in the runner area. Previous studies have demonstrated the high accuracy and stability of the detached eddy simulation (DES) turbulence model in solving large separation problems of fluid machinery [18]. The DES model has the advantages of both the large eddy simulation (LES) model and the Reynolds-averaged Navier–Stokes (RANS) model; thus, it can solve both large eddy structures far away from the wall using the LES model and the turbulent flow near the wall using the RANS model. Thus, the DES turbulence model was used in the present numerical simulation. The numerical model of unsteady flow was initially established and modified on the basis of the governing equations of steady flow. The time-averaged continuity equation and the Reynolds equation of turbulent flow are as follows:

$$\frac{\partial \rho}{\partial t} + \frac{\partial}{\partial x_i}(\rho u_i) = 0 \tag{1}$$

$$\frac{\partial}{\partial t}(\rho u_i) + \frac{\partial}{\partial x_j}(\rho u_i u_j) = -\frac{\partial p}{\partial x_i} + \frac{\partial}{\partial x_j}\{\mu \frac{\partial u_i}{\partial x_j} - \rho \overline{u'_i u'_j}\} + S_i \tag{2}$$

$$\frac{\partial}{\partial t}(\rho \varphi) + \frac{\partial}{\partial x_j}(\rho u_j \varphi) = \frac{\partial}{\partial x_j}\{\Gamma \frac{\partial \varphi}{\partial x_j} - \rho \overline{u'_j \varphi}\} + S \tag{3}$$

$$\frac{\partial}{\partial t}(\rho k) + \frac{\partial}{\partial x_j}(\rho k u_i) = \frac{\partial}{\partial x_j}[(\mu + \frac{\mu_t}{\sigma_k})\frac{\partial k}{\partial x_j}] + u_t(\frac{\partial u_i}{\partial x_j} + \frac{\partial u_j}{\partial x_i})\frac{\partial u_i}{\partial x_j} - \rho C_D \frac{k^{3/2}}{l} \tag{4}$$

where ρ is the fluid density, u is the fluid velocity, μ is the fluid dynamic viscosity, Γ is the fluid parameters, and S is the source term.

For the boundary conditions of the numerical model, the spiral case inlet was set as the mass flow inlet, and the draft tube outlet was set as the pressure outlet. The SIMPLEC algorithm was used to couple the pressure and velocity. The standard discrete format was used to describe the pressure term. The second-order upwind format was used to describe the momentum and turbulent kinetic energy dissipation rate. The solid wall was set to the condition of a no-slip wall, and the convergence accuracy was set to 10^{-5}. The calculation for a steady flow was firstly carried out. Then, according to the obtained stable macro parameters, the result of the steady-flow calculation was taken as the initial condition of the unsteady-flow calculation. It was verified in the calculation that the convergence accuracy could be met in 30 steps once the iterative convergence of the unsteady-flow calculation was stable. In order to ensure the numerical accuracy and save calculation time, the maximum number of iterations in each time step was set to 30, and the time step was set to 0.0005 s (rotate 3° for each time step). The total calculation time was 10 s with 166 cycles.

2.4. Selection of Working Conditions at Calculation Points

In this paper, the flow field in the first quadrant of the pump turbine was studied. The selected working points are shown in Table 2. The water head in the table was 32 m. The wave disturbance at the outlet of the draft tube used the wave data of an island in the East China Sea from the last ten years. In line with Yu et al. [26], the waveform was a sine wave, the maximum effective wave height was 10 m, and the period was 6 s. According to micro-amplitude wave theory [27], the expression of pressure at any point in the wave field can be expressed as

$$p = \rho g(\eta K_Z - Z) \tag{5}$$

where p is the pressure, ρ is the fluid density, η is the instantaneous vertical distance of the free surface above the average water surface, K_z is the pressure sensitivity coefficient, which decreases with the increase in the distance between the particle position and the still water surface, and z is the elevation of the free surface.

Table 2. Working conditions at the calculation points.

Working Condition	Relative Opening of Guide Vane is α_0 (Degrees)	Unit Speed n_{11} (r/min)	Unit Flow Q_{11} (m^3/s)	Outlet Pressure, Pout (Pa), of Draft Tube
Hydraulic turbine operating condition (Without wave disturbance)	6	40	0.19542	0
	12	40	0.41603	0
	24	40	0.65730	0
Hydraulic turbine operating condition (wave disturbance)	6	40	0.19542	10,000 sin $\frac{\sqrt{10}\pi t}{3}$
	12	40	0.41603	10,000 sin $\frac{\sqrt{10}\pi t}{3}$
	24	40	0.65730	10,000 sin $\frac{\sqrt{10}\pi t}{3}$

Therefore, the variation in hydrostatic pressure and hydrodynamic pressure can be regarded as a sine or cosine function. For small-amplitude waves, we assume that the hydrostatic pressure and hydrodynamic pressure are related to the water depth. According to [28], the formulas for the wave scale are as follows:

$$\lambda_{\mathrm{p}} = \lambda_k = \frac{l_p}{l_m} \tag{6}$$

$$\lambda_{\mathrm{f}} = {\lambda_k}^{-\frac{1}{2}} \tag{7}$$

where λ_k is the model length scale, l_p is the prototype length, l_m is the model length, λ_p is the pressure scale, and λ_f is the frequency scale.

In this paper, the ratio between the prototype and real pump turbine was one to ten. The downstream water level fluctuation was $\Delta H = 1 \times \sin\frac{2\pi t}{\mathrm{T}}$. The draft tube outlet pressure fluctuation $\Delta P = 10{,}000 \sin\frac{\sqrt{10}\pi t}{3}$ was taken as the wave disturbance condition. Without wave disturbance, the data points in 10 rotating periods (0.6 s) of the runner were analyzed. With wave disturbance, the data points in a sinusoidal period (about 1.9 s) of the wave were analyzed.

3. Validation of Simulation Method

In order to obtain the evolution of the pressure pulsation in the draft tube of the model pump turbine, three operating points of the guide vane opening were simulated. Figure 3 compares the experimental results and numerical simulation results for the case of no wave disturbance. The unit rotating speed and unit discharge are given by Equations (8) and (9), respectively [29]:

$$n_{11} = \frac{nD_2}{\sqrt{H}} \tag{8}$$

$$Q_{11} = \frac{Q}{{D_2}^2\sqrt{H}} \tag{9}$$

where n_{11} is the unit rotating speed, Q_{11} is the unit discharge, D_2 is the diameter of the runner outlet, and H is the hydraulic turbine head.

It can be seen that the simulation results fit fairly well with the experimental results. The errors of both the unit speed n_{11} and unit flow Q_{11} are within 3%. This indicates that the numerical simulation method used in this study is effective and reliable.

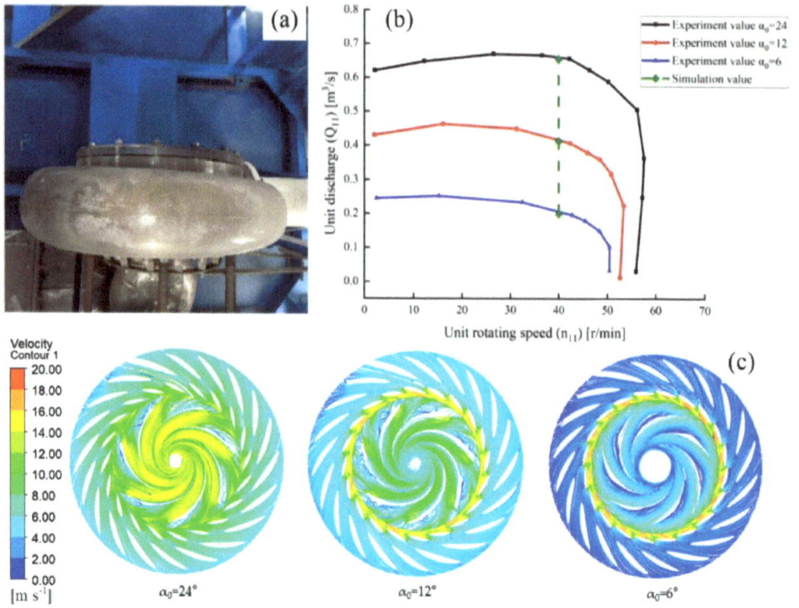

Figure 3. Comparison of simulated and experimental data: (**a**) the testbed of the Francis pump turbine; (**b**) the variation in the unit discharge against the unit rotating speed; (**c**) the velocity field of the pump turbine with different guide vane openings.

4. Results and Discussion

4.1. Pressure and Velocity Distribution in Draft Tube

As shown in Figure 4, for several operating conditions without wave disturbance, a large range of negative pressure is observed in the central area of the straight cone section of the draft tube. At $\alpha_0 = 6°$ and $\alpha_0 = 12°$, the fluid speed of the draft tube inlet is unbalanced. In addition, an obvious pressure gradient is observed in the central area of the straight cone section of the draft tube, and the highest flow rate is 13.4 m/s at $\alpha_0 = 6°$. With the increase in the flow rate, the circumferential velocity component produced by the outlet speed of the runner decreases at $\alpha_0 = 24°$, with the maximum flow rate reducing to 6.66 m/s. With the increase in the loss of steady water, the pressure gradient at the inlet of the draft tube disappears, the negative pressure area in the draft tube decreases, and the maximum pressure appears at the elbow of the draft tube.

Figure 5 shows the static pressure distribution on the axial surface of the draft tube in a wave cycle under the running condition of $\alpha_0 = 6°$. Due to the influence of wave disturbance, the pressure of the indicated section varies a lot at different times, with the maximum pressure at T/4 and the minimum pressure at 3T/4. Due to the combined effect of the swirl at the runner outlet and the low flow rate with a small opening, there is still a low-pressure zone in the center of the straight cone section of the draft tube, and the pressure value changes in line with the sinusoidal variation law of wave disturbance. However, the pressure distribution law of the indicated section is generally the same as that without wave disturbance, and the effect induced by wave disturbance is relatively small.

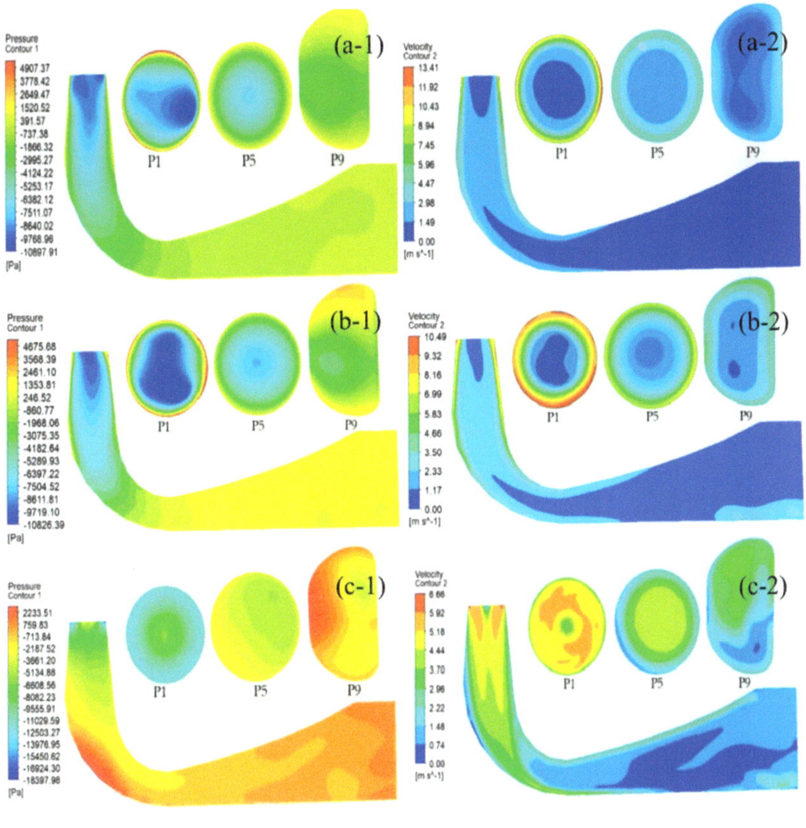

Figure 4. Static pressure and velocity distribution of draft tube with different opening without wave disturbance: (**a-1**) static pressure distribution of draft tube with $\alpha_0 = 6°$, (**a-2**) velocity distribution of draft tube with $\alpha_0 = 6°$, (**b-1**) static pressure distribution of draft tube with $\alpha_0 = 12°$, (**b-2**) velocity distribution of draft tube with $\alpha_0 = 12°$, (**c-1**) static pressure distribution of draft tube with $\alpha_0 = 24°$, (**c-2**) velocity distribution of draft tube with $\alpha_0 = 24°$.

As shown in Figure 6, under the running condition of $\alpha_0 = 12°$, with the increase in the opening of the guide vane, the inlet velocity of the draft tube gradually becomes even and stable, and the pressure gradient in the central area of the straight cone section slows down. Similar to the running condition of $\alpha_0 = 6°$, the maximum pressure of the draft tube appears in the area near the side walls of the draft tube inlet. However, a new vortex area appears at the end of the draft tube diffusion section. At the same time, the pressure value is smaller than that at $\alpha_0 = 6°$, the variation in pressure follows the sinusoidal variation law of wave disturbance, and the pressure distribution law affected by the wave disturbance is relatively small.

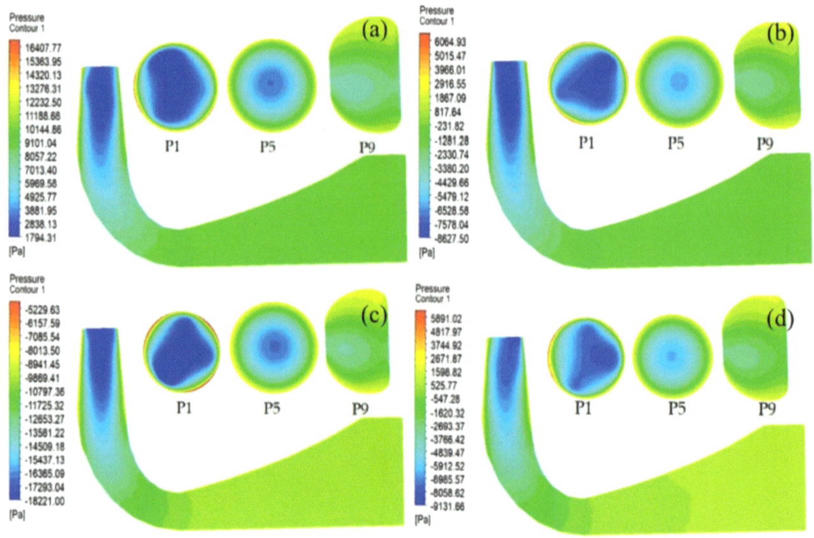

Figure 5. Static pressure distribution of the draft tube with α_0 = 6° under wave disturbance: (**a**) T/4, (**b**) T/2, (**c**) 3T/4, and (**d**) T.

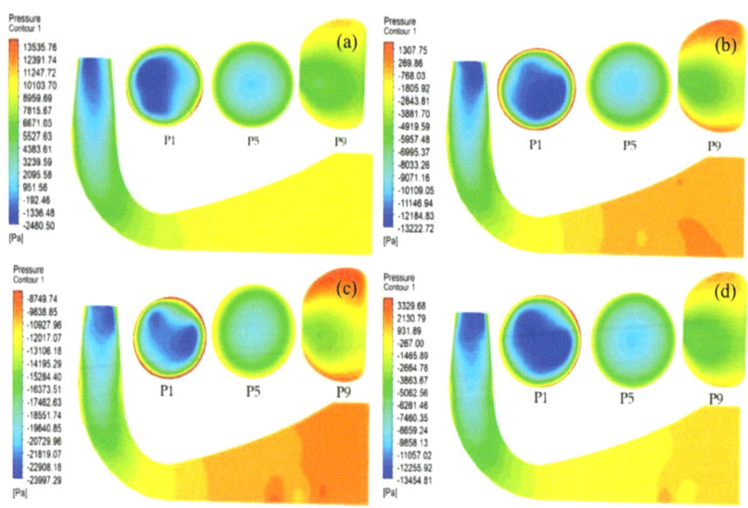

Figure 6. Static pressure distribution of the draft tube with α_0 = 12° under wave disturbance: (**a**) T/4, (**b**) T/2, (**c**) 3T/4, and (**d**) T.

As shown in Figure 7, for the situation with wave disturbance, under the running condition of α_0 = 24°, with the increase in the opening of the guide vane, there is almost no pressure gradient in the center of the straight cone section of the draft tube. At the inlet of the draft tube, due to the large water flow rate and the elbow blocks of the water flow, a local high-pressure area appears at the elbow of the draft tube. In other areas, the pressure increases gradually from the inlet of the draft tube to the outlet of the draft tube. The variation in the pressure fits well with the sinusoidal variation law, and the pressure distribution law affected by the wave disturbance is relatively small. In the cloud chart of the pressure in the draft tube, an obvious pressure gradient can be observed in the center of the straight cone section of the draft tube under the conditions of $\alpha_0 = 6°$ and $\alpha_0 = 12°$. In order to further explore the influence of wave disturbance on the draft tube pressure pulsation, it is necessary to analyze the pressure pulsation of each monitoring point in the frequency domain.

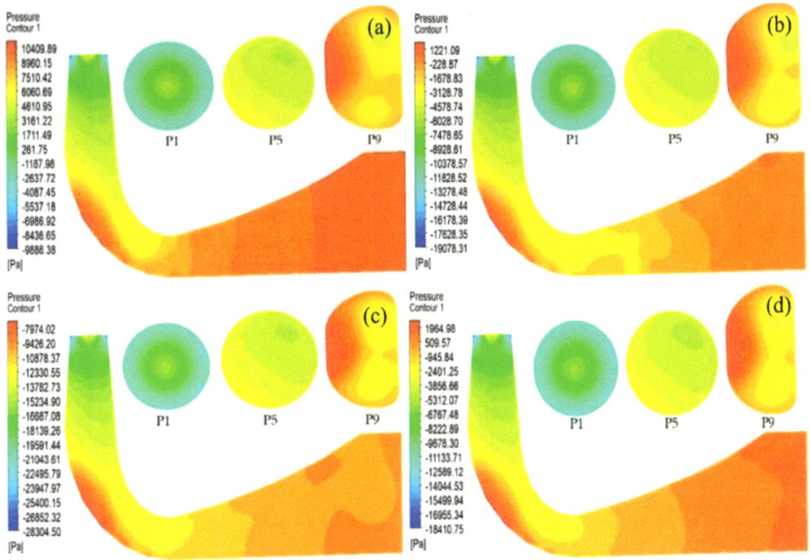

Figure 7. Static pressure distribution of the draft tube with $\alpha_0 = 24°$ under wave disturbance: (**a**) T/4, (**b**) T/2, (**c**) 3T/4, and (**d**) T.

4.2. The Peak Value of Pressure Pulsation

In order to describe the influence of the guide vane opening and wave disturbance on the draft tube pressure pulsation more intuitively and quantify the main characteristics of the draft tube pressure pulsation, the pressure pulsation coefficient peak $\Delta H'$ and pressure pulsation coefficient C_p are introduced [30]:

$$\Delta H' = \frac{\Delta H}{H} = \frac{P_{\text{imax}} - P_{\text{imin}}}{\rho g \text{H}} \times 100\% \tag{10}$$

$$\text{C}_\text{p} = \frac{P_\text{i} - \overline{P}}{\rho g \text{H}} \times 100\% \tag{11}$$

where ΔH is the peak value of the pressure pulsation, H is the water head, P_{imax} and P_{imin} are the corresponding maximum and minimum pressures at point i, C_p is the dimensionless pressure pulsation coefficient, P_i is the corresponding pressure at point i, and $\overline{P}$ is the average pressure over time.

In addition, the Q criterion is adopted to identify the transient vortex structure [31], and the definition of this criterion is shown in Equations (12)–(14).

The governing equation of the velocity gradient tensor is as follows:

$$\lambda^3 - P\lambda^2 + Q\lambda - R = 0 \tag{12}$$

where λ is the eigenvalue of the characteristic equation, and P, Q, and R are the three invariants of the velocity gradient tensor. Q is the second invariant, defined as

$$Q = \frac{1}{2}(\Omega^2 - S^2) \tag{13}$$

where Ω is the vorticity tensor, and S is the strain rate tensor. $Q > 0$ indicates that vorticity is dominant in the flow field region, and $Q < 0$ indicates that the strain rate or viscous stress is dominant in the flow field region. In the Cartesian coordinate system, Q can be written as

$$Q = -\frac{1}{2}\left[\left(\frac{\partial u}{\partial x}\right)^2 + \left(\frac{\partial v}{\partial y}\right)^2 + \left(\frac{\partial w}{\partial z}\right)^2\right] - \left(\frac{\partial u}{\partial y}\right)\left(\frac{\partial v}{\partial x}\right) - \left(\frac{\partial u}{\partial z}\right)\left(\frac{\partial w}{\partial x}\right) - \left(\frac{\partial v}{\partial z}\right)\left(\frac{\partial w}{\partial y}\right) \tag{14}$$

It can be seen from Figure 8a that when there is no wave disturbance, the amplitude of the pressure pulsation at each monitoring point decreases with the decrease in the opening. When the opening of the guide vane is $\alpha_0 = 24°$, the maximum pressure pulsation coefficient of the draft tube is obviously higher than that of $\alpha_0 = 6°$ and $\alpha_0 = 12°$. In addition, the difference between the maximum pressure pulsation coefficient at $\alpha_0 = 6°$ and that at $\alpha_0 = 12°$ is relatively small. The maximum of the pressure coefficient pulsation decreases slowly along the center line of the monitoring points in the straight cone section, which means that the pressure pulsation propagation has a small attenuation in the straight cone section of the draft tube. The maximum pressure pulsation coefficient is 1.22% at P1 under $\alpha_0 = 24°$, and the minimum pressure pulsation coefficient is 0.36% at P9 under $\alpha_0 = 6°$. It can be seen from the Q criterion vortex distribution diagram that, due to the combined effect of the swirl at the runner outlet and the low flow rate with a small opening, the vortex is obvious in the center of the straight cone section of the draft tube both at $\alpha_0 = 6°$ and $\alpha_0 = 12°$. The influence range of the eccentric vortex is about 0.85 D_2 at $\alpha_0 = 6°$, and about 0.65 D_2 at $\alpha_0 = 12°$. The vortex dissipates at $\alpha_0 = 24°$.

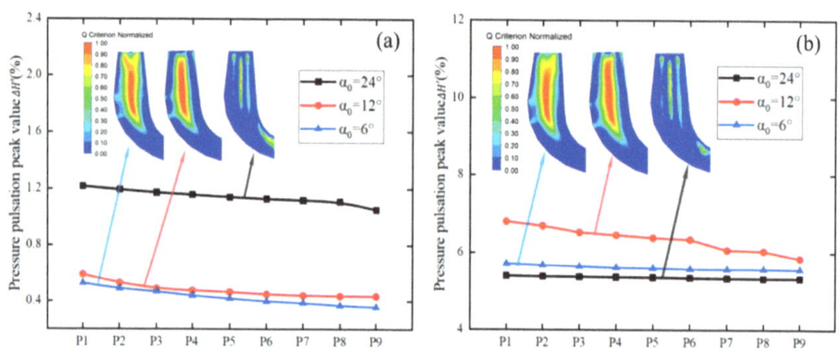

Figure 8. The maxima of pressure pulsation coefficient of monitoring points of the draft tube with different opening: (**a**) without wave disturbance, (**b**) wave disturbance.

It can be seen from Figure 8b that the wave disturbance has a negligible influence on the vortex distribution in the straight cone section of the draft tube; however, it has a significant influence on the amplitude of the pressure pulsation at each monitoring point. Under the running condition of $\alpha_0 = 12°$, the peak value of the maximum pressure pulsation coefficient at P1 is 6.82%. Under the running condition of $\alpha_0 = 24°$, the peak value of the minimum pressure pulsation coefficient at P9 is 5.33%. This means that the draft tube, as an important overcurrent component in the pump turbine, is too strong. In particular, the peak value of the maximum pulse coefficient with wave disturbance is more than five times larger than that without disturbance, which is a potential safety problem for the safe operation of the pump turbine.

4.3. Characteristics of Draft Tube Pressure Pulsation

Figure 9a,c,e show the frequency-domain results of the pressure pulsation at each measuring point without wave disturbance, where f is the frequency, and f_n is the frequency conversion of the runner. It can be seen that when the opening is small, the pressure amplitude is low, due to the low flow rate. The internal flow pattern of the pump turbine is complicated in the case where the pump is running under a small opening; thus, there are more vortices and backflows, which may lead to more low-frequency components in the pressure pulsation under a small opening. When the opening of the guide vane reaches $\alpha_0 = 24°$, the vortex gradually dissipates, but the pressure amplitude still increases with the increase in the flow rate. There is a low-frequency pulsation of 1.95 Hz in the central basin of the straight cone section of the draft tube with different opening degrees, which is the main frequency of the draft tube pressure pulsation. Overall, the pressure coefficient C_p increases obviously with the increase in the guide vane opening. In addition, under the same opening degree, the frequency components of all monitoring points are basically the same.

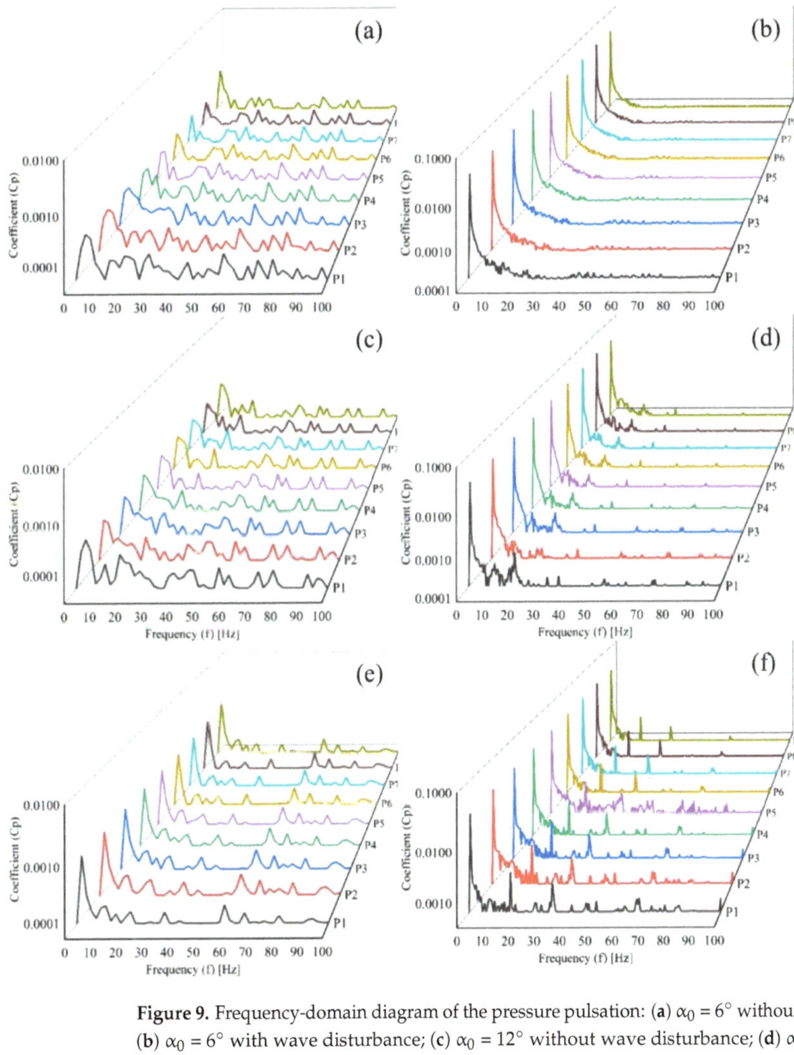

Figure 9. Frequency-domain diagram of the pressure pulsation: (**a**) $\alpha_0 = 6°$ without wave disturbance; (**b**) $\alpha_0 = 6°$ with wave disturbance; (**c**) $\alpha_0 = 12°$ without wave disturbance; (**d**) $\alpha_0 = 12°$ with wave disturbance; (**e**) $\alpha_0 = 24°$ without wave disturbance; (**f**) $\alpha_0 = 24°$ with wave disturbance.

Figure 9b,d,f show the frequency domain of the pressure pulsation at different measuring points for the situation with wave disturbance. Compared to the situation without wave disturbance, the pressure amplitude increases sharply, and the amplitude of the original main frequency of 1.95 Hz at each monitoring point is no longer obvious. It is replaced by a low-frequency pulsation with a frequency of 0.5 Hz, which equals the frequency of the wave disturbance. The frequency spectrum of the pressure pulsation under each opening is different, and no significant regularity is observed. The frequency of the pressure pulsation under each opening changes, but the characteristics of the low frequency and high amplitude of the draft tube pressure pulsation remain.

4.4. Classification of Major Frequencies

As can be seen from Figure 10a,c,e, for the situation with no wave disturbance, with the increase in the guide vane opening, the overall pressure amplitude in the low-frequency region increases, and the pressure amplitude is 0.11 f_n under all running conditions. As a similar unsteady pressure phenomenon, Yu et al. also confirmed that the main frequency of the pressure pulsation in the draft tube of a pump turbine unit was about 0.1 f_n [32]. Under the running condition of $\alpha_0 = 6°$, the frequency corresponding to the maximum pressure amplitude of many upstream monitoring points becomes 0.23 f_n. Under the running condition of $\alpha_0 = 12°$, although the frequency corresponding to the maximum pressure amplitude of P1 is only 0.23 f_n, other monitoring points also have higher amplitudes at 0.23 f_n. When $\alpha_0 = 24°$, the pressure amplitude corresponding to 0.23 f_n of each monitoring point is obviously smaller. According to [21], the rotating frequency of the vortex rope of a pump turbine is generally 1/3~1/5 of the rotating frequency, which is the main frequency of the tail race vortex rope. The rotating frequency of the runner is 1.0 f_n, which affects the whole draft tube. There is no change in the amplitude of the rotating frequency under each opening. With the increase in the guide vane opening, the amplitude of the pressure pulsation of the blade frequency of 7.0 f_n decreases gradually. Both the rotating frequency and the blade frequency come from the dynamic and static interference between the runner and the guide vane. For other frequencies between 3.5 f_n and 2.0 f_n, these frequencies generally come from the vortex and secondary flow, and their corresponding amplitudes are small.

As can be seen from Figure 10b,d,f, when there is wave disturbance and the guide vane opening is $\alpha_0 = 6°$ and $\alpha_0 = 12°$, there is still a high pressure amplitude at the main frequency of 0.23 f_n of the vortex rope. With the increase in the opening of the guide vane, the high-amplitude pressure pulsation of the main frequency of 0.23 f_n of the vortex rope gradually disappears. Because the monitoring point P1 is close to the runner, it is greatly affected by the dynamic and static interference, and has a high pressure amplitude at the rotating frequency of 1.0 f_n of the runner. Overall, the maximum pressure amplitude of each monitoring point generally increases by 5~15 times, and the main frequency of the pressure pulsation is close to 0.5 Hz, which is the frequency of wave disturbance itself. In addition, there are several 0.5 Hz frequency-doubled, high-amplitude pressure pulsations, which indicates that the pressure waves generated by the waves propagate from the downstream side to the inside of the draft tube, and the superposition of the pressure pulsation signals produces more low-frequency, high-amplitude pressure pulsations.

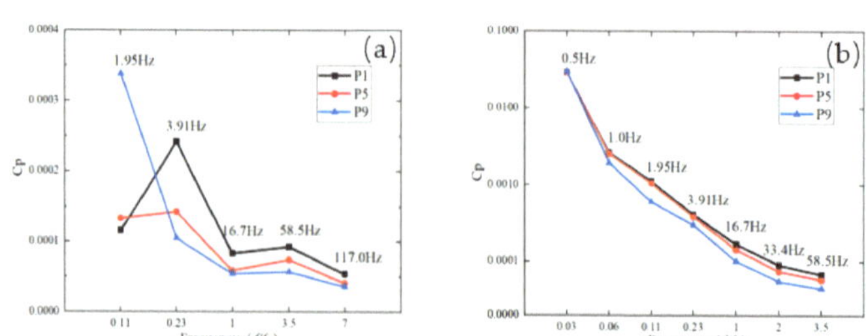

Figure 10. *Cont.*

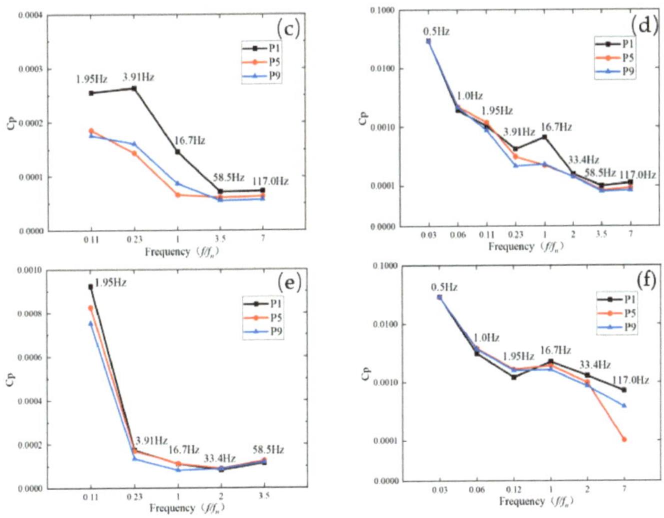

Figure 10. Main frequency types under different guide vane openings: (**a**) α_0 = 6° without wave disturbance; (**b**) α_0 = 6° with wave disturbance; (**c**) α_0 = 12° without wave disturbance; (**d**) α_0 = 12° with wave disturbance; (**e**) α_0 = 24° without wave disturbance; (**f**) α_0 = 24° with wave disturbance.

5. Conclusions

In this paper, we analyzed the transient hydraulic characteristics of the draft tube of a pump turbine under wave disturbance and discussed the influence of wave disturbance on the draft tube pressure pulsation under different guide vane openings. The concluding remarks are as follows.

Wave disturbance affected the pressure value in the draft tube. Its influence on the eccentric vortex rope in the straight cone section and the pressure distribution law in the draft tube was relatively small. For the situation without wave disturbance, the maximum pressure pulsation coefficient peak of the monitoring points appeared at α_0 = 24° with a value of 1.22%. For the situation with wave disturbance, the monitoring points' maximum pressure pulsation coefficient peak appeared at α_0 = 12° with a value of 6.82%. The maximum pressure pulsation at all monitoring points universally increased 5~15 times.

For the situation without wave disturbance, the central pressure amplitude of the straight cone section of the draft tube was small, and an obvious eccentric vortex appeared under the running condition of a small opening. With the increase in the opening, the eccentric vortex gradually dissipated. The flow rate was the main factor that affected the pressure pulsation. For the situation with wave disturbance, the central pressure amplitude of the straight cone section of the draft tube increased sharply. The wave disturbance and eccentric vortex were the main factors affecting the pressure pulsation.

For the situation without wave disturbance, the main frequency of the draft tube under each opening was 0.11 f_n. The main frequency of the pressure pulsation in the eccentric vortex was 0.23 f_n, showing the typical characteristics of a low frequency and high amplitude. For the situation with wave disturbance, the main frequency of the draft tube became 0.5 Hz, equaling the frequency of the wave disturbance. In addition, there were several 0.5 Hz frequency-doubled, high-amplitude pressure pulsations. The pressure pulsation frequency spectrum varied a lot, and the main frequency characteristics still existed.

The flow law and pressure pulsation propagation law of the pump turbine found in this study can provide a reference for the optimal design of seawater pumped storage units. In addition, it would be interesting to investigate the effect of wave disturbance on the flow field and pressure pulsation of pump turbines under more complex operation conditions in further research.

Author Contributions: Conceptualization, J.H. and Z.M.; methodology, J.H., Q.W. and H.S. (Hongge Song); validation, B.C., Q.W. and H.S. (Hongge Song); formal analysis, J.H.; writing—original draft, B.C., J.H. and Q.W.; writing—review and editing, Z.M., Q.W. and H.S. (Hui Shen). All authors have read and agreed to the published version of the manuscript.

Funding: This research was funded by the Joint Funds of the Zhejiang Provincial Natural Science Foundation (LZJWZ22E090004).

Institutional Review Board Statement: Not applicable.

Informed Consent Statement: Not applicable.

Data Availability Statement: The numerical simulation data and experimental data used to support the findings of this study are available from the corresponding author upon request.

Conflicts of Interest: The authors declare no conflict of interest.

Nomenclature

C_p	Coefficient of pressure fluctuation
C_D	DES turbulence model constant
D_1	Inlet diameter of runner (m)
D_2	Outlet diameter of runner (m)
f	Frequency (Hz)
f_n	Rotation frequency (Hz)
g	Acceleration of gravity (m/s^2)
H	Head (m)
K	Production term of turbulent kinetic energy
K_z	Pressure sensitivity coefficient
l_p	Prototype length (m)
l_m	Model length (m)
n	Rotational speed (r/min)
n_{11}	Specific speed (r/min)
p	Static pressure (Pa)
Q_{11}	Specific flow rate (m^3/s)
S	Generalized source term of momentum equation
S_i	Invariant measure of the strain rate
Z	Elevation of the free surface (m)
α_0	Guide vane opening (degree)
σ_k	Constant of the turbulence model
ρ	Density (kg/m^3)
μ	Dynamic viscosity (Pa s)
φ	Velocity potential
Γ	Fluid parameters
Ω	Vorticity tensor
η	Efficiency (%)
λ	Eigenvalue of the characteristic equation
λ_k	Scale length of wave model
λ_p	Scale of wave pressure
λ_f	Scale of wave frequency
ΔH	Peak value of pressure pulsation (m)
$\Delta H'$	Relative pressure pulsation amplitude (%)
CFD	Computational fluid dynamics
DES	Detached eddy simulation
FFT	Fast Fourier transform
LES	Large eddy simulation
RANS	Reynolds-averaged Navier–Stokes
SST	Shear stress transport
SIMPLEC	Semi-Implicit Method for Pressure-Linked Equations-Consistent

References

1. *2017/68*; Resource Census Results of Seawater Pumped Storage Hydropower Plant. National Energy Administration: Beijing, China, 2017.
2. Zhang, X.; Zhang, P.; Chen, X. Development and application of seawater pumped storage hydropower plant. *Mech. Electr. Tech. Hydropower Plant* **2019**, *42*, 66–70. [CrossRef]
3. He, R.F.; Wang, F.; Zhang, H. Technical route and key parameters selection of seawater variable speed pumped storage units. *Water Res. Hydro. Eng.* **2020**, *51*, 184–189. [CrossRef]
4. Timoshevskiy, M.V.; Churkin, S.A.; Kravtsova, A.Y.; Pervunin, S.K.; Dmitriy, M.M. Cavitating flow around a scaled-down model of guide vanes of a high-pressure turbine. *Int. J. Multiphase Flow* **2016**, *78*, 75–87. [CrossRef]
5. Shi, W.H.; Cha, H.; Luo, H.; Wang, K.Y. Preliminary study on development of seawater pumped storage hydropower plant in China. *China Energy* **2015**, *37*, 36–40.
6. Ramos, H.M.; Amaral, M.P.; Covas, D.I.C. Pumped-Storage Solution towards Energy Efficiency and Sustainability: Portugal Contribution and Real Case Studies. *J. Water Resour. Prot.* **2014**, *06*, 1099–1111. [CrossRef]
7. Fujihara, T.; Imano, H.; Oshima, K. Development of Pump Turbine for seawater Pumped-storage hydropower Plant. *J. Am. Soc. Civ. Eng.* **1998**, *47*, 199–202.
8. Wang, X.; Liu, D.; Liu, X.; Wen, T.; Lianchen, X. Analysis on flow structures and pressure pulsation in vaneless space of reversible pump turbine. *J. Hydrol. Eng.* **2021**, *40*, 59–72. [CrossRef]
9. Goyal, R.; Trivedi, C.; Kumar Gandhi, B.; Cervantes, M.J. Numerical simulation and validation of a high head model Francis turbine at part load operating condition. *J. Inst. Eng.* **2018**, *99*, 557–570. [CrossRef]
10. Guo, T.; Xu, L.H.; Luo, Z.M. Influence of Upstream Disturbance on Vortex Rope Evolution and Pressure Fluctuation in Draft Tube. *J. Trans. Chin. Soc. Agric. Mach.* **2022**, *53*, 192–570.
11. Wang, T.T.; Zhang, C.B.; Xie, T.T.; Cao, W.Z. Analysis on vortex ropes and pressure pulsations in draft tube of variable-speed Francis turbine. *J. Hydrol. Eng.* **2021**, *40*, 95–101. [CrossRef]
12. Zhang, Y.N.; Liu, K.H.; Li, J.W.; Xian, H.Z.; Du, X.Z. Analysis of the vortices in the inner flow of reversible pump turbine with the new omega vortex identification method. *J. Hydro.* **2018**, *30*, 463–469. [CrossRef]
13. Zhou, L.J.; Wang, Z.W.; Huang, Y.F. Draft tube surge and runner outlet flow patterns. *J. Tsinghua Univ. Sci. Technol.* **2002**, *42*, 1670–1673. [CrossRef]
14. Ji, J.T.; Zheng, X.B.; Luo, X.Q. Analysis of pressure fluctuation in draft tube based on simulation of unsteady flow in whole passage of axial flow turbine. *J. Hydrol. Eng.* **2009**, *28*, 147–151.
15. Gao, X. Study of flow characteristics of draft tube of tidal power station. *Eng. J. Wuhan Univ.* **2019**, *52*, 594–599. [CrossRef]
16. Ji, L.; Xu, L.; Peng, Y.; Zhao, X.; Li, Z.; Tang, W.; Liu, D.; Liu, X. Experimental and Numerical Simulation Study on the Flow Characteristics of the Draft Tube in Francis Turbine. *Machines* **2022**, *10*, 230. [CrossRef]
17. Wang, Z.; Zhu, B.; Wang, X.; Qin, D. Pressure Fluctuations in the S-Shaped Region of a Reversible Pump-Turbine. *Energies* **2017**, *10*, 96. [CrossRef]
18. Cavazzini, G.; Covi, A.; Pavesi, G.; Ardizzon, G. Analysis of the unstable behavior of a pump-turbine in turbine mode: Fluid-dynamical and spectral characterization of the S-shape characteristic. *J. Fluids Eng.* **2016**, *138*, 021105. [CrossRef]
19. Yang, J.; Lv, Y.; Liu, D.; Wang, Z. Pressure Analysis in the Draft Tube of a Pump-Turbine under Steady and Transient Conditions. *Energies* **2021**, *14*, 4732. [CrossRef]
20. Guo, T.; Xu, L.; Wang, W. Influence of Upstream Disturbances on the Vortex Structure of Francis Turbine Based on the Criteria of Identification of Various Vortexes. *Energies* **2021**, *14*, 7626. [CrossRef]
21. Chen, T.; Zheng, X.H.; Zhang, Y.N.; Li, S.C. Influence of upstream disturbance on the draft-tube flow of Francis turbine under part-load conditions. *J. Hydrodyn.* **2018**, *30*, 131–139. [CrossRef]
22. Dewan, Y.; Custer, C.; Ivashchenko, A. Simulation of the Francis-99 Hydro Turbine During Steady and Transient Operation. *J. Phys. Confer. Ser.* **2017**, *782*, 12003. [CrossRef]
23. Nam, M.C.; Dechun, B.; Yue, X.G.; Jin, M.R. Design optimization of hydraulic turbine draft tube based on CFD and DOE method. *Earth. Environ. Sci.* **2018**, *136*, 012019. [CrossRef]
24. Zheng, Y.; Jiang, W.Q.; Chen, Y.J.; Sun, O.R. Investigation on low frequency pulsating and draft tube vortex of tubular turbine. *J. Trans. Chin. Soc. Agric. Mach.* **2018**, *49*, 165–171. [CrossRef]
25. Peng, S.Y.; Li, Z.G.; Li, X.R.; Chai, X.; Liu, D.Y.; Zhao, Y.Z.; Liu, X.B.; Wang, W.J. Flow analysis of tubular turbine draft tube based on vortex analysis. *Eng. J. Wuhan Univ.* **2020**, *53*, 679–685. [CrossRef]
26. Yu, J.M.; Ji, Q.Y.; Zheng, H.; Yang, L.Y.; Wang, R.Q.; Chen, C.J. The investigation of wave energy in Zhoushan coastal area based on SWAN model. *Mar. Fore.* **2020**, *37*, 38–49. [CrossRef]
27. Huang, D.B. *Fundamentals of Water Wave Theory*; National Defense Industry Press: Beijing, China, 2011; ISBN 978-71-1807-651-6.
28. *JTJ/T 234-2001*; Wave Model Test Regulation. Nanjing Hydraulic Research Institute: Nanjing, China, 2002.
29. Zheng, Y.; Chen, D.X. *Hydraulic Turbine*; China Water and Power Press: Beijing, China, 2011; ISBN 978-75-0848-368-9.
30. Xu, L.C.; Peng, Y.J.; Tang, W.; Liu, D.M.; Liu, X.B. Flow characteristics and pressure pulsation in the S characteristic area of model pump turbine. *Chin. J. Hydro.* **2022**, *37*, 213–225. [CrossRef]

31. Lu, J.; Qian, Z.; Lee, Y.H. Numerical investigation of unsteady characteristics of a pump turbine under runaway condition. *Renew. Energy* **2021**, *169*, 905–924. [CrossRef]
32. Yu, A.; Wang, Y.; Tang, Q.; Lv, R.; Yang, Z. Investigation of the vortex evolution and hydraulic excitation in a pump-turbine operating at different conditions. *Renew. Energy* **2021**, *171*, 462–478. [CrossRef]

sustainability

MDPI

Article

Finite Element Analysis of Different Infill Patterns for 3D Printed Tidal Turbine Blade

Glecelyn M. Galvez, Karl Andrie M. Olivar, Francis Rey G. Tolentino, Louis Angelo M. Danao * and Binoe E. Abuan

Department of Mechanical Engineering, University of the Philippines Diliman, Quezon City 1101, Philippines
* Correspondence: louisdanao@up.edu.ph

Abstract: The fabrication route for tidal turbine blades has been compounded with the appearance of additive manufacturing; with the use of infill patterns, improvement of mechanical strength and material reduction for 3D printed parts can be obtained. Through finite element analysis and three-point bend tests, the optimal infill lattice pattern, and the viability of the shell–infill turbine blade model as an alternative to the conventional shell-spar model was determined. Out of a selection of infills, the best infill pattern was determined as the hexagonal infill pattern oriented in-plane. A representative volume element was modeled in ANSYS Material Designer, resulting in the homogenized properties of the in-plane hexagonal lattice. After validation, the homogenized properties were applied to the tidal turbine blade. The shell–infill model was based on the volume of the final shell-spar model which had a blade deflection of 9.720% of the blade length. The difference in the deflection between the homogenized infill and the spar cross-section was 0.00125% with a maximum stress of 170.3 MPa which was within the tensile strength and flexure strength of the carbon fiber with onyx base material. Conclusively, the homogenized infill was determined as a suitable alternative to the spar cross-section. The best orientation of the infill relative to the horizontal orientation of the blade was 0 degrees; however, the lack of trend made it inconclusive whether 0 degrees was the absolute optimal infill orientation.

Keywords: finite element analysis; additive manufacturing; infill patterns; horizontal axis tidal turbine blade

Citation: Galvez, G.M.; Olivar, K.A.M.; Tolentino, F.R.G.; Danao, L.A.M.; Abuan, B.E. Finite Element Analysis of Different Infill Patterns for 3D Printed Tidal Turbine Blade. *Sustainability* **2023**, *15*, 713. https://doi.org/10.3390/su15010713

Academic Editors: Bin Huang, Ling Zhou, Yuquan Zhang and Jianhua Zhang

Received: 15 November 2022
Revised: 20 December 2022
Accepted: 27 December 2022
Published: 30 December 2022

1. Introduction

The Department of Energy (DoE) of the Philippines developed the National Renewable Energy Program (NREP) that projected the country's renewable energy capacity to be 15,304 MW [1] by year 2030. Experts of renewable energy are encouraged to design technologies to exploit the untapped energy potential for the country's benefit. With the appearance of additive manufacturing in the present, the fabrication route for makers to produce horizontal axis wind and tidal turbines has compounded considering the reduced cost of production and assembling procedure it offers.

According to ISO/ASTM 52900:2021 [2], additive manufacturing (AM) is the process of manufacturing a physical object using a 3D model, slicing the model, and then fabricating it by combining or adding parts together, generally, layer by layer [3]. AM is mostly used in rapid prototyping of products rather than producing actual products for the market, since there are a lot of manufacturing inconsistencies in AM [4,5]. However, recent developments in AM suggest that the manufacturing of actual products are viable [5–7]. Researchers are drawn to investigate the potential of AM processes in meeting the strength capabilities that traditionally manufactured products are favored with. Utilizing the ability of AM to produce complex geometries, lattice structures are incorporated into AM products with the primary goal to improve its mechanical strength while reducing material volume [4,5,7].

Engineering design iterations, performed on computer aided design (CAD) models of AM products, are subjected to finite element analysis (FEA) to predict product properties. However, inaccuracies in finite element modeling of 3D printed products appear due to manufacturing errors and inconsistencies of AM Products produced by different machines [5,6,8]. Meanwhile, recent research endeavors have validated the method of FEA on 3D printed products using experimental results [9], these studies reclaimed the validity of FEA on AM products. Their method can then be extended to predict the properties of complex geometric properties such as turbine blades [10,11].

The objective of this study was to fill the knowledge gap of applying additive manufacturing to horizontal axis tidal turbines (HATT), and to integrate infill patterns in an already existing HATT blade design that was tested using finite element analysis (FEA). The study investigated the effects of the infill pattern and its orientation on the turbine blade design by comparing the blade deflection and maximum stress results of the models using a homogenized modelling method. This study is first of its kind in showing the applicability of a representative volume element to simplify the modelling of complex internal structures, such as infill patterns commonly employed in 3D printing technologies to structural design and analysis of tidal turbine blades.

2. Review of Related Literature

2.1. Additive Manufacturing

Additive manufacturing (AM) is categorized based on the types of processes performed [5,7]. The common AM processes are: (1) fused deposition modelling (FDM) which fabricates products by layering thermal plastic based on a specified pattern, (2) laminated object manufacturing (LOM) that cuts sheets of material layer by layer using a laser, (3) stereolithography (SLA) which produced parts by curing material layer by layer with UV light and, lastly, (4) selective laser sintering (SLS) that heats up powdered material with a laser just before the sintering point to manufacture a part. These processes can work with different materials such as polymers, ceramics, metals, biomaterials, etc., allowing a wide selection of material without retrofitting the manufacturing device.

AM can manufacture complex geometries such as thin-walled products which are hard to produce with traditional manufacturing [4,5]. However, AM products are also characterized as having uncertain mechanical properties, poor surface quality, and dimensional accuracy [5,8] making it difficult to simulate 3D printed products' properties through finite element analysis. Recent researchers proposed models that can only be applied to specific machines; hence, an optimal approach towards manufacturing the 3D printed part must be taken to accurately predict the model properties [5,6,8].

2.2. Infill and Lattice Structures

Each layer in the printing process consists of a shell and the infill. The shell is the outline perimeter of the part that can be modified to the desired thickness while the infill is printed inside the shell. The infill is usually represented in percentages which correspond to the amount of material that a solid model must be filled. Infills vary with geometry and percentage which dictates the 3D printed parts' strength (Figure 1), manufacturing time and material [12].

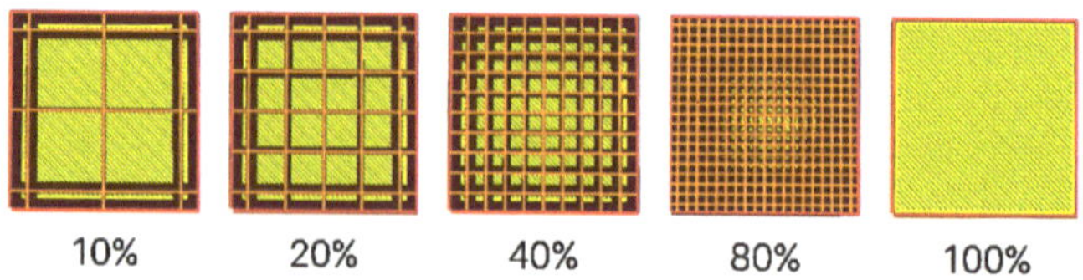

Figure 1. Grid infill at different infill densities.

Cho et al. [13] found out that layer thickness had higher influence than infill patterns on the structural resilience of the part, this is supported by Luzanin et al. [14]. However, increasing the thickness of layers beyond the recommended setting can affect the resolution and fidelity of the 3D printed part which may lead to additional post processing steps. This can simply be avoided by modifying the infill pattern or increasing the infill density of the 3D printed part.

A study by Gurrala and Regalla showed that optimal strength occurs at the highest infill density [15]. Similarly, Alvarez et al. recommended a solid (100%) infill to achieve high mechanical resistance and fast printing compared to infill values between 50% to 98% [16]. Dudescu and Racz showed that the increase in specimen strength and infill percentage was non-linear, the same result was obtained when elastic modulus was tested against infill percentage [17]. The optimal infill percentage for maximizing the stiffness-to-mass ratio of 3D printed structural members is around 10 to 20% and that infill orientation has a negligible effect on flexural rigidity [18].

The ultimate strength, strains and failure behavior of the 3D printed part are functions of the infill pattern (Figure 2) [19]. Research showed that the triangle infill pattern gives the highest mechanical strength and ultimate tensile strength-to-weight ratio when compared to grid, zig-zag, rectilinear and hexagonal infill patterns with the same amount of material [13,20]. Meanwhile, the hexagonal infill pattern proved to have superior characteristics in terms of high ultimate strength and decent ability to resist fracture than specimens of rectangular and diamond infill patterns [13,19]. The improved strength at the strut joints of the hexagonal infill also contributes to its advantage [21].

In other studies, circle packing algorithms were used to generate a honeycomb-shape infill structure with small and dense hexagon patterns in a high-stress area and large and coarse hexagon patterns in a low-stress area for the design of horizontal axis wind turbine (HAWT) blades [22].

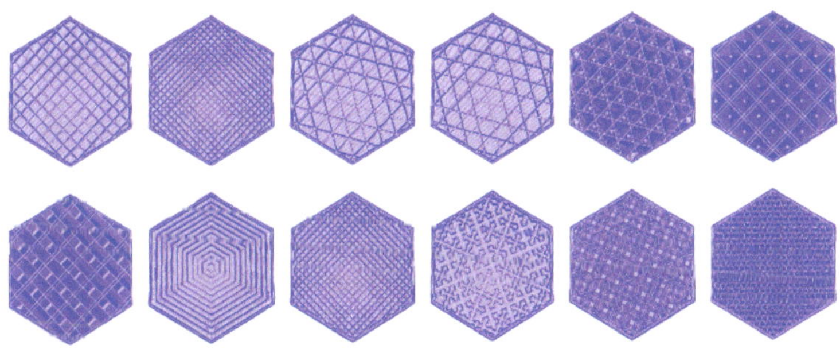

Figure 2. Examples of infills used in the 3D-printing industry. From left to right, Row 1: Grid, Lines, Triangles, Tri-hexagon, Cubic, Cubic (subdivision), Row 2: Octet, Quarter Cubic, Concentric, Zig-Zag, Cross, Cross 3D, and Gyroid Infill, adapted from ([23], Figure 1).

2.3. Finite Element Analysis of 3D Printed Parts

Despite the inaccuracies in parts fabricated through additive manufacturing, multiple studies have attempted to predict the material properties of 3D printed products incorporated with lattice structures. Their FEA results coincided with experimental results which validated their FEA methods.

Fernandez et al. assumed that the properties of the model were isotropic to simplify the determination of the elastic modulus of FDM products [9]. Their FEA results coincided with the experimental results, even though the anisotropic nature of the part was disregarded. This demonstrated the feasibility of performing FEA simulations in AM. From the concept of uniformity of the model, the method of homogenization was adapted by researchers, Bhandari and Lopez-Anido [24], to decrease the simulation time of AM products impregnated with lattice structures. The concept of homogenization is about modelling a volume-based lattice finite element model, in conjunction with the experimental data from the 3D printed samples, to generate the homogenized properties of the lattice which consists of Poisson's ratio and Young's modulus in different orientations. The homogenized properties are then used to determine the part's properties.

In the study of Xia Zhou et al., the bending behavior of hybrid sandwich composite structures with 3D printed lattice cores were observed by experimentation and simulation [25]. They created an FEA model of the sandwich with an adhesive using the material properties from the tensile test and subjected it to a simulated three-point bend test. The FEA results followed the same trend as the experimental results. They were also able to obtain properties of the other lattice structures through FEA without performing further experimentations since they had already validated the FEA results and experimental results of the model. This process could be applied to further studies in predicting various 3D printed lattice behavior.

2.4. Horizontal Axis Tidal Turbines

A horizontal axis tidal turbine (HATT) is a tidal current machine that is oriented parallel to the direction of the tidal current. Turbine blades are considered the most critical component of modern wind and tidal turbines. Tidal turbine blades tend to be shorter and stronger compared to wind turbine blades since they are made to withstand higher hydrodynamic forces [26]. Failure analysis must be carried out to determine if the blade can withstand the static and dynamic load it experiences during its design service life [27].

The conventional design cross-section of a turbine blade is made up of an aeroshell and internal webs and the most widely used composite for long wind turbines is resin infusion technology [28]. A 3D printing application on an airfoil design was already tested in the study of Kalyani [10] and Kim [11,22]. A design methodology was included in Kim's study where the density of the infill pattern was concentrated based on the stress profile from the computation of structural analysis. The optimized infill pattern density distribution in the airfoil was then subjected to static loading test by using different loads at different distances from the blade roots [10]. The same methodology was also applied to a wind turbine blade [29].

Bhandari and Lopez-Anido [24] conducted finite element modeling of 3D printed parts with cellular structures or infill pattern via the use of homogenized continuum finite element model to decrease the simulation time. The concept of homogenization was about representing a volume-based lattice finite element model which would be used to be able to determine the product properties from 3D printed parts. The volume-based lattice finite element was made using experimental data of 3D printed samples to obtain the Poisson's ratio and Young's modulus; these values were taken from past experiments. The FEM results were in accordance with the experimental results.

The geometric specifications of the HATT blade used in this study were based on the study of the design of a tidal turbine blade by Danao et al. [26]. The blade performance (Figure 3) and structural response due to hydrodynamic loading were tested using Qblade and ANSYS, respectively. The blade forces used were based on blade element momentum (BEM) simulation for 2 m/s and 5 m/s flow. The forces calculated were then used on the structural analysis of ANSYS to determine its stress and deformation.

The reviewed literature matrix (Table 1) summarizes the similarities and differences of this study to past research based on the given criteria. The row contains the studies while the column contains the criteria. The second column determines whether the study is about a HAWT/ HATT blade, the third whether the study involves 3D printing, the fourth whether the study involves infill pattern, the fifth whether it includes experimental bend test, and the last column whether it involves simulation of the blade. The study of Kim, S. et al. [22] involves topology optimization and simulation of a blade but it did not involve an experimental bend test of the specimen.

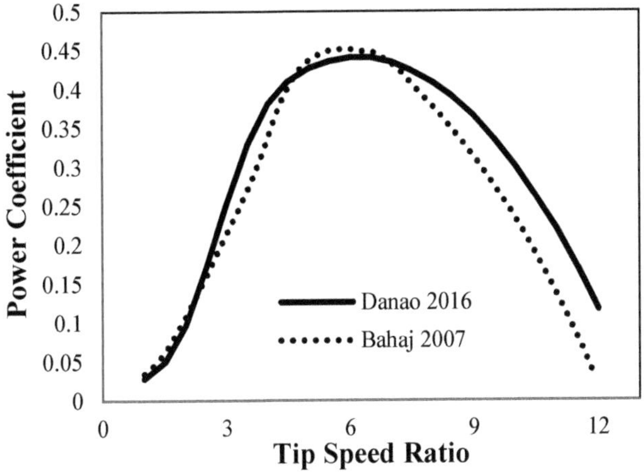

Figure 3. BEM predicted power coefficient for the proposed blade design and old design adapted from [26,30].

Table 1. Reviewed literature matrix.

Authors	HAWT/HATT	Additive Manufacturing	Infill Pattern	Flexure Testing	Simulation
Kim, S. et al. [22]	HAWT	Yes	Yes	No	Yes
Noronha, N. Munishamaiah, K. [31]	HAWT	No	No	No	Yes
Cho, P. et al. [13]	N/A	Yes	Yes	Yes	Yes
Kim, S. et al. [11]	N/A	Yes	Yes	Yes	Yes
Kalyani, K. et al. [10]	N/A	Yes	No	No	Yes
Bassett et al. [32]	HAWT	Yes	Yes	No	No
Danao et al. [26]	HATT	No	No	No	Yes
Bhandari, S. and Lopez-Anido, R [24]	N/A	Yes	Yes	Yes	Yes

For this study, the conventional spar web cross-section of a HATT blade was replaced by an optimized infill. Different infill pattern geometries and orientations were also tested since each pattern had a different mechanical property which was orthotropic. The parameters of the infill used were based on previous studies which recommended an infill percentage within the range of 10–20% to maximize the value of mass-to-stiffness ratio of 3D printed parts. The utilization of hexagonal and triangular infill patterns was based on their better strength and stiffness characteristics relative to other infill patterns.

Due to time constraints and resource availability, finite element analysis with minimal physical experiments was adapted to apply infill patterns to horizontal tidal turbine blades which was also not done on the previous studies in Table 1. The concept of homogenized continuum finite element model was applied to the infill pattern to speed up the process of simulation which produced orthotropic properties. The tensile test was used to validate the FEA simulation results and to obtain the mechanical properties of the polylactic acid (PLA) material used in the simulation while the bend test was utilized for validating the homogenized 3D printed sample. These concepts were combined to be able to produce an FEA model of a turbine blade with varying lattice structures while significantly reducing simulation time.

3. Methodology

3.1. PLA Material Properties

Polylactic acid (PLA) plastic was chosen as the material for this study due to its price and availability. The material properties of 3D printed PLA were determined using ASTM D638-16 type V standard (Table 2, Figure 4) from the tensile test for simulating the elastic behavior of the PLA material [33]. The printed specimens were subjected to post-processing to refine their edges. The specimens were printed on a 45-degree angle with respect to the loading axis since it was an effective estimate of the anisotropy and material property of 3D printed plastics [34]. The specimens were then set up on the Instron 3366 electronic universal material testing machine for tensile testing (Figure 5). Based on the experimental results, the elastic modulus of the PLA material was calculated and used as a reference for validating a simulated PLA sample.

Table 2. ASTMD638 type V standard.

Dimensions	Type V (mm)	Tolerances
W—Width of narrow section	3.18	±0.5
L—Length of narrow section	9.53	±0.5
WO—Width overall, min	9.53	+3.18
LO—Length overall, min	63.5	No max
G—Gage length	7.62	0.25
D—Distance between grips	25.4	±5

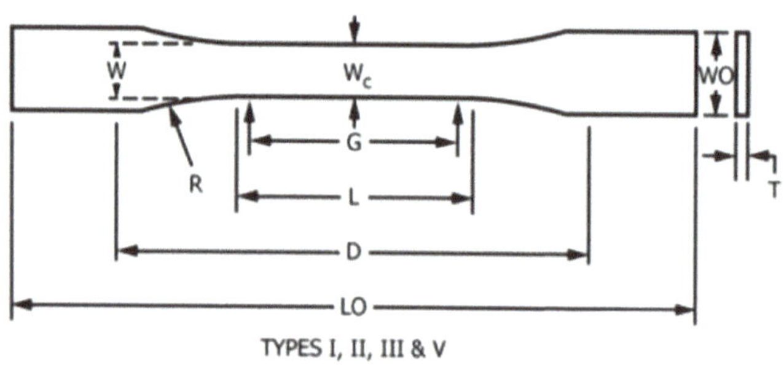

Figure 4. Specimen drawing: ASTMD638 type V standard [33].

Figure 5. Tensile test experiment (**a**) before test, (**b**) failure after test.

The simulation of the tensile test was then created using ANSYS static structural simulation, in which the CAD model of the specimen (Figure 6a) was given the PLA material properties from the experimental results and boundary conditions to simulate the necessary tensile loads. The simulation was limited to the elastic response of the model. The elastic modulus of PLA was fitted with the experimental data and was found to be 577.11 MPa. The comparison of tensile test and simulation results for the stress–strain curves of the 3D printed PLA samples were compared (Figure 6b). It was found that the simulated tensile stress–strain curve followed the trend of the experimental data quite well; the proposed model for 3D printed PLA can describe the tensile stress–strain behavior with good accuracy.

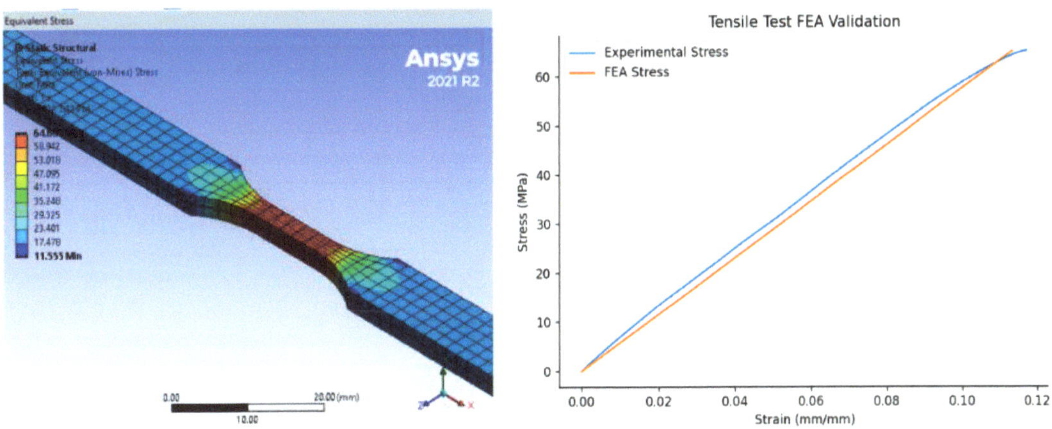

(a) (b)

Figure 6. (**a**) Simulated tensile test experiment and (**b**) comparison of results.

3.2. Lattice Core Design

The lattice core designs were based on the triangle infill and hexagonal infill since the triangle infill pattern yielded the highest mechanical strength while the hexagonal had the highest elastic modulus [13,20]. The design specifications for the infill patterns were calculated using Li's method for designing 3D printed lattices [35]. The design was constrained to have 0.8 mm wall width and the relative density of the lattice structures was fixed to 15%, since it was recommended that the relative density of the lattice structures be 10–20% in maximizing the stiffness-to-mass ratio [18]. From the constraints, it was calculated that the side length of the triangle for the triangle infill was 7 mm while the hexagon had a side length of 4 mm. To determine the optimal lattice unit-cell configuration for the bending performance of 3D printed PLA infill pattern, eight sandwich models (Figure 7) with varying lattice orientations were prepared. The sample specimens were designed according to the ASTM C393 standards [36].

3.3. Lattice Sample Preparation and Mechanical Testing

For each of the tests conducted, three samples were prepared for each sandwich model. They were 3D printed with the same printer and filament used in the tensile tests. The Instron 3366 electronic universal testing machine (UTM) was then set up with the configuration specified by ASTM C393-16 test standards (Figure 8) and Instron software to gather the necessary data from the experiment. A clamp velocity of 5 mm/min in the downward direction was used in the bending test to avoid the viscoelastic response of the plastic material. The results were recorded for validating the infill via behavioral bending which was based on Xia Zhou and their group's research; their simulated results had good correlation with their experimental results which were the foundation for simulating other infill patterns [25].

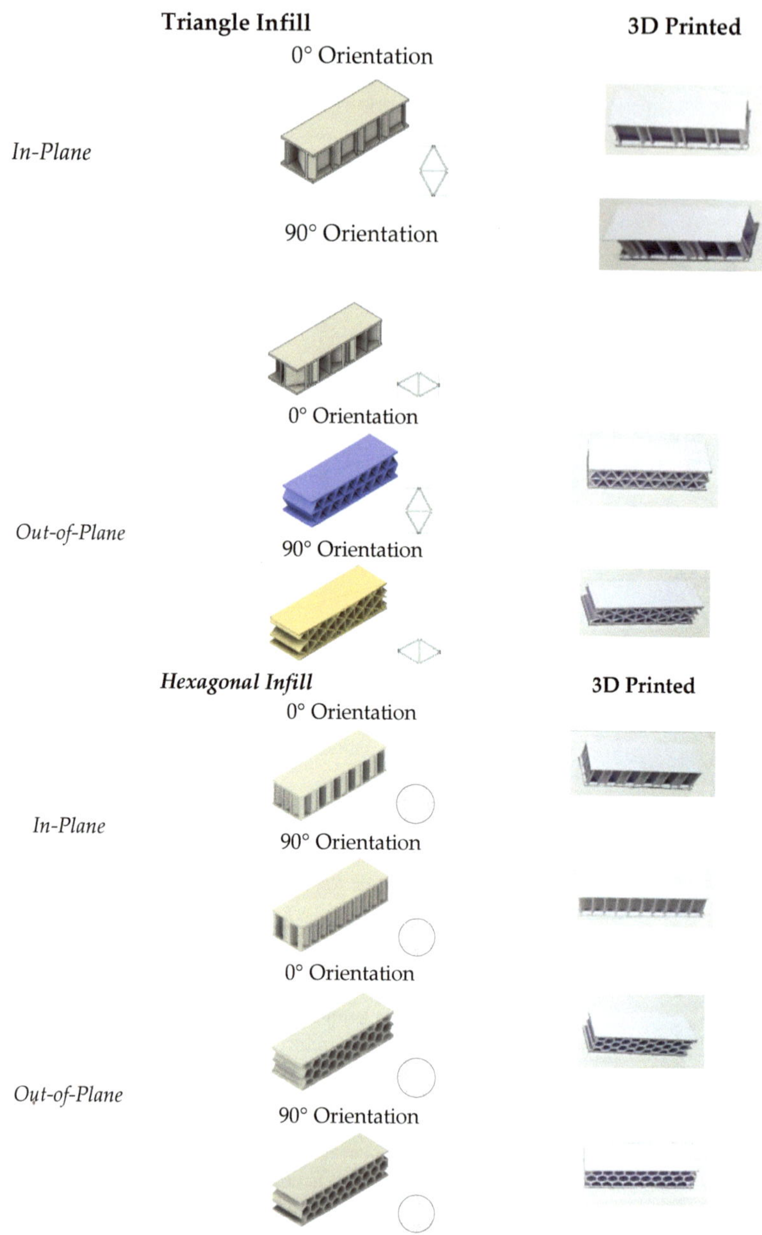

Figure 7. Sandwich models and printed samples with varying lattice structure configuration for 3-point bend test.

Figure 8. Experimental bending test setup.

3.4. Bend Test Simulation and Homogenization Validation

To ease the computational process of simulating infill patterns for the turbine blade, the best lattice core design was homogenized as a representative volume element (RVE) for application to models with large volume [24]. ANSYS Material Designer was utilized to make the RVE of the chosen infill pattern. The mesh sizes used in the homogenization of the hexagonal lattice structure were successively refined for convergence. Based on the outputs of the Material Designer in ANSYS, the resulting mechanical properties have converged under the 0.5 mm mesh size within 5% error. The ASTM C393-16 sandwich model was simulated in ANSYS and was given the constant PLA material properties. The orthotropic results of the RVE were applied as material properties to the infill of the ASTM C393 model to make a homogenized model of the sandwich. Various mesh sizes were tested for the sandwich infill for mesh independence. It was subjected to the same bend test setup by applying the same load and specifications taken from the bend test experiment. A fixed support was applied to the contact region between the right pin and the beam, remote displacement was applied to the pins, and load was applied on the upper pin. The load vs. displacement data from the bend test simulation were recorded.

The CAD model of the chosen infill pattern was also subjected to the simulated bend test. (Figure 9). A fixed support was applied on the contact region between the right pin and the beam, remote displacement was applied to the pins, and load was applied on the upper pin. This led to three datasets for the bend test: the experimental bend results, CAD model results, and the homogenized CAD model results. The three datasets were compared to validate the results of calculated homogenization and simulation properties [24,25].

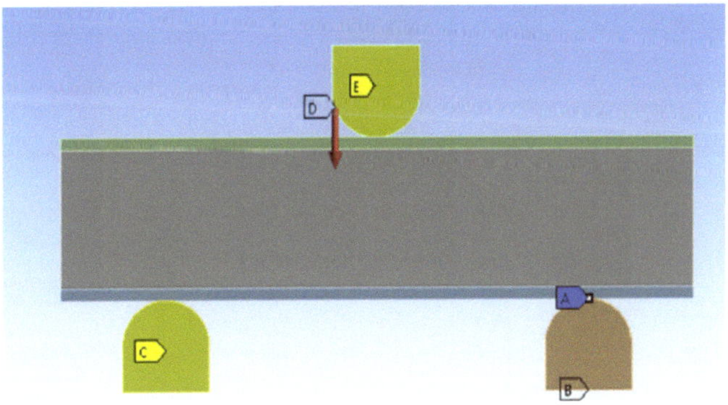

Figure 9. FEA bending test setup where point A is a pin connection to avoid rigid body motion, B and C are pin supports, E as the load pin with D as the remote force.

3.5. Turbine Blade Models and Simulation Setup

After obtaining the homogenized properties, it was implemented into the proposed blade design by Danao et al. [26] with the specifications stated (Table 3). The rotor design of their study had a good overall coefficient of performance at 0.44–0.45 at a tip speed ratio of 6.

Table 3. Turbine blade specifications [26].

Radial Position (m)	Chord Length (m)	Twist (°)	Foil Profile
0.4	0.250	20	NACA 4424
0.6	0.2312	14.5	NACA 4420
0.8	0.2126	11.1	NACA 4418
1	0.1938	8.9	NACA 4417
1.2	0.175	7.4	NACA 4416
1.4	0.1562	6.5	NACA 4415
1.6	0.1376	5.9	NACA 4414
1.8	0.1188	5.4	NACA 4413
2	0.1	5	NACA 4412

The turbine blade design was simulated with the same boundary conditions as the study of Danao et al. [26]. A fixed support was applied to the blade root of the turbine blade and remote forces, which intersected the axis of twist, were applied on the cross-sections of the blade (Figure 10). The remote forces used on each section of the blade were derived from the tangential and normal forces computed using the blade force data of the BEM simulation (Table 4). The forces used were the load experienced by the blade at 2 m/s fluid speed. In addition to that, a rotational velocity of 6 rad/s was imposed to introduce the centrifugal effects in the simulation. The turbine blade should not exceed a maximum deflection of 10% of the blade span which was 0.16 m.

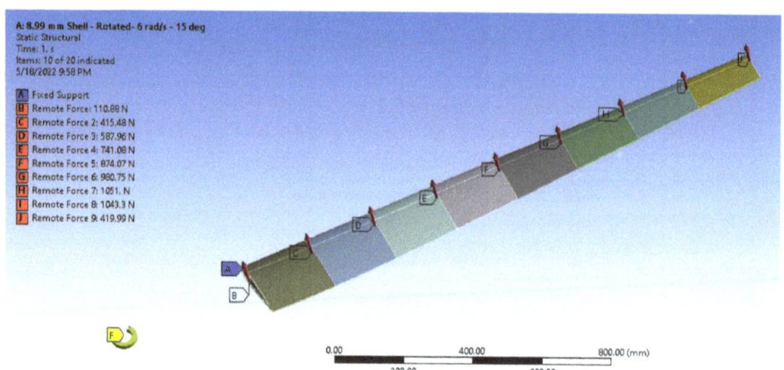

Figure 10. Simulation setup showing the blade with applied remote forces and fixed support.

Table 4. Blade force data from BEM Simulation at 2 m/s flow speed adapted from [26].

Radial Position (m)	Tangential Force (N)	Normal Force (N)	Resultant Force (N)
2	36.072	418.441	419.933
1.8	121.475	1036.20	1043.30
1.6	142.404	1041.26	1050.95
1.4	151.03	969.055	980.754
1.2	154.613	860.287	874.07
1	155.63	724.558	741.084
0.8	151.657	568.066	587.962
0.6	134.533	393.101	415.485
0.4	40.06	103.387	110.877

Three models representing the turbine blade design were created using the coordinate of each of the section's hydrofoil and using the loft function to seamlessly connect each section to the adjacent ones to produce a continuous blade surface. These models were the solid blade model, shell-spar model, and the shell–infill model.

The first blade model, the solid blade model, was designed to have no space within the shell (Figure 11). It was a preliminary model that served as the basis of the suitability of the material used in the blade design to determine the minimum possible deflection of the material used. With the simulation of this model, it was found that PLA was not a suitable material for turbine blade design due to blade tip deflections exceeding 10% of the blade length.

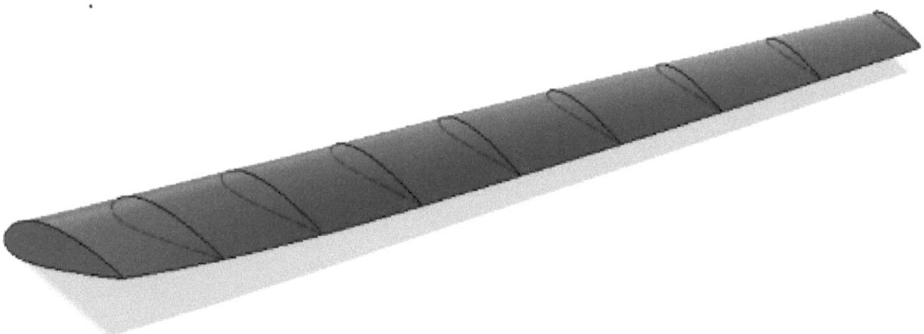

Figure 11. Solid blade model.

The second model, shell-spar model, was created with a shell of the turbine blade and a spar (Figure 12) based on the original design. The original design used in the numerical simulations of Danao et al. [26] had a constant shell thickness of 2% of the highest chord length while the spar thickness was 8% of the highest chord length. Adjustments were made to the thickness of the shell to accommodate for the material used for simulating the blade loads. The final design iteration of the shell-spar blade was chosen on the criteria of a maximum blade tip deflection that was less than 0.16 m or 10% of the blade length. The maximum stress was calculated from the simulation of the shell-spar model.

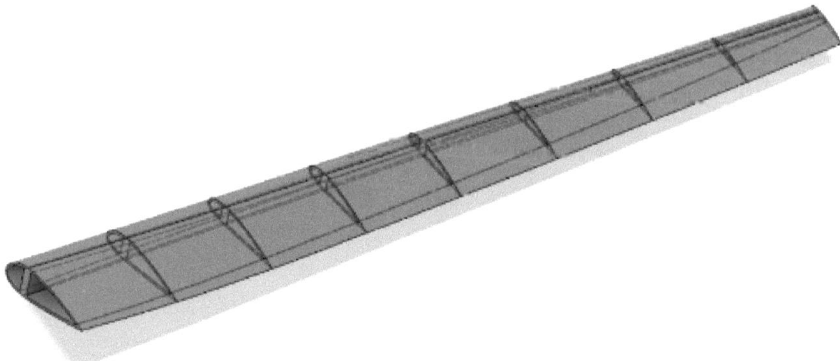

Figure 12. Shell-Spar blade model.

The third model, the shell–infill model, was created with a shell of the turbine blade and was filled with an infill (Figure 13). The shell and infill dimensions were specified based on the volume of the second model. The volume of the third model was set to have an equal material volume with the second model to compare the effects of the infill pattern to the blade model; using the same amount of material, the differences in the geometry was expected to affect the structural behavior of the blade. The dimensions of the final shell–infill model were calculated iteratively and verified by inspecting the volume of both CAD models. Additionally, the maximum stress was calculated from the simulation.

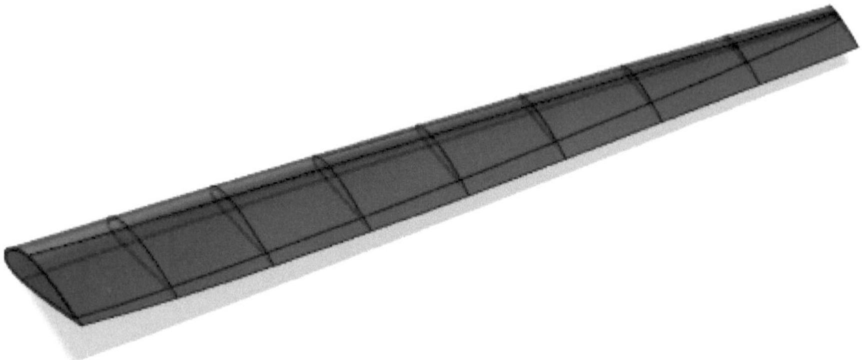

Figure 13. Shell–infill blade model.

3.6. Replacement of PLA with Carbon Fiber with Onyx Base

It was found that PLA material was not suitable for the loadings specified in the previous section. It was replaced with carbon fiber with an onyx base taken from the composite material data sheet from Markforged [37].

The onyx base is a material with a high quality surface finish, chemical resistivity, and heat tolerance while the carbon fiber reinforcement has a high strength-to-weight ratio and was used to replace material that used machined aluminum (Table 5). Together, the materials formed a composite with the Markforged 3D printer which could be laid on the printing platform together; this resulted in the carbon fiber being parallel to the build plate [37]. The solid blade model was simulated with the new material and yielded a result that was lower than 10% of the blade length.

Table 5. Material properties of carbon fiber with onyx base [37].

Property	Material Properties
Young's modulus	60 GPa
Poisson's ratio	0.3061
Tensile strength	760 MPa
Flexural strength	540 MPa

3.7. Turbine Blade Infill Orientation

The infill has orthotropic properties which could affect the structural performance of the turbine blade based on its orientation. The orientation of the homogenized infill of the shell–infill blade model was set with respect to the horizontal (Figure 14). A variation study of infill orientation was conducted to investigate the effects of infill orientation relative to the blade. A unique global coordinate system for the infill body was created to rotate the infill. The infill body was then rotated along the y-axis of the global coordinate system created. Afterwards, the blade was subjected to the same blade loading conditions. The blade tip deflection was observed from $-15°$ to $25°$ infill orientation. The zero degree orientation is shown in Figure 15.

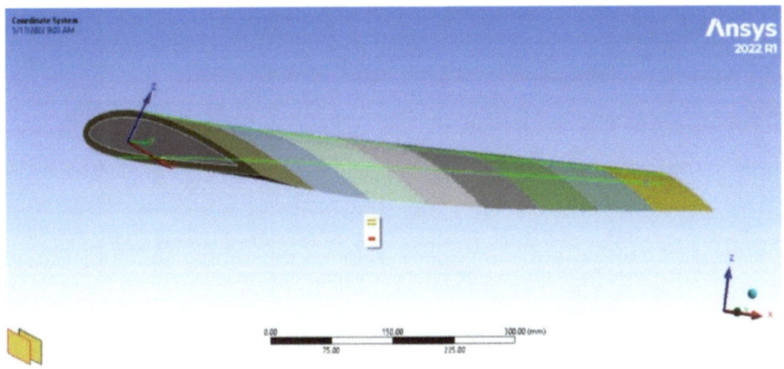

Figure 14. Turbine blade with infill orientation in ANSYS.

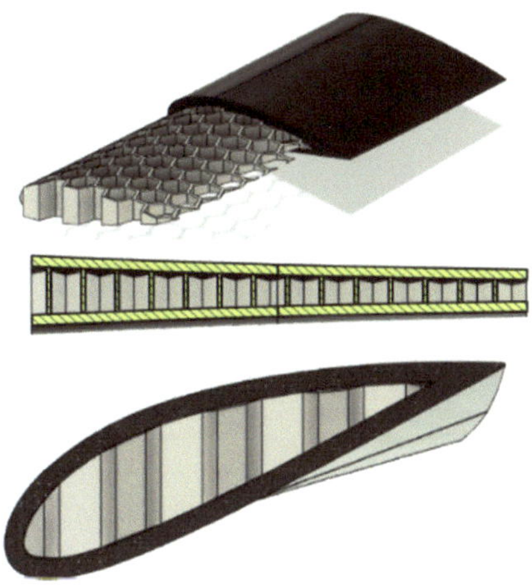

Figure 15. Visual representation of blade turbine sections with 0-degree orientation.

4. Results and Discussion

4.1. Infill Pattern Selection

The experimental data from the bend tests were extracted from the Instron software and post processed. The load-deflection curves from the bend tests, of the triangle and hexagonal infills with each having four different configurations, are plotted in Figure 16. These experimental data were the basis for the selection of the optimal infill configuration that was applied to the tidal turbine blade.

Based on these plots, the infill pattern and configuration that has the highest yield point was the in-plane hexagonal infill, oriented 90 degrees from the horizontal. Similar findings were found by Harpool's study where the hexagonal infill configuration was proven to have superior characteristics in terms of high ultimate strength [12]. The in-plane direction of the infill was also supported by Eryildiz [21], reasoning that the strut joints of hexagonal infills contribute to the strength of the infill. Furthermore, the slight difference in performance of 0° and 90° oriented hexagonal infills can be related to the aspect ratio and orientation of the lattice core. Since the in-plane hexagonal infill (90-degree orientation) has the highest yield point, it was used as the basis for simulating the infill pattern.

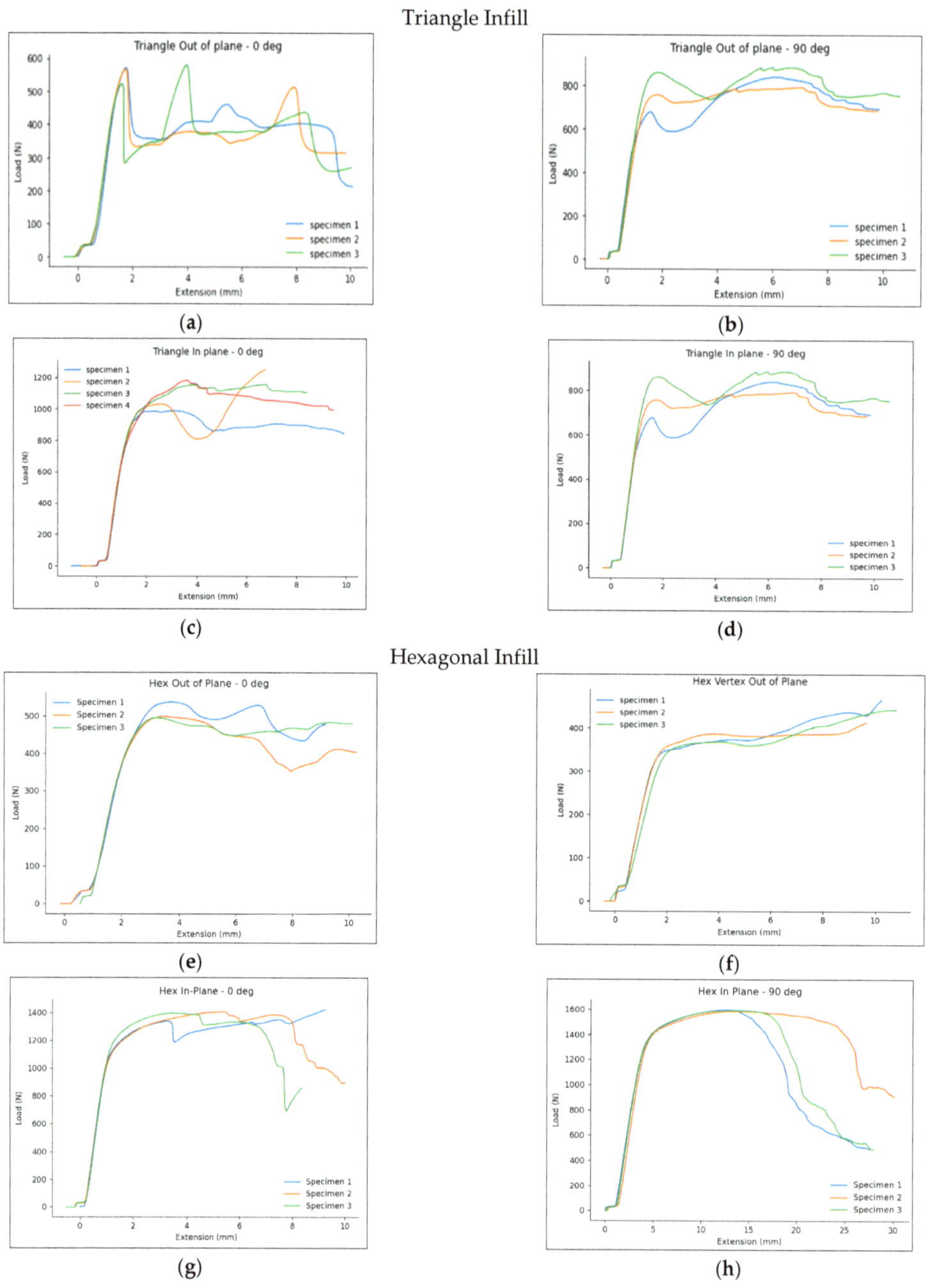

Figure 16. Load deflection curves from bend test. Triangle infill (**a**) out of plane, 0°, (**b**) out of plane, 90°, (**c**) in plane, 0°, (**d**) in plane, 90°. Hexagonal infill (**e**) out of plane, 0°, (**f**) out of plane, 90°, (**g**) in plane, 0°, (**h**) in plane, 90°.

A notable behavior in the load-displacement curve can be observed, particularly in Figure 16a, where there is an apparent fluctuation in the value of the load following the increase in displacement. This is the result of the layer-by-layer failure of the lattice structure, which results in a significant drop in load during the crushing of a particular layer. After the localized failure, the structure behaves as one elastic beam again, thus showing a gradual increase in the load as displacement increases.

4.2. Homogenized PLA Lattice

The RVE of the chosen hexagonal infill was designed in Material Designer from ANSYS with the following hexagonal pattern parameters: side length of 4 mm, an infill thickness of 0.8 mm, and relative density of 15%. The thickness of the infill was oriented perpendicular to the Z axis while two sides of the hexagon were oriented perpendicular to the X axis.

From the homogenized results, it was observed that the highest Young's modulus was oriented in the Z direction while the moduli in the X and Y directions were approximately equal (Table 6). In essence, enough bearing load must be applied along the Z-axis to counter the maximum modulus. This agreed with the results of the flexure tests, as hexagonal infills oriented in-plane demonstrated a stiffer behavior than infills oriented out-of-plane.

Table 6. Homogenized properties of the PLA lattice.

Property and Direction	Material Properties
Young's Modulus X	3.88631 MPa
Young's Modulus Y	3.88627 MPa
Young's Modulus Z	86.568 MPa
Poisson's Ratio XY	0.91915
Poisson's Ratio YZ	0.014726
Poisson's Ratio XZ	0.014725
Shear Modulus XY	1.0205 MPa
Shear Modulus YZ	17.484 MPa
Shear Modulus XZ	17.484 MPa

4.3. Validation Study of Homogenized Lattice Structure

The simulations performed were limited to the elastic behavior of the bend tests since the experimental data of the material were also limited to the elastic behavior of the PLA. In consideration of the blade performance, it was crucial for the turbine blade to deform within the elastic region given that any plastic deformation of the turbine blade would gravely affect its hydrodynamic performance. A study by Park, S. [38] stated that a deformation of about 14 percent of the blade length decreased the power coefficient of the blade by 1.9%, which is important for accurate power predictions at the off-design stage. For this reason, the plasticity of the material was out of the scope of this research.

The elastic region of the bend test was plotted against each other and compared. The homogenized CAD model were simulated in three mesh sizes: default mesh size, 2 mm and 3 mm element size. The mesh independence study demonstrated that the homogenized properties converged at the default mesh size. Based on the plot in Figure 17, the CAD model and the homogenized CAD model results were almost identical, meaning that the FEA simulations were conclusive on their own. Simulations were validated and solidified by experimental results; however, the experimental load-deflection curve appears to be slightly deviated from the lattice and homogenized simulation load-deflection curve in Figure 17.

As can be seen from Figure 17, the convergence of results occurred in the default size of mesh. The standard deviation of the experimental, default size (0.5 mm), and 3 mm size mesh bend test was 449.79, 287.21, and 270.81. Hence, the standard deviation percentage difference of the experimental bend test with the default size and 3 mm size is 36.14% and 39.79%, respectively. The deviation was believed to be associated to the inconsistencies of AM, anisotropy, porosity and imperfections attributed to the 3D printed sandwich models, and the limitations of the splicing program used. These aspects affect the accuracy of the volume fraction and distribution of printed materials that directly contribute to the deflection behavior of the sandwich models [35]. The deviation was also believed to be associated to the lack of the elasto-mechanical constitutive characteristics of the 3D printed PLA materials which may be obtained using the Modified Mazars damage model that Zhou employed in his study [25].

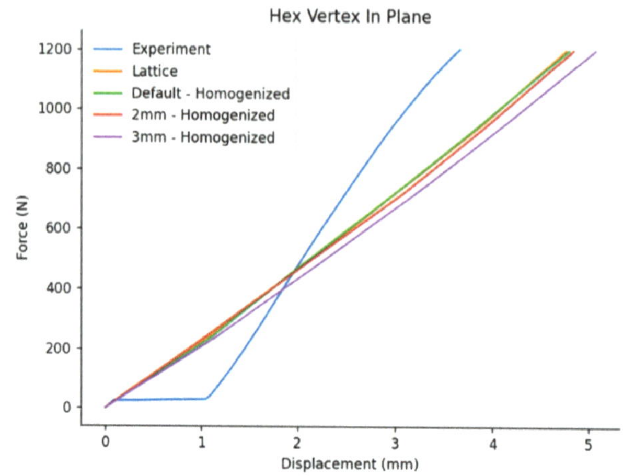

Figure 17. Load Deflection Curves from Bend Test.

Despite the slight deviation, the plot of the load–extension curves from the different modes were comparable. This showed that the developed FEA simulations for homogenization could be used to describe the linear elastic behavior of the sandwich structure in three-point bending tests with reasonable accuracy.

4.4. 3D Printing Material Replacement

A solid model of the tidal turbine blade was simulated with the PLA material and loading condition based on the research of Danao et al. [26]. From Figure 18, the computed maximum deflection occurred at the tip of the blade. The reported deflection was larger than the blade span itself apart from the fact that the solution did not converge. PLA was not suitable for the blade loading conditions subjected to it.

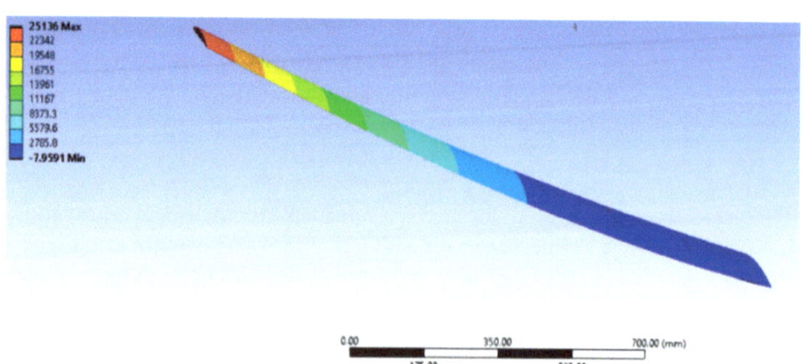

Figure 18. Solid blade model simulation with PLA.

The carbon fiber with onyx base was chosen to replace PLA for the simulation of the tidal turbine blade. The Young's modulus of the new material was 60 GPa while its Poisson's ratio was 0.3061 [37]. The homogenization process was conducted again with the same geometric specifications of the hexagonal infill pattern that were used from the PLA setup. Table 7 shows the homogenized properties of the carbon fiber lattice.

Table 7. Homogenized Properties of the Carbon Fiber lattice.

Property and Direction	Material Properties
Young's Modulus X	0.37195 GPa
Young's Modulus Y	0.37199 GPa
Young's Modulus Z	9 GPa
Poisson's Ratio XY	0.92441
Poisson's Ratio YZ	0.012399
Poisson's Ratio XZ	0.012398
Shear Modulus XY	0.0966 GPa
Shear Modulus YZ	1.8531 GPa
Shear Modulus XZ	1.853 GPa

The material had the highest values for the Young's modulus in the Z direction, a low Poisson's ratio on both the YZ and XZ direction, and high shear modulus also in the YZ and XZ direction. The results from the homogenized properties were in line with the results from the actual bending test of the beam where the material deformation has the lowest values in the in-plane orientations. The Z axis of the of the homogenized infill was the basis of the orientation of the infill in the turbine blade since it was the axis that had the highest value of the modulus of elasticity.

4.5. Turbine Blade Models Using Carbon Fiber with Onyx Base

In this section, the carbon fiber with onyx base material properties and homogenized properties were used for the blade shell and infill, respectively. A similar simulation setup was adopted from the previous PLA simulations performed. In Figure 19, the deformation of the solid blade model was reported to be 148.61 mm; this served as the minimum tip deflection that a carbon fiber with onyx base turbine can achieve.

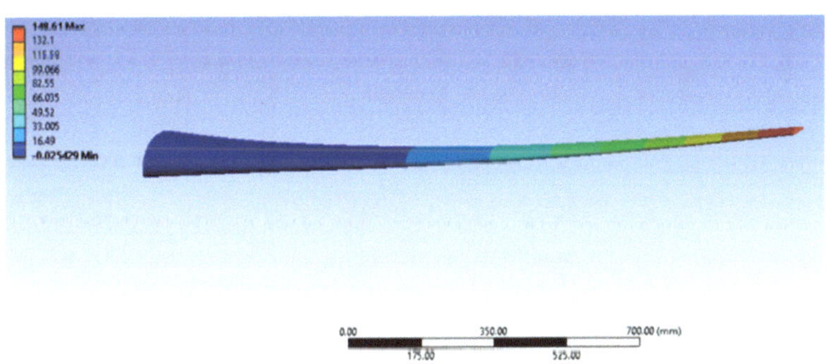

Figure 19. Solid blade model (carbon fiber with onyx base material).

The FEA results from the iterations of blade shell thickness for the shell-spar model are summarized in Table 8 which were based on the maximum chord length (MCL) of the blade. These simulations were conducted to determine the tip deflection for the corresponding blade thickness. The spar thickness used in all the shell-spar models was equal to 8% MCL (20 mm).

Table 8. Turbine blade configurations and deflection.

Turbine Blade Model	Description	Deflection (mm)	Deflection (%)
Solid Blade model	Solid Body	148.61	9.288125
Shell-Spar model	2% of MCL Shell Thickness (5 mm)	188.52	11.7825
Shell-Spar model	3% of MCL Shell Thickness (7.5 mm)	161.67	10.11
Shell-Spar model	3.5% of MCL (8.75 mm)	155.51	9.719375
Shell–infill model	3.596% of MCL Shell Thickness (8.99 mm) + 15% Infill	155.53	9.720625

The original design for the shell-spar model has a tip deflection of 188.52 mm which was greater than the recommended tip deflection for tidal turbine blades. From several iterations, the final shell-spar model was set to a shell thickness equal to 3.5% of the maximum chord length. It had a deformation of 9.719375% which was less than 10% of the total length of the blade (Figure 20), and it became the basis for the geometric specification of the turbine blade with the homogenized infill by adjusting the shell thickness and homogenized infill volume of the turbine blade. The volume of the turbine blade shell-spar model with 3.5% MCL (7.5 mm) shell thickness was matched with the volume of the turbine blade with the homogenized infill. The resulting deformation of the shell–infill model was observed to be 9.720625% of the span (Figure 21), still well within 10% of the blade length.

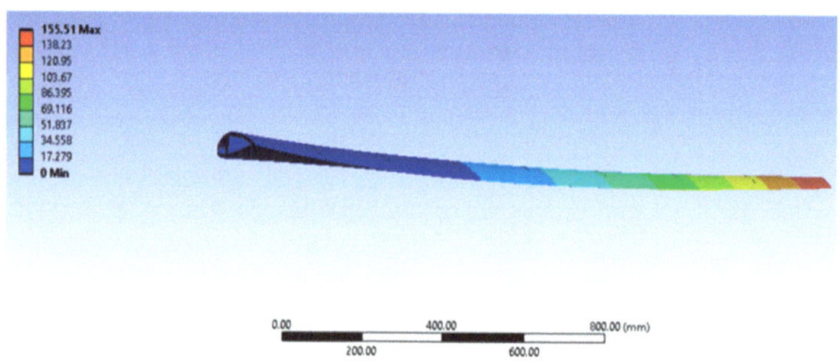

Figure 20. Directional deformation (mm) along the Z axis of the final shell-spar model (carbon fiber with onyx base material).

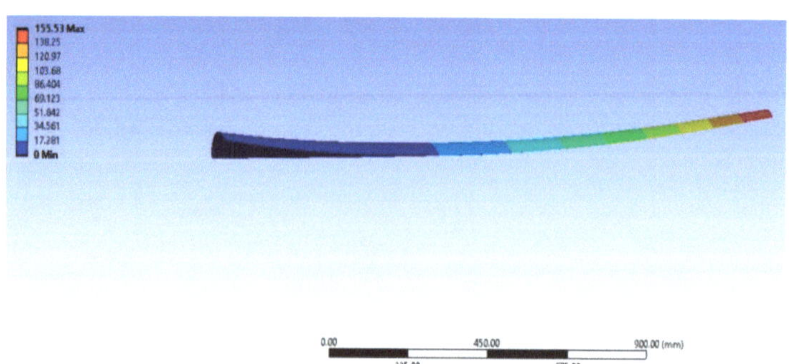

Figure 21. Directional deformation (mm) along the Z axis of shell–infill model (carbon fiber with onyx base material).

The maximum stress for the shell-spar model was found to be 166.96 MPa (Figure 22) while the maximum stress for the shell–infill model was found to be 170.3 MPa (Figure 23). Both maximum stresses were located a third of the blade span with respect to the blade root which was similar to the results of Danao et al. [26], and the locations of the maximum stresses were attributed to the rapid change in blade thickness [26]. The maximum stress value was well within the tensile strength (760 MPa) and flexural strength of the material (540 MPa). This meant that the turbine blade would not fail under the given loads. The datasheet lacked a value for the yield point and thus it was inconclusive if the stress was still within the elastic region.

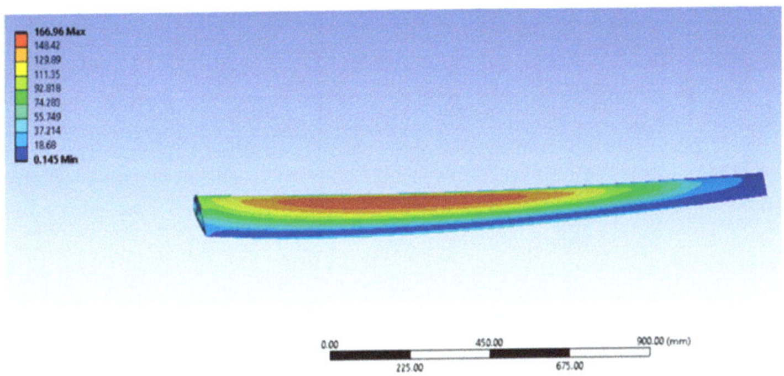

Figure 22. Von Mises stress (MPa) of shell-spar model (carbon fiber with onyx base material).

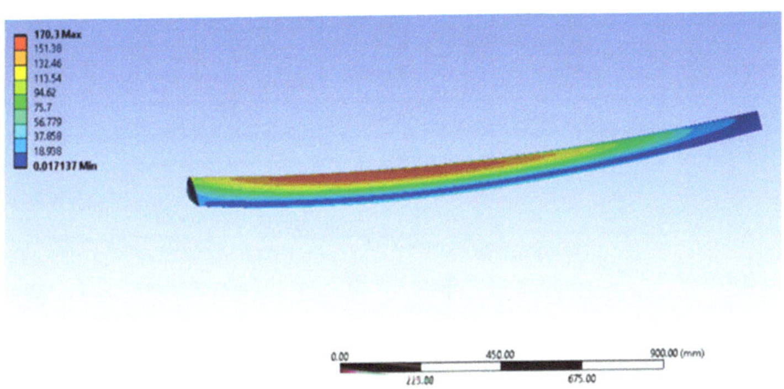

Figure 23. Von Mises stress (MPa) of shell-infill model (carbon fiber with onyx base material).

Based on the deflection along Z axis results and the equivalent stress results, the use of homogenized infill pattern can be used as an alternative to the spar cross-section of the blade due to the 0.00125% difference in the deflection and 2% difference in the maximum stress values.

4.6. Homogenized Infill Orientation

From the results of the infill orientation simulations, a dip in the deflection values was seen in the 0-degree infill orientation demonstrated in Figure 24a. At first glance, there was no discernable trend in varying the orientation of the infill pattern with respect to the horizontal. However, the variation of the deflection results relative to the blade span was minuscule, which ranges from 9.721–10.18%, that can easily be observed in Figure 24b. There is only a 0.46% difference between the maximum and the minimum value of the deflections for the different orientations.

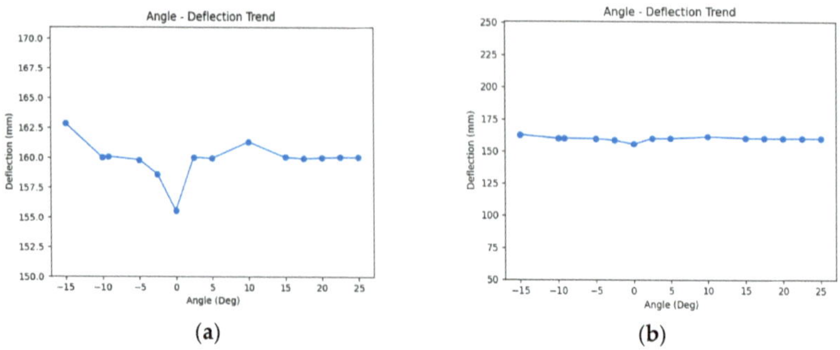

Figure 24. Angle vs. deflection trend of the turbine blade with different infill orientations. (**a**) Reduced y-axis range plot and (**b**) enlarged y-axis range plot.

The results are not sufficient to say that the 0-degree orientation was the most optimal orientation to achieve the lowest tip deflection. This was attributed to the complex geometry of the turbine blade model. The default orientation (0-degree) of the infill pattern was the best configuration of the different infill angles. Assuming that the turbine blade was a simple beam with the blade span parallel to the length of the beam, the 0-degree orientation of the blade has a flatwise configuration defined in the study of Rajkumar. By rotating the blade with respect to the y-axis, it approached an edgewise configuration which had the lowest stiffness. The findings of Rajkumar's study corresponded to the results of this study; as the infill orientation was rotated away from the flatwise configuration, the blade experienced lower modulus of elasticity and flexural strength [23].

5. Conclusions

This study addressed the knowledge gap of applying additive manufacturing to horizontal axis tidal turbines (HATT) through a numerical approach. The structural influence of infill patterns on tidal turbine blade designs was investigated which proved the viability of integrating infill patterns in tidal turbine blades as an alternative to the traditional design which was characterized by the blade shell and spar model.

Out of a selection of infills, the optimal infill pattern determined through the experiment was in-plane hexagonal infill, oriented 90° from the horizontal. The homogenized properties of the in-plane hexagonal infill (90°) impregnated with 3D printed materials were modeled and validated through finite element modelling and ASTM C393 bend test results. The slight deviations that occurred between the experimental results and the homogenization simulations were associated to the discrepancies of the simulation model and the 3D printed parts that were prone to anisotropy, porosity and imperfections. Despite these, the homogenized properties were in accordance with the experimental results and can be used with reasonable accuracy. The validated homogenized properties were then utilized in the static structural analysis of the tidal turbine blades with infill.

The initial findings of the FEA simulations showed that PLA was not qualified to be used in tidal turbine blades as it exceeded the blade deflection limit. The material was replaced with carbon fiber with onyx base which has significantly higher modulus of elasticity than PLA. The difference in the deflection between the homogenized infill and the spar cross-section was 0.00125% with a maximum stress of 170.3 MPa which was within tensile strength and flexure strength of the carbon fiber with onyx base material. This meant that the turbine blade would not fracture under the loading conditions. Due to the application of homogenized lattice structure, it is conclusive that the homogenized infill was considered as an alternative to the spar cross-section.

The variation study for the infill orientation revealed that the 0 degrees with respect to the horizontal orientation was the best infill orientation, resulting in the lowest deflection.

Owing to the lack of observed trend among the infill orientation tested, it was not conclusive if the 0-degree configuration was the absolute optimal infill orientation.

Author Contributions: Conceptualization, G.M.G., K.A.M.O., F.R.G.T., L.A.M.D. and B.E.A.; methodology, G.M.G., K.A.M.O., F.R.G.T., L.A.M.D. and B.E.A.; software, G.M.G., K.A.M.O. and F.R.G.T.; validation, G.M.G., K.A.M.O. and F.R.G.T.; formal analysis, G.M.G., K.A.M.O., F.R.G.T. and B.E.A.; writing—original draft, G.M.G., K.A.M.O. and F.R.G.T.; writing—review and editing, L.A.M.D. and B.E.A.; visualization, G.M.G., K.A.M.O. and F.R.G.T.; supervision, L.A.M.D. and B.E.A.; funding acquisition, L.A.M.D. and B.E.A. All authors have read and agreed to the published version of the manuscript.

Funding: The APC was funded by the Department of Science and Technology through the Engineering Research and Development for Technology Program.

Institutional Review Board Statement: Not applicable.

Informed Consent Statement: Not applicable.

Data Availability Statement: Not applicable.

Conflicts of Interest: The authors declare no conflict of interest.

Nomenclature

3D	three dimensional
HAWT	horizontal axis wind turbine
AM	additive manufacturing
LOM	laminated object manufacturing
ASTM	American Society for Testing and Materials
MCL	maximum chord length
BEM	blade element momentum
PLA	polylactic acid
CAD	computer aided design
RVE	representative volume element
FDM	fused deposition modelling
SLA	stereolithography
FEA	finite element analysis
SLS	selective laser sintering
FEM	finite element method
UTM	universal testing machine
HATT	horizontal axis tidal turbine
UV	ultraviolet

References

1. Department of Energy. National Renewable Energy Program. Republic of the Philippines—Department of Energy. 2017. Available online: https://doe.gov.ph/national-renewable-energy-program (accessed on 24 April 2022).
2. ISO/ASTM 52900:2021. 2021. Available online: https://www.iso.org/obp/ui/#iso:std:iso-astm:52900:ed-2:v1:en:ref:4 (accessed on 20 October 2021).
3. Rouway, M.; Nachtane, M.; Tarfaoui, M.; Chakhchaoui, N.; Omari, L.E.H.; Fraija, F.; Cherkaoui, O. 3D printing: Rapid manufacturing of a new small-scale tidal turbine blade. *Int. J. Adv. Manuf. Technol.* **2021**, *115*, 61–76. [CrossRef]
4. Pereira, T.; Kennedy, J.V.; Potgieter, J. A comparison of traditional manufacturing vs additive manufacturing, the best method for the job. *Procedia Manuf.* **2019**, *30*, 11–18. [CrossRef]
5. Kumar, M.B.; Sathiya, P. Methods and materials for additive manufacturing: A critical review on advancements and challenges. *Thin-Walled Struct.* **2020**, *159*, 107228. [CrossRef]
6. Abdulhameed, O.; Al-Ahmari, A.; Ameen, W.; Mian, S.H. Additive manufacturing: Challenges, trends, and applications. *Adv. Mech. Eng.* **2019**, *11*, 1687814018822880.
7. Gibson, I.; Rosen, D.W.; Stucker, B. Development of additive manufacturing technologies. In *Additive Manufacturing Technologies: Rapid Prototyping to Direct Digital Manufacturing*; Springer: Berlin/Heidelberg, Germany, 2021; pp. 17–40.
8. Bikas, H.; Stavropoulos, P.; Chryssolouris, G. Additive manufacturing methods and modelling approaches: A critical review. *Int. J. Adv. Manuf. Technol.* **2016**, *83*, 389–405. [CrossRef]

9. Fernández, V.; García, A.; Camacho, A.M.; Claver, J.; Rodríguez, A.; Sebastián, M.A. Methodology to estimate the modulus of elasticity of parts manufactured by FFF/FDM combining finite element simulations and experimental tests. *IOP Conf. Series: Mater. Sci. Eng.* **2021**, *1193*, 012103. [CrossRef]
10. Kalyani, K.; Shafee, S.; Rajasri, A.; SaiManichandra, V. Examination of Aerofoil Blade Fabricated by Additive Manufacturing Process. *Int. J. Recent Technol. Eng.* **2019**, *2S11*, 1–3.
11. Kim, S.; Chen, X.; Dreifus, G.; Lindahl, J.; Kang, I.; Kim, J.H.; Selim, M.; Nuttall, D.; Messing, A.; Nycz, A.; et al. *An Integrated Design Approach for Infill Patterning of Fused Deposition Modeling Ang Its Application to an Airfoil*; Oak Ridge National Lab. (ORNL): Oak Ridge, TN, USA, 2017; pp. 2–5.
12. Harpool, T.D. Observing the Effect of Infill Shapes on the Tensile Characteristics of 3D Printed Plastic Parts. Master's Thesis, Wichita State University, Wichita, KS, USA, 2016; pp. 64–77.
13. Cho, E.; Hein, H.; Lynn, Z.; Hla, S.; Tran, T. Investigation on Influence of Infill Pattern and Layer Thickness on Mechanical Strength of PLA Material in 3D Printing Technology. *J. Eng. Sci. Res.* **2019**, *3*, 27–37. [CrossRef]
14. Luzanin, O.; Movrin, D.; Plancak, M. Effect of layer thickness, deposition angle, and infill on maximum flexural force in FDM-built specimens. *J. Technol. Plast.* **2014**, *39*, 52–56.
15. Gurrala, P.K.; Regalla, S.P. Multi-objective optimisation of strength and volumetric shrinkage of FDM parts. *Virtual Phys. Prototyp.* **2014**, *9*, 127–138. [CrossRef]
16. Álvarez, K.; Lagos, R.F.; Aizpun, M. Investigating the influence of infill percentage on the mechanical properties of fused deposition modelled ABS parts. *Ing. E Investig.* **2016**, *36*, 110. [CrossRef]
17. Dudescu, C.; Racz, L. Effects of Raster Orientation, Infill Rate and Infill Pattern on the Mechanical Properties of 3D Printed Materials. *Acta Univ. Cibiniensis* **2017**, *69*, 23–30. [CrossRef]
18. Porter, J.H.; Cain, T.M.; Fox, S.L.; Harvey, P.S. Influence of infill properties on flexural rigidity of 3D-printed structural members. *Virtual Phys. Prototyp.* **2018**, *14*, 148–159. [CrossRef]
19. Adams, D.; Turner, C.J. An implicit slicing method for additive manufacturing processes. *Virtual Phys. Prototyp.* **2017**, *13*, 2–7. [CrossRef]
20. XZhou, X.; Hsieh, S.-J.; Ting, C.-C. Modelling and estimation of tensile behaviour of polylactic acid parts manufactured by fused deposition modelling using finite element analysis and knowledge-based library. *Virtual Phys. Prototyp.* **2018**, *13*, 177–190. [CrossRef]
21. Eryildiz, M. The effects of infill patterns on the mechanical properties of 3D printed PLA parts fabricated by FDM. *Ukr. J. Mech. Eng. Mater. Sci.* **2021**, *7*, 1–8. [CrossRef]
22. Kim, S.; Dreifus, G.; Beard, B.; Glick, A.; Messing, A.; Hassen, A.A.; Lindahl, J.; Liu, P.; Smith, T.C.; Failla, J.; et al. Graded Infill Structure of Wind Turbine Blade Core Accounting For Internal Stress in Big Area Additive Manufacturing. In Proceedings of the CAMX Composite and Advanced Materials Expo, Dallas, TX, USA, 15–18 October 2018; pp. 5–6.
23. Tanveer, Q.; Mishra, G.; Mishra, S.; Sharma, R. Effect of infill pattern and infill density on mechanical behaviour of FDM 3D printed Parts-a current review. *Mater. Today: Proc.* **2022**, *62*, 100–108. [CrossRef]
24. Bhandari, S.; Lopez-Anido, R. Finite element modeling of 3D-printed part with cellular internal structure using homogenized properties. *Prog. Addit. Manuf.* **2018**, *4*, 143–154. [CrossRef]
25. Zhou, X.; Li, J.; Qu, C.; Bu, W.; Liu, Z.; Fan, Y.; Bao, G. Bending behavior of hybrid sandwich composite structures containing 3D printed PLA lattice cores and magnesium alloy face sheets. *J. Adhes.* **2021**, *98*, 1–19. [CrossRef]
26. Danao, L.; Abuan, B.; Howell, R. Design Analysis of a Horizontal Axis Tidal Turbine. In Proceedings of the Asian Wave and Tidal Conference 2016, Singapore, 24–28 October 2016; pp. 4–7.
27. Malhotra, P.; Hyers, R.W.; Manwell, J.F.; McGowan, J.G. A review and Design Study of blade testing systems for Utitulity-scale wind turbines. *Renew. Sustain. Energy* **2012**, *16*, 284–292. [CrossRef]
28. Kong, C.; Sugiyama, Y.; Lee, J.; Soutis, C. Full Scale Structural Experimental Investigation of an E-glass/ Epoxy Composite Wind Turbine Blade. *Adv. Compos. Lett.* **2002**, *11*, 096369350201100502. [CrossRef]
29. Mishnaevsky, L.; Branner, K.; Petersen, H.N.; Beauson, J.; McGugan, M.; Sorensen, B.F. Materials for Wind Turbine: An overview. *Materials* **2017**, *10*, 1285. [CrossRef] [PubMed]
30. Bahaj, A.S.; Batten, W.M.J.; McCann, G. Experimental verifications of numerical predictions for the hydrodynamic performance of horizontal axis marine current turbines. *Renew. Energy* **2007**, *32*, 2479–2490. [CrossRef]
31. Noronha, N.; Munishamaiah, K. Design and analysis of micro horizontal wind turbine using MATLAB and Qblade. *Int. J. Adv. Sci. Technol.* **2020**, *29*, 8881–8882.
32. Bassett, K.; Carriveau, R.; Ting, D.-K. 3D printed wind turbines part 1: Design considerations and rapid manufacture potential. *Sustain. Energy Technol. Assessments* **2015**, *11*, 186–193. [CrossRef]
33. *D638-14*; Standard Test Method for Tensile Properties of Plastics. ASTM: West Conshohocken, PA, USA, 2020.
34. Sarvestani, H.Y.; Akbarzadeh, A.; Niknam, H.; Hermenean, K. 3D printed architected polymeric sandwich panels: Energy absorption and structural performance. *Compos. Struct.* **2018**, *200*, 886–909. [CrossRef]
35. Li, T.; Wang, L. Bending behavior of sandwich composite structures with tunable 3D-printed core materials. *Compos. Struct.* **2017**, *175*, 46–57. [CrossRef]
36. *C393/C393M-16*; Standard Test Method for Flexural Properties of Sandwich Constructions. ASTM: West Conshohocken, PA, USA, 2016.

37. Markforged. Material Data Sheet Composites. In *Carbon Fiber with Onyx Base Data Sheet, pg 2, REV 5.0*; Markforged, Inc.: Watertown, MA, USA, 2021. Available online: https://static.markforged.com/downloads/composites-data-sheet.pdf (accessed on 26 December 2022).
38. Park, S.; Park, S.; Rhee, S. Influence of Blade Deformation and Yawed inflow on performance of a horizontal axis tidal turbine blade. *Renew. Energy* **2016**, *92*, 327–331. [CrossRef]

sustainability

Article

Design and Test for a New Type of Permanent Magnet Synchronous Generator Applied in Tidal Current Energy System

Yuxiang Zhao [1,2], Caixia Mo [1,2], Wanqiang Zhu [1,2,*], Jianmei Chen [1,2], Baigong Wu [3], Xiao Zhang [4], Xueming Zhang [1,2] and Liwei Chen [5]

1 School of Physics, Northeast Normal University, 5268 Renmin Street, Changchun 130024, China
2 Jilin Provincial Key Laboratory of Advanced Energy Development and Application Innovation, 5268 Renmin Street, Changchun 130024, China
3 Marine Equipment and Technology Institute, Jiangsu University of Science and Technology, Zhenjiang 212000, China
4 School of Energy and Power Engineering, Changchun Institute of Technology, Changchun 130012, China
5 Hangzhou Jianghe Hydro-Electric Science & Technology Co., Ltd., Hangzhou 310012, China
* Correspondence: zhuwq773@nenu.edu.cn

Abstract: This article takes a 120 W direct-drive permanent magnet synchronous generator (PMSG) for tidal current energy as the research object, with the aim of starting at low flow velocity, low voltage regulation, and high sealing performance. The research was conducted on the design method of the generator, and the parameter calculation formula is modified based on the structural characteristics of the generator. The design scheme is simulated using the finite element method, and the voltage regulation is optimized using evolutionary algorithm. A prototype was manufactured based on the optimized scheme, and drag experiments and sealing tests were conducted. After that, the generator and impeller were assembled, and water tank drag experiments were conducted. The experimental results verified the reliability of the motor design method in this paper.

Keywords: tidal current energy; permanent magnet synchronous generator; finite element simulation; genetic algorithm; sealing test

Citation: Zhao, Y.; Mo, C.; Zhu, W.; Chen, J.; Wu, B.; Zhang, X.; Zhang, X.; Chen, L. Design and Test for a New Type of Permanent Magnet Synchronous Generator Applied in Tidal Current Energy System. *Sustainability* **2023**, *15*, 7378. https://doi.org/10.3390/su15097378

Academic Editor: Grigorios L. Kyriakopoulos

Received: 6 March 2023
Revised: 9 April 2023
Accepted: 25 April 2023
Published: 28 April 2023

1. Introduction

As fossil fuels continue to deplete and environmental issues gain more attention, there is a growing global focus on finding clean and renewable energy sources to replace traditional fossil fuels [1]. Tidal current energy is a kind of kinetic energy produced by the periodic horizontal movement of seawater, which is caused by the gravitational pull of the sun and moon. China possesses a significant amount of tidal current energy resources, which has become a topic of interest for many domestic research institutions [2].

The power generation cost of tidal power generation devices decreases with the increase of installed capacity. Research has shown that, when the installed capacity reaches 100 MW, the cost of tidal power generation will be highly competitive compared to traditional power generation models. This power generation method will become an effective supplement to traditional power generation methods and will ultimately partially replace traditional power generation.

Tidal current power generation was first researched in coastal countries in Europe. In 2015, the 1 MW project prototype designed by the Italian National Institute of Ocean Engineering was installed in Canada's Bay of Fundy [3]. In 2017, the HS1500 horizontal shaft unit (1.5 MW) designed by Atlantis Resources UK was installed in relevant sea areas and realized grid connection operation [4]. China has increased investment in tidal current power generation technology since the beginning of the 21st century. In April 2002, the "Wanxiang I" tidal current power generation device (70 kW) was researched and built by Harbin Engineering University; it was then installed in Guishan waterway for a test operation [5]. In 2013, Northeast Normal University began to research and design a 300 kW horizontal-axis self-variable pitch project prototype, which was the first time that passive automatic pitch angles adjustment technology had been applied to tidal power generation [6]. In 2017, Zhejiang University designed the 650 kW horizontal-axis tidal current power generation device, which was successfully connected to the grid in Zhoushan [7].

In terms of generator design, Moury et al. proposed a permanent magnet synchronous generator with axial permanent magnet structure, which is complex in structure and high in cost [8]. Lindh et al. compared the difference between the surface-mounted and the embedded permanent magnet structures, then proposed a type of new embedded structure, but the new structure was complex and inconvenient to install [9]. Fizari et al. proposed a method to design a 1.5 kW PMSG that required low-speed starting performance and high reliability. Maxwell design software was used to obtain the parameter design results with 20 poles and 30 slots [10]. Nikbakhsh et al. first introduced the analysis, calculation, and design process of the PMSG, and used the ant colony optimization algorithm to optimize the design. The optimized design improves the efficiency and reduces the volume [11]. Quintal Palomo et al. conducted a finite element analysis of a surface-mounted PMSG. To obtain the lowest total harmonic distortion and cogging torque under the highest induced voltage, several optimization tests were conducted [12]. Faiz et al. introduced the design process of a three-phase PMSG for wind power generation. The AlNiCo permanent magnetic material with low coercivity was used to replace the ordinary and expensive NdFeB permanent magnetic material. To reduce the cogging torque and improve the generator performance, the optimal number of cogging and magnetic poles was determined [13].

After analyzing various power generation devices and design methods, it becomes clear that there is currently no research dedicated to underwater generators specifically designed for low-speed tidal current. This paper presents a design model for an underwater generator with large air gap that takes into consideration the low-speed start, large torque and insulation sealing characteristics applied in tidal current energy system. The design is evaluated through simulation analysis, and a prototype is manufactured based on the evolutionary algorithms optimization results for reduce the voltage regulation, a kind of underwater centrifugal sealing technology is used to seal the generator. A pool drag experiment is conducted to test the prototype's ability to verify the low-speed starting performance of the generator. The experimental results demonstrate the effectiveness of the design and sealing methods for tidal current energy PMSG presented in this paper.

2. Underwater Generator Design

2.1. Generator Specifications

The primary dimensions of the generator include the outer diameter (D_{il}) of the armature near the air gap and the effective length (L_{ef}) of the iron core. To determine the main dimensions of the generator, it is necessary to first identify the rated parameters [14]. The rated parameters of the generator are shown in Table 1.

Table 1. Rated parameters of the generator.

Rated Power	Rated Voltage	Rated Current	Rated Rotation Speed
120 W	36 V	1.93 A	28 r/min

2.2. Generator Primary Dimensions

The generator adopts an embedded magnetic pole structure on the surface of the outer rotor. To preliminarily determine the primary dimensions of the generator, the calculated power P' shall be calculated according to Formula (1) based on the rated power P_N.

$$P' = \frac{K_E P_N}{\cos \varphi_N} \tag{1}$$

where K_E is ratio of induced potential and terminal voltage at rated load, $\cos\varphi_N$ is power factor at rated load.

Because the underwater sealing performance of the generator needs to be considered, the thickness of the sealant layer on the stator surface should be reserved [15] when designing the air gap length. By referencing the primary dimension formula used for the traditional motor, the primary dimension formula of the tidal current energy PMSG is modified as follows:

$$\frac{{D_{il}}^2 L_{ef} n}{P'} = \frac{\alpha_{xz} 6.1 \times 10^4}{\alpha'_p K_{Nm} K_{dp} A B_\delta} = C_A \tag{2}$$

where n is generator rotation speed, α'_p is calculated pole arc coefficient, K_{Nm} is air gap magnetic field waveform coefficient, K_{dp} is armature winding coefficient, A is line load, B_δ is maximum value of air gap magnetic density. α_{xz} is air gap correction coefficient, which is used to correct the influence of the sealant layer on the air gap length.

The variation range of K_{Nm} and K_{dp} is small, the electromagnetic loads A and B_δ are selected according to experience, and the approximate value ${D_{il}}^2 L_{ef}$ of the effective volume can be determined by Formula (2). However, the same value of the effective volume could have different shapes. In order to reflect the geometric shape of the generator, the main dimension ratio λ is introduced. The expression of λ is Formula (3).

$$\lambda = \frac{L_{ef}}{\tau} \tag{3}$$

where τ is polar distance. Select the right λ and the armature outer diameter D_{il}, value then the effective length of iron core L_{ef} could be preliminarily determined.

2.3. Size of Permanent Magnets

The permanent magnet size includes the axial length L_M, radial magnetization length h_M, and area S_M. L_M is slightly smaller than the axial length of the generator core, so only the radial magnetization length h_M and area S_M need to be designed. These two parameters can be obtained from Formulas (4) and (5) [16].

$$h_M = \frac{2k_\alpha k_\delta b_{m0} \delta}{(1 - b_{m0})\sigma_0} \tag{4}$$

$$S_M = \frac{2\sigma B_\delta S_\delta}{B_r} \tag{5}$$

where k_α is the rotor structure coefficient, k_δ is the saturation coefficient, and b_{m0} is the estimated value of permanent magnet no-load working point, δ is air gap length, σ_0 is the no-load magnetic leakage coefficient, σ is the magnetic leakage coefficient, B_δ is the air gap magnetic density, S_δ is the air gap area, and B_r is the remanence.

2.4. Number of Stator Slot

Once the number of poles and phases are established, the number of stator slots is determined by the number of slots q for per pole and phase. Tidal current energy PMSG operates at low speed, so it requires multi-pole structure, which influences the parameter q. However, choosing an integer value for q could cause an increase in the cogging torque and electromagnetic loss. Therefore, the generator adopts fractional slot winding design, and the number of slots q per pole and phase can be found using Formula (6).

$$q = \frac{Z}{2pm} = b + \frac{c}{d} = b + \frac{e}{2p} \tag{6}$$

where Z is the number of stator slots, p is the number of pole pairs, m is the number of phases, b is an integer.

2.5. Winding Design

The induced voltage can be calculated using Formula (7):

$$E = 4.44Nf\phi_g \tag{7}$$

where N is the number of turns, f is the frequency of generator, Φ_g is the magnetic flux of each pole, which can be obtained from Formula (8):

$$\phi_g = \frac{2\pi R_g L_{ef} B_g}{2p} \tag{8}$$

where R_g is the average air gap radius and B_g is the air gap magnetic flux density.

According to the selected current density J and rated current I, the conductor area and diameter can be calculated using Formulas (9) and (10):

$$A_\omega = \frac{I}{J} \tag{9}$$

$$d_\omega = \sqrt{\frac{4A_\omega}{\pi}} \tag{10}$$

2.6. Generator Design Parameters

The basic parameters of the generator calculated according to the above formulas and experience are shown in Table 2:

Table 2. Generator design parameters.

Pole Number	Slot Number	Rotor External Diameter	Stator External Diameter	Stator Bore	Air Gap Length	Core Length	Permanent Magnet Thickness	Pole Arc Coefficient
40	72	620 mm	590 mm	476 mm	2.5 mm	25 mm	5 mm	0.94

3. Finite Element Simulation Analysis

3.1. Simulation Modeling

Ansoft Maxwell software adopts finite element discrete form to transform electromagnetic field calculation in engineering into huge matrix solutions [17]. This paper uses the RMxprt module in Ansoft Maxwell to establish a PMSG model for simulation analysis and imports the prototype model into the Maxwell 2D environment [18]. The established generator quarter model is shown in Figure 1, and the winding arrangement of the generator is shown in Figure 2.

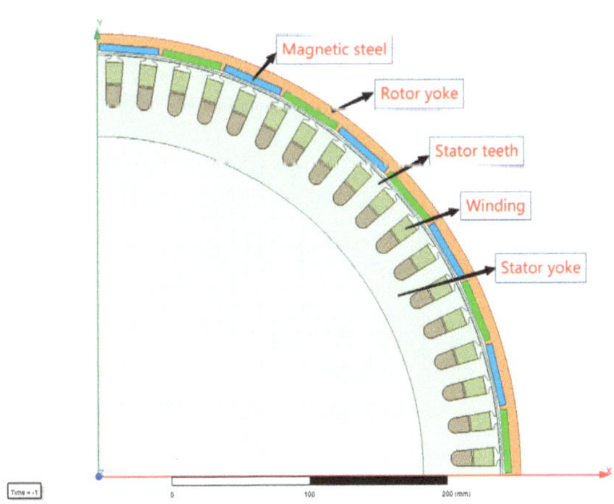

Figure 1. Generator quarter model.

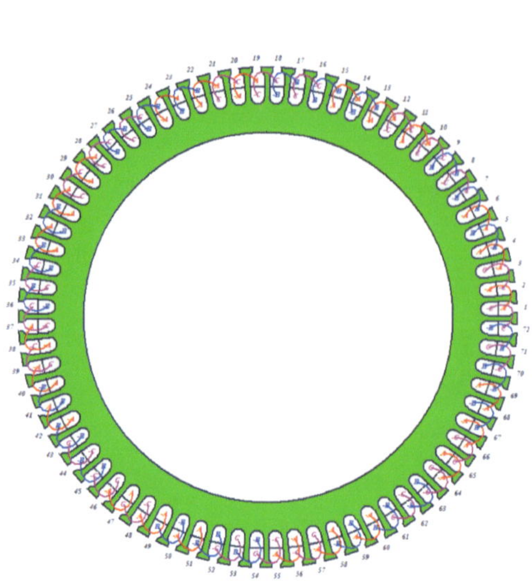

Figure 2. Winding arrangement diagram.

3.2. No-Load Simulation Analysis

The simplest operating mode of the PMSG is the no-load mode, where the air gap magnetic field is provided solely by the permanent magnet. Analyzing the no-load operation is a crucial step in verifying whether the generator performance meets standards and the magnetic circuit design is reasonable. For conducting no-load analysis, the model excitation source is selected as current source, and the current value is set to 0 A. The simulation results are presented In Figure 3. As shown in Figure 3, the magnetic density distribution is reasonable when the generator operates under no load, and there are no oversaturation points. Additionally, the distribution of magnetic field lines is uniform, and the magnetic leakage of the permanent magnet is small. As shown in Figure 4, the no-load voltage waveform is close to a sine wave, and the maximum phase voltage $U_{\max}$ is 30.6 V. According to Formula (11):

$$U_e = 1.225U_{\max} \tag{11}$$

where U_e is the rated voltage, the no-load simulation line voltage is 37.5 V, which is close to the designed rated voltage of 36 V.

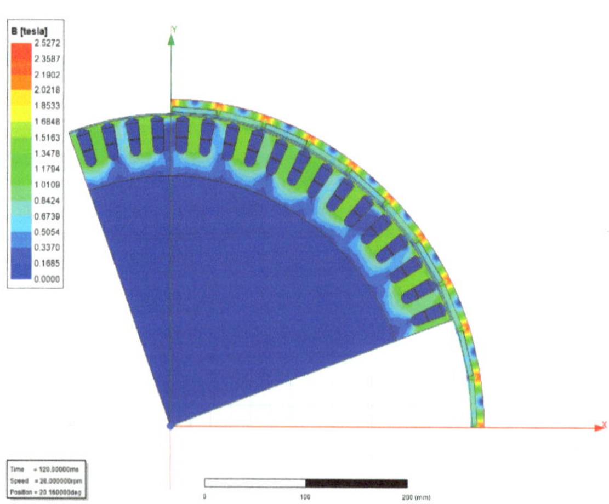

Figure 3. No load magnetic cloud diagram of generator.

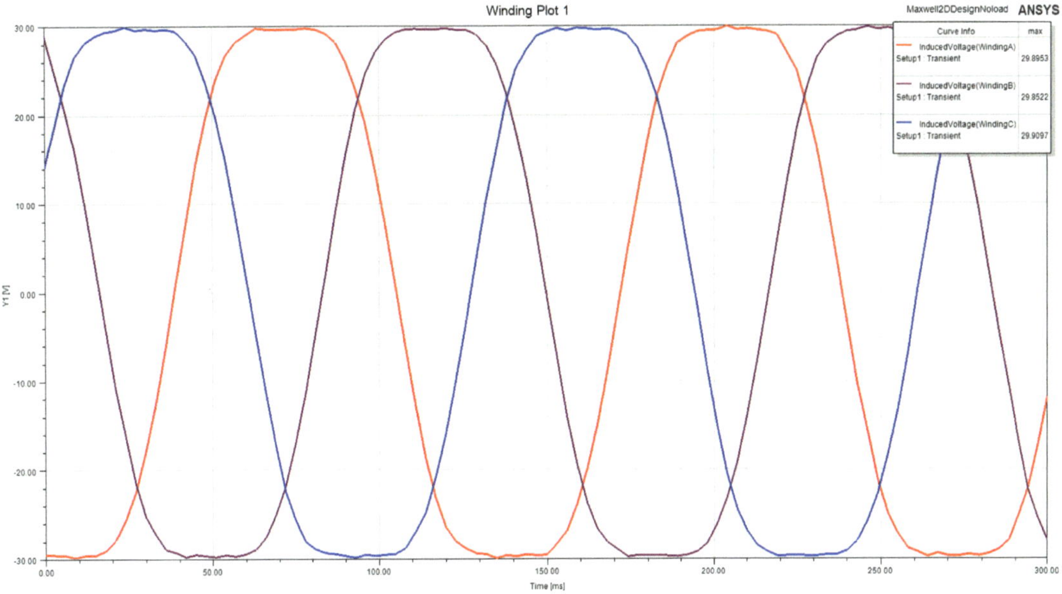

Figure 4. No load induced voltage of generator.

It can be seen from Figure 5 that the maximum cogging torque of the generator is 0.84 N·m, but that it is only 2% of the rated torque, which can meet the design requirements.

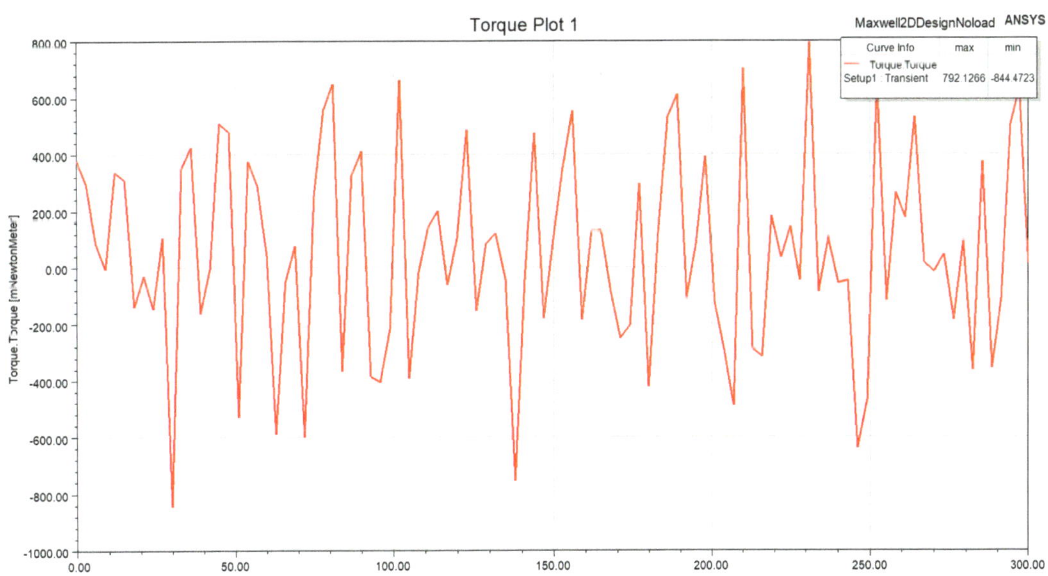

Figure 5. No load cogging torque of generator.

3.3. Load Simulation Analysis

By constructing the external circuit model of the generator, it is possible to simulate and analyze the generator at rated load. The load is a pure resistive load of 22.8 Ω. The results of the load simulation are presented in Figures 6–9. Figure 6 shows that there are no oversaturation points in the magnetic density of the generator during load operation. Figure 7 shows that the maximum load-induced phase voltage of the generator is 29.8 V, and according to Formula (11), the load simulation line voltage is 36.5 V, satisfying the design requirement of 36 V rated voltage. Figure 8 shows that the maximum load-induced current of the generator is 2.63 A, which matches the design requirement of 1.93 A rated current. Figure 9 shows that the load torque of the generator is 42 N·m, according to Formula (12):

$$T = \frac{9550P}{n} \tag{12}$$

where T is torque, P is rated power, and n is rated rotation speed. The simulation result is close to the calculated value of 41 N·m.

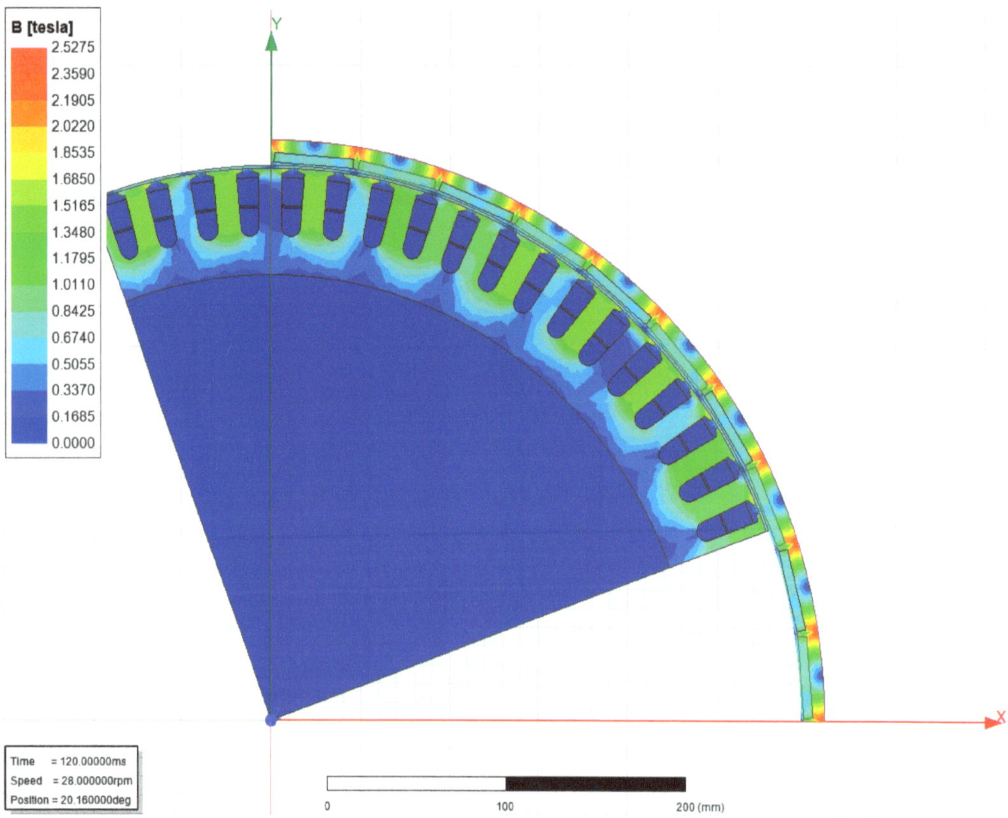

Figure 6. Load magnetic cloud diagram of generator.

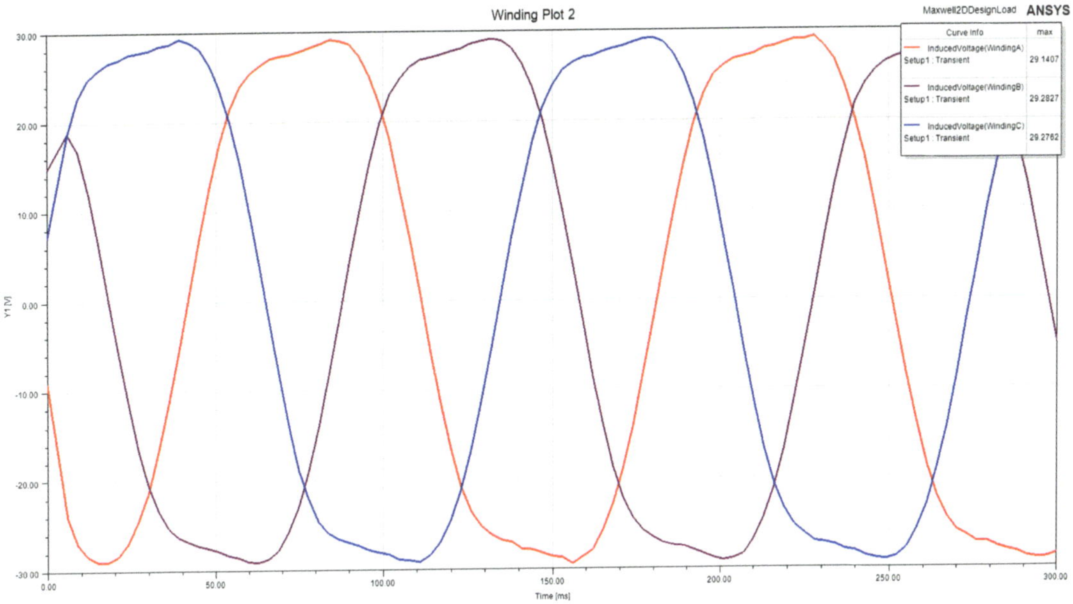

Figure 7. Load induced voltage of generator.

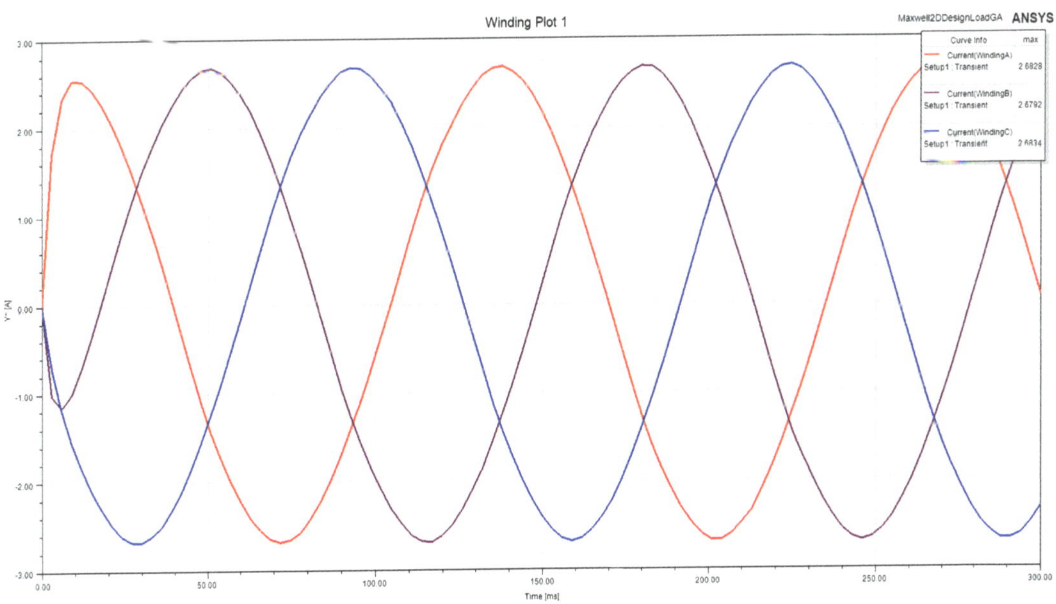

Figure 8. Line current of generator.

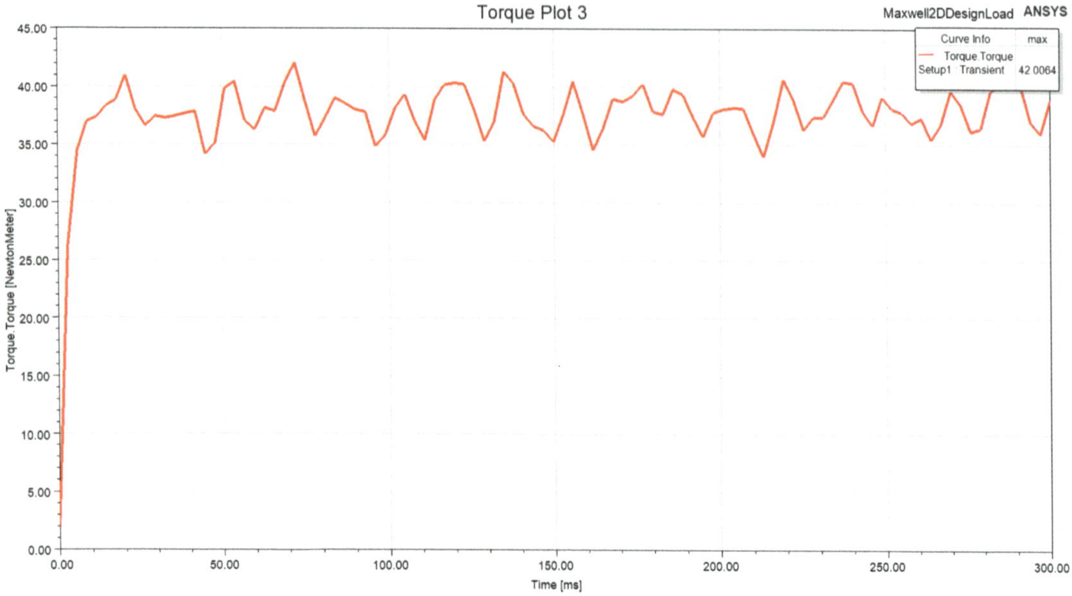

Figure 9. Load cogging torque of generator.

4. Genetic Algorithm Optimization of Generator Parameters

4.1. Analysis of the Voltage Regulation

Voltage regulation ΔU represents the change of output voltage when the load changes. Its value depends entirely on the basic characteristics of the generator itself, and its expression is:

$$\Delta U = \frac{E_0 - U}{U_N} \times 100\% \tag{13}$$

where E_0 is the no-load induced voltage, U is the output voltage, and U_N is the rated voltage.

Since it is difficult to adjust the air gap magnetic field after the PMSG is made, it is necessary to consider how to reduce ΔU at the design stage. The motor parameters, such as synchronous reactance X_q and X_d, armature winding resistance R_1, and leakage reactance X_1 must be reasonably designed at each phase. The specific factors that affect the above parameters are air gap length, iron core length, permanent magnet magnetization direction length, and stator slot size. Therefore, the influence of each parameter on the voltage regulation is analyzed below [19].

4.1.1. Effect of Air Gap Length

The approximate calculation formula of quadrature axis armature reaction reactance X_{aq} of PMSG is:

$$X_{aq} = \frac{E_0 F_{aq}}{I_N F_\delta} K_{aq} \frac{1}{1 + \frac{b_M}{2\mu_r \delta K_\delta}} \tag{14}$$

where F_{aq} is quadrature axis armature magnetomotive force, F_δ is air gap magnetic potential difference, K_{aq} is the conversion coefficient of the quadrature axis armature magnetomotive force, b_M is permanent magnet width, μ_r is the relative recovery permeability of the permanent magnet, δ is air gap length, and K_δ is air gap coefficient. The air gap length can be determined by the above formula.

4.1.2. Effect of Core Length

Under the rated load, the PMSG's direct axis electromotive force (E_d) and quadrature axis armature reactance (X_{aq}) are both correlated to axial length (L_M) of the permanent magnet and calculated length (L_{ef}) of the armature. The expressions are:

$$E_d = 4.44 f_1 N K_{dp} K_\phi \Phi_{\delta N} \tag{15}$$

$$\phi_{\delta N} = (b_{mN} - h_{mN}\lambda_\sigma) B_r A_m \times 10^{-4} \tag{16}$$

$$X_{aq} = \frac{2}{\pi} L_{ef} \tau K_{dp} N \frac{B_{aq1}}{I_q} \tag{17}$$

$$L_{ef} = L_1 + 2\delta \tag{18}$$

where A_m = $b_M L_M$, L_1 is the length of stator core. The axial length L_M of the permanent magnet is equal to L_1. It can be seen that L_1 can directly affect E_d and X_{aq}.

4.1.3. Effect of Permanent Magnet Magnetization Direction Length

When the PMSG is under rated load, the relationship between E_d and permanent magnet thickness h_{Mp} is as follows:

$$b_{mN} = \varphi_{mN} = \frac{\lambda_n (1 - f')}{\lambda_n + 1} \tag{19}$$

$$h_{mN} = f_{mN} = \frac{\lambda_n f' + 1}{\lambda_n + 1} \tag{20}$$

$$f' = \frac{F_{ad}}{\sigma_0 H_c h_{Mp} \times 10^{-2}} \tag{21}$$

where λ_N is the composite magnetic guide unit value of the external magnetic circuit, σ_0 is no-load magnetic leakage coefficient, and F_{ad} is direct axis armature magnetomotive force. It can be seen that h_{Mp} can change the voltage regulation by influencing the direct axis electromotive force E_d.

4.1.4. Effect of Stator Slot Size

The calculation formula of each phase armature winding leakage reactance X_1 is as follows:

$$X_1 = 15.5 \frac{f}{100} \left(\frac{N}{100}\right)^2 \frac{L_1}{pq} \sum \lambda \times 10^{-2} \tag{22}$$

where λ is the total magnetic leakage permeability, and its expression is:

$$\sum \lambda = \lambda_t + \lambda_d + \lambda_E + \lambda_s \tag{23}$$

where λ_t is tooth-tip-specific magnetic leakage conductance, λ_d is differential leakage permeance, λ_E is the specific leakage magnetic conductivity at the end, λ_S is slot-specific magnetic permeability [20]. The slot-specific magnetic leakage conductance is mainly affected by the stator slot type, and its expression is:

$$\lambda_s = \frac{1}{4}\left[0.31 + \frac{2h_1}{3(d_2 + b)} + \frac{h_2}{b} + \frac{k_1 h_3}{b + d_1} + k_2\left(0.985 + \frac{h_5}{b_0}\right)\right] \tag{24}$$

The specific meaning of the parameter is shown in Figure 10.

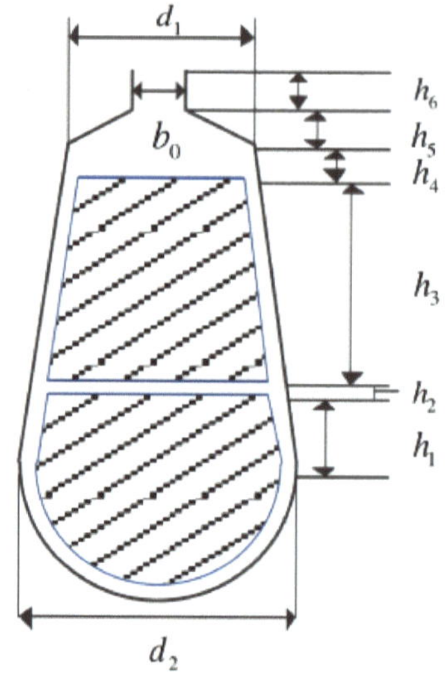

Figure 10. Stator slot size parameters.

It can be seen from the above formulas that the stator slot size changes the voltage regulation by affecting the winding leakage reactance X_1.

4.2. Voltage Regulation Genetic Algorithm Optimization

Select air gap length δ, core length L_1, permanent magnet thickness h_{Mp}, slot width b_0, slot body width d_2 and slot depth h as optimization variables, and the objective function expression to be optimized is:

$$f(x) = Min[\Delta U(\sigma, L_1, h_{Mp}, b_0, d_2, h)] \tag{25}$$

The genetic algorithm parameters are shown in Table 3, the range of optimization variables are shown in Table 4, and the optimization results are shown in Table 5. The effect of air gap length, iron core length, and permanent magnet thickness on load voltage is shown in Figure 11, and the effect of slot width, slot body width, and slot depth on no-load voltage is shown in Figure 12. The generator performances before and after optimization are shown in Table 6.

Table 3. Genetic algorithm parameters.

Population Size	Maximum Genetic Algebra	Individual Length	Generation Gap	Cross Probability	Probability of Variation
40	50	40	0.95	0.7	0.01

Table 4. Value range of optimization variable.

Variable	δ/mm	L_1/mm	h_{Mp}/mm	b_0/mm	d_2/mm	h/mm
min	2.0	20.0	4.0	4.0	8.0	30.0
max	3.0	30.0	6.0	6.0	12.0	45.0

Table 5. Optimization results.

Variable	δ/mm	L_1/mm	h_{Mp}/mm	b_0/mm	d_2/mm	h/mm
Before optimization	2.5	25.0	5.0	5.0	10.0	35.0
After optimization	2.8	24.3	4.5	5.3	11.2	33.5

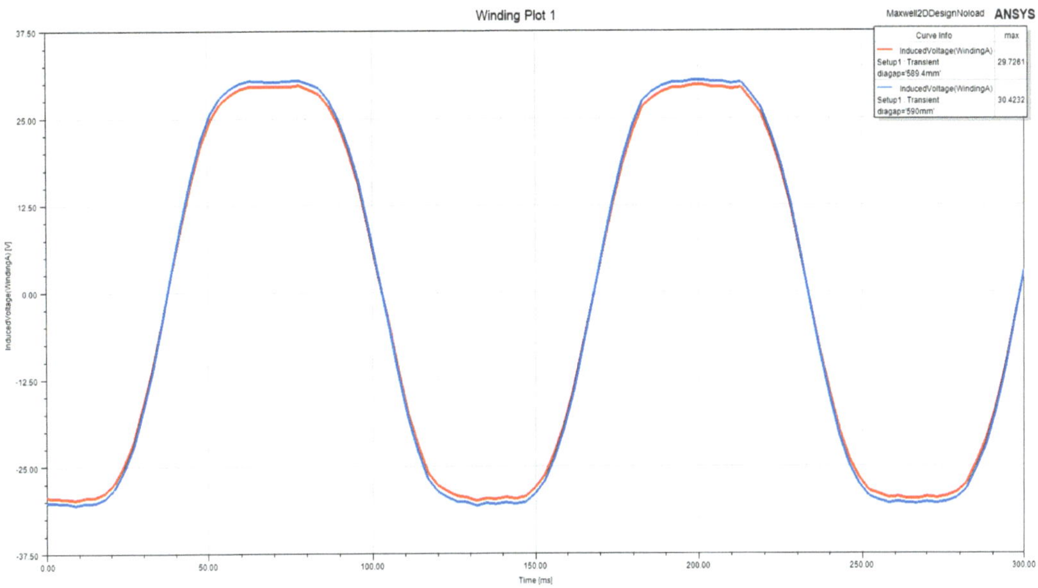

Figure 11. Effect of air gap length, iron core length, and permanent magnet thickness on load voltage.

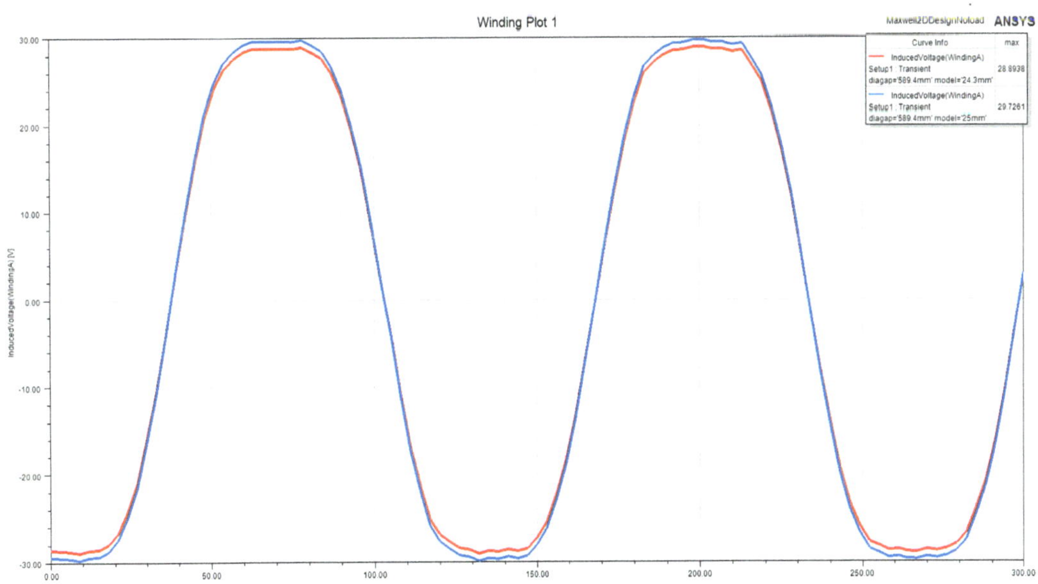

Figure 12. Influence of slot width, slot body width, and slot depth on no-load voltage.

Table 6. The motor performance before and after optimization.

Performance Index	Voltage Regulation Rate/%	Stator Tooth Magnetic Density/T	Stator Yoke Magnetic Density/T	Air Gap Magnetic Density/T
Before optimization	7.93	1.62	0.96	0.73
After optimization	6.85	1.82	0.98	0.84

Table 5 demonstrates that the air gap length, slot width, slot body width, and slot depth of the generator have all increased through optimization, while the iron core length and permanent magnet thickness have decreased. Figures 11 and 12 show that the load voltage and no-load voltage of the optimized generator have been reduced. Table 6 shows that the optimized generator stator tooth magnetic density, stator yoke magnetic density, and air gap magnetic density have all increased. The optimized generator voltage regulation has decreased 13.6% compared to the previous model, which confirms the effectiveness of the genetic optimization algorithm. Based on these optimization results, a prototype was produced for further drag experiments.

5. Actual Test of Underwater Generator

5.1. Generator Dragging Experiment

The generator dragging experiment took place at Hangzhou Jianghe Factory with the test temperature of 32.5 °C and the rate load of 20 Ω for the test system. Figure 13 displays a photo of the generator dragging experiment. As demonstrated in Figure 14, the generator began operating at a speed close to 5 rpm, and at the rated rotation speed of 28 rpm, the output power reached 121.3 W, which fulfilled the design requirements for rated power. The voltage regulation of the prototype obtained through no-load and rated load experiments is 7.05%, which is similar to the optimized simulation results.

Figure 13. Photo of the motor dragging experiment.

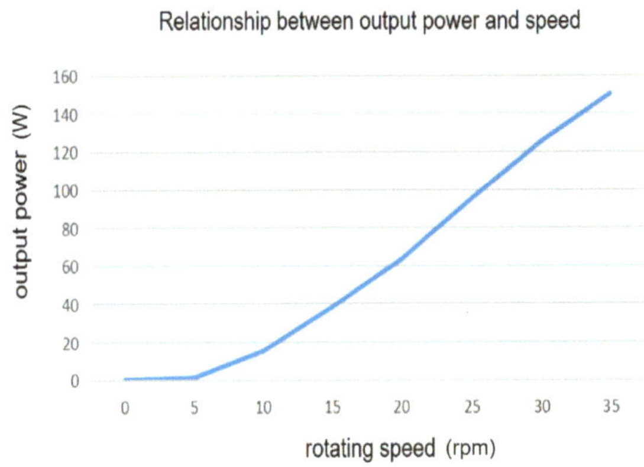

Figure 14. Measured output power.

5.2. Generator Sealing Experiment

The horizontal-axis tidal current generator operates in a seawater immersion environment, so its insulation and sealing performance determine the reliability of the generator's operation. The traditional horizontal-axis tidal current generator adopts a sealed cabin structure for sealing, and only a dynamic sealing structure is designed at the transmission shaft to ensure power transmission and underwater sealing of the generator. However, as the operating time of the generator increases, it is difficult for this method to ensure the reliability of the generator due to the inherent principle defects of dynamic sealing [21]. The heat generated during the full power operation of the generator causes the internal pressure in the sealed compartment of the generator to be greater than the external pressure, while during low-speed operation, the cooling of the generator causes the external pressure in the sealed compartment to be greater than the internal pressure, resulting in alternating large changes in pressure at the dynamic sealing area, causing seawater leakage and damaging the sealing reliability of the generator [22]. The prototype used polyurethane elastic resin for overall seals, and a certain thickness of polyurethane colloid is formed on the surface of the stator and rotor. The thickness of the stator outer surface adhesive layer is 1.0 to 2.0 mm, the inner surface adhesive layer of the stator is 5 mm, and the thickness of the adhesive layer at both ends of the stator is 10 mm, the surface adhesive layer of the stator is smooth and dense. After the sealing is completed (Figure 15), the underwater pressure test (Figure 16) was carried out on the generator. The stator and rotor of the generator were placed in an underwater pressure environment of 20 m, and the pressure tank was maintained at 0.3 MPa. After 52 h, the insulation resistance was greater than 5 M Ω, which meets the operational requirements for the horizontal-shaft tidal current energy underwater generator.

Figure 15. Photo of the generator after sealing.

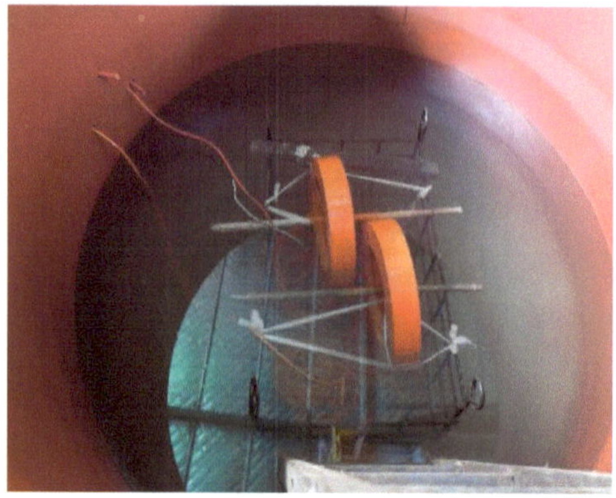

Figure 16. Photo of the generator sealing experiment.

5.3. Water Tank Dragging Experiment

After the sealing test, the generator and impeller have been assembled, and the water tank dragging experiment was conducted to simulate different water flow speeds by varying the forward speed of the trailer. Figures 17 and 18 depict the experimental setup and photos, respectively. The structure of the traction system mainly includes the traction platform, tidal current power generation device, circuit breaker, three-phase rectifier filter, programmable electronic load, and laptop monitoring and recording software. The circuit breaker is used to connect the front-end energy capture and generation system with the back-end energy conversion monitoring system; The three-phase sorting filter rectifies and filters the three-phase alternating current into a smoother direct current; programmable electronic loads are used to control the load output characteristics of power generation systems at different flow rates; laptops are used for real-time monitoring and recording of drag test data. The electronic load used for this test is the Feisi FT6300 series single programmable DC electronic load, which is set to constant resistance mode. Based on the calculation of the design parameters of the generator, the resistance values are set to 30 Ω, 35 Ω, and 40 Ω, with drag flow rates ranging from 0.2 m/s to 1.0 m/s and flow rate steps of 0.05 m/s. The cables for the power generation device and current meter are connected to the status monitoring system on shore. The generator output is connected to the battery charging and discharging module, which can monitor the generator's speed and electrical parameters during the charging and discharging process. The monitoring system hardware mainly consists of rectification, speed measurement module, power monitoring module, voltage reduction module, power supply module, battery, processing control unit, 4G communication transmission module, etc. The processing control unit obtains information such as speed, generator output voltage, current, frequency, temperature, depth, flow rate, flow direction, and underwater sound signals, and uploads this information to the cloud server through a 4G communication module. The monitoring data returned by the monitoring module is viewed on the server side of the monitoring system. The results of the water tank dragging experiment are shown in Table 7. When the flow rate was 0.25 m/s and the resistance was kept constant at 30 Ω, the system was started, and the generator's average output power was 3.5 W. At the flow rate of 0.8 m/s, the average output power of the generator was 122.31 W. The output voltage changed slightly when the load was changed, demonstrating the generator's suitability for low flow rate tidal current energy resources in China.

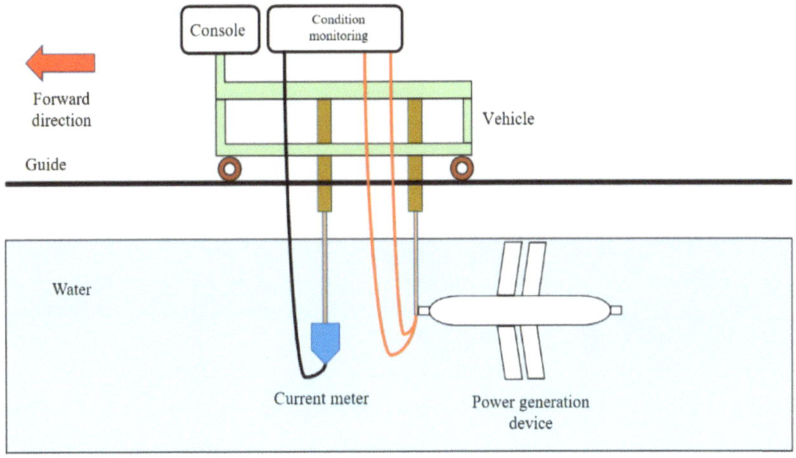

Figure 17. Schematic diagram of the dragging experiment.

Figure 18. Photo of the dragging experiment.

Table 7. Water tank dragging experiment results.

Load/Ω	Start Flow Rate/m/s	Starting Power/W	Rated Flow Rate/m/s	Output Power/W	Output Voltage/V
30	0.25	3.65	0.88	123.85	38.74
35	0.25	3.56	0.90	123.18	37.21
40	0.26	3.49	0.91	122.31	36.52

6. Conclusions

This paper designed a 120 W, 28 rpm PMSG that can be used specifically for tidal current energy. The design formula is corrected based on the operating conditions. The design scheme is validated using finite element simulation. Further optimization is conducted using the genetic optimization algorithm to reduce the voltage regulation. After completing the design, a prototype is manufactured for a generator dragging experiment, underwater pressure test, and water tank dragging experiment. The experimental results show that the prototype meets the design requirements, verifying the effectiveness of the generator design method proposed in this study. This provides a reference for the design of low-speed direct-drive PMSG for tidal current energy and lays a technical foundation for the long-term reliable operation of power generation devices using tidal current energy.

Author Contributions: Conceptualization, C.M.; Formal analysis, X.Z. (Xueming Zhang); Investigation, L.C.; Methodology, J.C.; Software, B.W.; Validation, X.Z. (Xiao Zhang); Writing—original draft, Y.Z.; Writing—review & editing, W.Z. All authors have read and agreed to the published version of the manuscript.

Funding: This research was funded by the international cooperation project of the Ministry of Science and Technology, grant number 2021YFE0107600; this research was also funded by Key R&D projects of Jilin Provincial Department of Science and Technology, grant number 20210203104SF, 20210402071GH.

Institutional Review Board Statement: Not applicable.

Informed Consent Statement: Not applicable.

Data Availability Statement: Some or all data, models, or code generated or used during the study are available from the corresponding author by request.

Conflicts of Interest: The authors declare no conflict of interest.

References

1. Tian, J.; Yu, L.; Xue, R.; Zhuang, S.; Shan, Y. Global low-carbon energy transition in the post-COVID-19 era. *Appl. Energy* **2022**, *307*, 118205. [CrossRef] [PubMed]
2. Jouffray, J.B.; Blasiak, R.; Norström, A.V.; Österblom, H.; Nyström, M. The blue acceleration: The trajectory of human expansion into the ocean. *One Earth* **2020**, *2*, 43–54. [CrossRef]
3. Khan, N.; Kalair, A.; Abas, N.; Haider, A. Review of ocean tidal, wave and thermal energy technologies. *Renew. Sustain. Energy Rev.* **2017**, *72*, 590–604. [CrossRef]
4. Eng, B.A.M.B. *Combined Experimental Testing and Numerical Modelling of a Novel Vertical Axis Tidal Turbine*; National University of Ireland: Galway, Ireland, 2018.
5. Zhang, Y.-L.; Lin, Z.; Liu, Q.-L. Marine renewable energy in China: Current status and perspectives. *Water Sci. Eng.* **2014**, *7*, 288–305.
6. Li, B.; Li, L.; Yang, L.; Yang, X.; Xiao, Y. Utilization and research status of tidal current energy. *Solar Energy* **2010**, *9*, 39–42.
7. Gu, Y. Study on the Key Technologies of the Grid-connected Marine Current Turbine's Control System. Ph.D. Thesis, Zhejiang University, Hangzhou, China, 2018.
8. Moury, S. *Design of Low Speed Axial Flux Permanent Magnet Generators for Marine Current Application*; Memorial University of Newfoundland: St. John's, NL, Canada, 2010.
9. Lindh, P.M.; Jussila, H.K.; Niemela, M.; Parviainen, A.; Pyrhonen, J. Comparison of concentrated winding permanent magnet motors with embedded and surface-mounted rotor magnets. *IEEE Trans. Magn.* **2009**, *45*, 2085–2089. [CrossRef]
10. Fizari, A.J.; Wijaya, F.D.; Cahyadi, A.I.; Adiyasa, I.W. Design and Simulation Permanent Magnet Synchronous Generator 1.5 kW for Ocean Current Turbine. In Proceedings of the 2019 IEEE Conference on Energy Conversion (CENCON), Yogyakarta, Indonesia, 16–17 October 2019; pp. 121–125.
11. Nikbakhsh, A.; Izadfar, H.; Beromi, Y. Design and optimization of permanent magnet synchronous generator for use in hydrodynamic renewable energy by applying ACO and FEA. *IIUM Eng. J.* **2017**, *18*, 158–176. [CrossRef]
12. Quintal-Palomo, R.; Dybkowski, M.; Gwoździewicz, M. Parametric analysis for the design of a 4 pole radial permanent magnet generator for small wind turbines. *Power Electron. Drives* **2016**, *1*, 175–186.
13. Faiz, J.; Valipour, Z.; Shokri-Kojouri, M.; Khan, M.A. Design of a Radial Flux Permanent Magnet Wind Generator with Low Coercive Force Magnets. In Proceedings of the 2016 2nd International Conference on Intelligent Energy and Power Systems (IEPS), Kyiv, Ukraine, 7–11 June 2016; pp. 1–7.
14. Hebala, A.; Ghoneim, W.A.; Ashour, H.A. Detailed Design Procedures for Low-Speed, Small-Scale, PMSG Direct-Driven by Wind Turbines. In Proceedings of the 2018 XIII International Conference on Electrical Machines (ICEM), Alexandroupoli, Greece, 3–6 September 2018; pp. 697–703.
15. Liwei, C.; Wenpeng, C.; Yinchao, L.; Shanguo, P.; Junsheng, W. Filling and sealing technology of ocean underwater generator. *Polyurethane Ind.* **2018**, *33*, 34–36.
16. Madani, N.; Cosic, A.; Sadarangani, C. A permanent magnet synchronous generator for a small scale vertical axis wind turbine. In Proceedings of the 2015 IEEE International Electric Machines & Drives Conference (IEMDC), Coeur d'Alene, ID, USA, 10–13 May 2015; pp. 48–52.

17. Madani, N. Design of a Permanent Magnet Synchronous Generator for a Vertical Axis Wind Turbine. Master's Thesis, Royal Institute of Technolgy, Stockholm, Sweden, December 2011.
18. Htet, T.Z.; Zhao, Z.; Gu, Q. Design Analysis of Direct-Driven PMSG in Wind Turbine Application. In Proceedings of the 2016 International Conference on System Reliability and Science (ICSRS), Paris, France, 15–18 November 2016; pp. 7–11.
19. Jeong, C.L.; Kim, Y.K.; Hur, J. Optimized design of PMSM with hybrid-type permanent magnet for improving performance and reliability. *IEEE Trans. Ind. Appl.* **2019**, *55*, 4692–4701. [CrossRef]
20. Li, W.; Cheng, M. Systems. Investigation of influence of winding structure on reliability of permanent magnet machines. *Trans. Electr. Mach. Syst.* **2020**, *4*, 87–95. [CrossRef]
21. Zhang, Z.; Zhang, Y.; Zheng, Y.; Zhang, J.; Fernandez-Rodriguez, E.; Zang, W.; Ji, R. Power fluctuation and wake characteristics of tidal stream turbine subjected to wave and current interaction. *Energy* **2023**, *264*, 126185. [CrossRef]
22. Zhang, Y.; Zang, W.; Zheng, J.; Cappietti, L.; Zhang, J.; Zheng, Y.; Fernandez-Rodriguez, E. The influence of waves propagating with the current on the wake of a tidal stream turbine. *Appl. Energy* **2021**, *290*, 116729. [CrossRef]

sustainability

Article

Comparative Evaluation of the Immersed-Solid Method for Simulating the Flow Field around Hydrofoil

Zilong Hu [1], Di Zhu [2,*], Ruofu Xiao [1,3] and Ran Tao [1,3]

1 College of Water Resources and Civil Engineering, China Agricultural University, Beijing 100083, China
2 College of Engineering, China Agricultural University, Beijing 100083, China
3 Beijing Engineering Research Center of Safety and Energy Saving Technology for Water Supply Network System, China Agricultural University, Beijing 100083, China
* Correspondence: zhu_di@cau.edu.cn

Abstract: The wall boundary is important in computational-fluid-dynamics simulations. If extremely small leakage, changing leakage or a moving body exists in the simulation case, the difficulty in meshing and solving near-wall flow increases. The immersed-solid method, which inserts a rigid, solid body into the entire fluid domain, was a choice to solve the wall-boundary-solution problems mentioned above, without considering mesh deformation. The purpose of this paper is to verify the effectiveness of the immersed-solid method in the simulation of extremely small leakage, changing leakage or a moving body, and to provide a theoretical basis for the use of the submerged-solid method in engineering. In this study, the NACA0015 hydrofoil was used to check the hydrodynamic characteristics in using the immersed-solid method. The comparative study was conducted at the incidence angle of 8 degrees and a Reynolds number of 5.0×10^5, by using the immersed-solid and traditional no-slip-wall boundary. The results show that the flow striking and separation with pressure rise and drop can be correctly captured using an immersed-solid setup with boundary tracking. However, the accuracy of pressure and velocity field using the immersed-solid method was insufficient. The turbulence-kinetic energy was much higher around the immersed-solid foil body. Generally, the immersed-solid method can qualitatively predict the correct hydrodynamic characteristics. Its convergence ability is better, and it can save approximately 20% of CPU time, even if the grid density is 4.39 times of the traditional no-slip wall. Therefore, the immersed-solid method can be a good choice for engineering-flow cases with complex wall problems.

Keywords: wall boundary; immersed-solid method; turbulent flow; hydrofoil

Citation: Hu, Z.; Zhu, D.; Xiao, R.; Tao, R. Comparative Evaluation of the Immersed-Solid Method for Simulating the Flow Field around Hydrofoil. *Sustainability* **2023**, *15*, 3493. https://doi.org/10.3390/su15043493

Academic Editors: Bin Huang, Ling Zhou, Yuquan Zhang and Jianhua Zhang

Received: 9 January 2023
Revised: 10 February 2023
Accepted: 13 February 2023
Published: 14 February 2023

1. Introduction

Computational fluid dynamics (CFD) is widely used in the performance evaluation [1] and optimization [2] of hydraulic machinery. The wall boundary is the most common boundary type in computational-fluid-dynamics (CFD) simulations, especially considering fluid flows around solid bodies [3,4]. Currently, an accurate solution of near-wall flow is no more difficult. The direct solving of the usage of different wall-function estimations and mixed formulations can be applicable to wall function insensitive to y+ [5], pipe flow [6] and high Reynolds-number flow [7]. However, some near-wall puzzles still exist to be solved in modern-engineering CFD simulations. For example, extremely small leakage is difficult to simulate when it is much smaller than the other regions [8]. Putting mesh topology inside extremely small leakage is a challenge, especially without compromising the accuracy with the simulation cost. Another example is variable leakage, such as leakage in piston-pump housing and gate housing. Blanco et al. [9] used dynamic-grid technology to simulate the flow of a piston pump, and Akoz et al. [10] compared eight different grid systems to obtain a grid that can accurately simulate the flow of a sluice gate. In this case, researchers need to change both the grid-topology distribution and the density. This will strongly affect the time cost of CFD's pre-process and the solution process. Moving a solid body in fluid is also a difficult problem [11]. If the solid body only translates, the mesh-re-building process can be simplified and well solved. If translation and rotation happen together [12,13], the mesh re-building becomes difficult, and even be unable to be made automatic. The immersed-solid method has become another wall-boundary choice in the commercial software ANSYS CFX, putting a solid domain into the entire fluid domain [14]. It provides a new way for the near-wall flow simulations to solve the puzzles above. Yoon et al. [15] simplified the calculation of small-clearance leakage flow in an external gear pump by the immersed-solid method, overcoming numerical difficulties and saving calculation time. The simulation results show that there is little difference between the numerical predictions and the experimental data. Zhang et al. [16,17] applied the immersed-solid method to the simulation of moving-boundary problems which were difficult to solve by other methods, and simulated the flow field inside the double-helical pump. At present, the immersed-solid method has been widely used in the simulation of extremely small leakage, changing leakage, and the moving body, but the gap between its accuracy and the traditional wall-boundary simulation is not clear, and a good mesh-matching scheme on the fluid–solid interface has not been found.

Hydrofoil, which is a typical, simplified solid body, is chosen as the studied object in this case, due to its standardized profile. It has been widely investigated in both numerical simulations and experiments [18–21]. For typical hydrofoils, the lift/drag forces, flow-striking and separation point, surface-pressure distribution, cavitation performance and structural-response characteristics are already clear. Therefore, the hydrodynamic characteristics can be well studied for comparing the immersed-solid method with the traditional wall-boundary type [22]. In this study, there are three main points to be solved. Firstly, it is necessary to find a way to mesh the fluid domain, solid domain and the fluid–solid interface when using the immersed-solid method. The mesh independency in using the immersed-solid method and using the traditional wall boundary are checked and compared. Secondly, the flow field should be examined in detail, including the pressure field, velocity field and turbulence field. This will be helpful for the application of the immersed-solid method in engineering flow cases. Thirdly, the convergence ability should be comparatively studied in the immersed-solid case and in the traditional wall-boundary case. The time cost, hardware cost and CFD pre-processing cost need a comprehensive comparison to provide guidance for engineering cases.

2. Numerical Methods

2.1. Turbulence Modeling

The numerical study adopts the Reynolds-averaged Navier–Stokes (RANS) equation. The RANS equation is the equation of average turbulent motion obtained by averaging the physical quantity of turbulence with the sum of an average value and a pulsation value. The eddy-viscosity turbulence model is used to close the averaged equations. It follows the Boussinesq eddy-viscosity hypothesis [23], which establishes the relationship between the eddy viscosity, μ_t, and the Reynolds stress, τ_{ij}. In this study, the SST turbulence model [24–26], which deals well with the strong freestream and improves the accuracy of adverse pressure gradients by combining the standard k-ε model and Wilcox's k-ω model, was used. The SST model was based on the turbulence-kinetic-energy, k, equation and the specific-dissipation-rate, ω, equation [27]:

$$\frac{\partial(\rho k)}{\partial t} + \frac{\partial(\rho u_j k)}{\partial x_j} = P_k - \beta^* \rho \omega k + \frac{\partial}{\partial x_j}\left[(\mu + \sigma_k \mu_t)\frac{\partial k}{\partial x_j}\right] \tag{1}$$

$$\frac{\partial(\rho \omega)}{\partial t} + \frac{\partial(\rho u_j \omega)}{\partial x_j} = \frac{\alpha \omega}{k} P_k - \beta \rho \omega^2 + \frac{\partial}{\partial x_j}\left[(\mu + \sigma_\omega \mu_t)\frac{\partial \omega}{\partial x_j}\right] + 2(1 - F_1)\frac{\rho \sigma_{\omega 2}}{\omega}\frac{\partial k}{\partial x_j}\frac{\partial \omega}{\partial x_j} \tag{2}$$

where the production term P_k can be expressed by:

$$P_k = \tau_{ij}\frac{\partial u_i}{\partial x_j} \tag{3}$$

$$\tau_{ij} = \mu_t\left(2S_{ij} - \frac{2}{3}\frac{\partial u_k}{\partial x_k}\delta_{ij}\right) - \frac{2}{3}\rho k\delta_{ij} \tag{4}$$

where δ_{ij} is the Kronecker number and S_{ij} is the shear-stress tensor, so that

$$S_{ij} = \frac{1}{2}\left(\frac{\partial u_i}{\partial x_j} + \frac{\partial u_j}{\partial x_i}\right) \tag{5}$$

and

$$\mu_t = \frac{1}{\max(a_1\omega, SF_2)} \tag{6}$$

F_1 is the 1st blending function for variants (for example, φ), by

$$\varphi = F_1\varphi_1 + (1 - F_1)\varphi_2 \tag{7}$$

where F_1 follows

$$F_1 = \tanh\left\{\left\{\min\left[\max\left(\frac{\sqrt{k}}{\beta^*\omega d_w}, \frac{500\nu}{d_w^2\omega}\right), \frac{4\rho\sigma_{\omega 2}k}{CD_{k\omega}d_w^2}\right]\right\}^4\right\} \tag{8}$$

where d_w is the distance to the nearest wall, ν is the kinematic viscosity, and $CD_{k\omega}$ is

$$CD_{k\omega} = \max\left(2\rho\sigma_{\omega 2}\frac{1}{\omega}\frac{\partial k}{\partial x_j}\frac{\partial \omega}{\partial x_j}, 10^{-20}\right) \tag{9}$$

F_2 is the 2nd blending function:

$$F_2 = \tanh\left\{\left[\max\left(2\frac{\sqrt{k}}{\beta^*\omega d}, \frac{500\nu}{d_w^2\omega}\right)\right]^2\right\} \tag{10}$$

In the above equations, ρ is the density. x, t, and u denote coordinate, time and velocity, respectively. The constants are $\sigma_{k1} = 0.85$, $\sigma_{\omega 1} = 0.65$, $\beta_1 = 0.075$ for the k-ω setup and $\sigma_{k2} = 1.00$, $\sigma_{\omega 2} = 0.856$, $\beta_2 = 0.0828$ for the k-ε setup. The other constants are $\beta^* = 0.09$, $a_1 = 0.31$ and $\alpha = 0.556$.

2.2. Wall Treatments

2.2.1. No-Slip Wall

The no-slip wall boundary has been popularly used in computational-fluid-dynamics (CFD) research [22]. It assumes that fluid molecules may not slip from a wall in an ideal situation [28,29]. Generally, a common approximation for fluid slip can be drawn as:

$$u - u_w = \beta_s\frac{\partial u}{\partial n} \tag{11}$$

where u is the fluid velocity, u_w is the velocity on the wall, β_s is the slip length so that $\beta_s = C_s \cdot l_{mfp}$ where l_{mfp} is the mean free path and C_s is the empirical coefficient [30]. For a no-slip wall-boundary condition, β_s is 0, which means $u = u_w$.

2.2.2. Automatic-Wall Functions

If the SST turbulence model was applied, the automatic-wall treatments [6] were also adopted, to empirically model the near-wall flow. As shown in Figure 1, the red line in the figure is the change curve of fluid velocity with the distance from the wall. The near-wall-flow region included 3 main parts: the viscous sublayer, buffer layer and logarithmic layer [31,32]. In the viscous sublayer, there is [33]:

$$u^+ = y^+ \tag{12}$$

where u^+ is the dimensionless velocity of t $u^+ = u/u_t$; u is the velocity, u_t is the friction velocity of $u_t = \sqrt{\tau_w/\rho}$., τ_w is the wall shear stress, and y^+ is the dimensionless off-wall distance of $y^+ = y \cdot u_t$, where y is the off-wall distance. In the logarithmic layer, there is [33]:

$$u^+ = \frac{1}{\kappa} \ln y^+ + C^+ \tag{13}$$

where κ is the von Karman constant and C^+ is the velocity-deviation constant. The automatic-wall functions blend the ω solution and the u_t solution by [34]

$$\omega = \sqrt{\omega_{vis}^2 + \omega_{log}^2} \tag{14}$$

$$u_t = \sqrt[4]{\left(u_t^{vis}\right)^4 + \left(u_t^{log}\right)^4} \tag{15}$$

where superscript/subscript *vis* represent the viscous sublayer and the buffer layer, respectively, and superscript/subscript *log* represent the logarithmic layer.

$$\omega_{vis} = \frac{6\nu}{\beta y^2}; \omega_{\log} = \frac{u_t}{C_\mu^{1/4} \kappa y} \tag{16}$$

$$u_t^{vis} = \frac{u}{y^+}; u_t^{log} = \frac{u}{\log E_u y^+} \tag{17}$$

where E_u is the model constant of 9.81, and C_μ is the eddy-viscosity coefficient. Here we define y as the mesh-element–centroid off-wall distance in the CFD simulation; if the y value is small, $1/y^2$ dominates, to emphasize the viscous characters. On the contrary, if the y value is large, $1/y$ dominates, being mainly logarithmic.

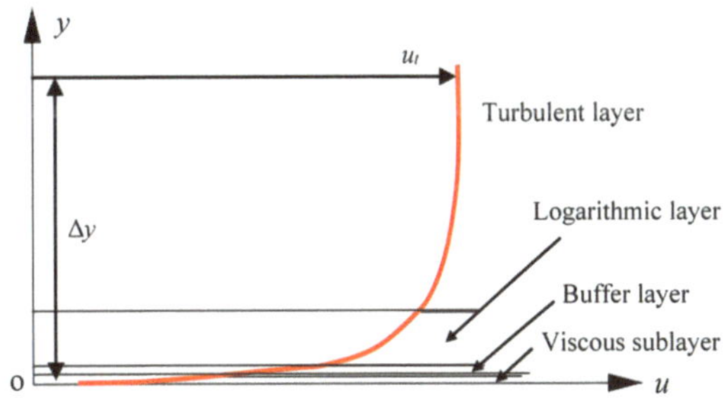

Figure 1. Law of the near-wall velocity distribution. u: velocity; y: off-wall distance.

2.2.3. Immersed-Solid Method

The immersed-solid method is enabled in the commercial software ANSYS CFX by immersing a rigid-solid domain into a fluid domain [14]. This immersed-solid domain was more like a region in which a momentum source is added to force the fluid to move with the solid. Figure 2 shows a schematic map of the mesh elements and nodes when using the immersed-solid method. Table 1 lists the acronyms defined for the immersed-solid method.

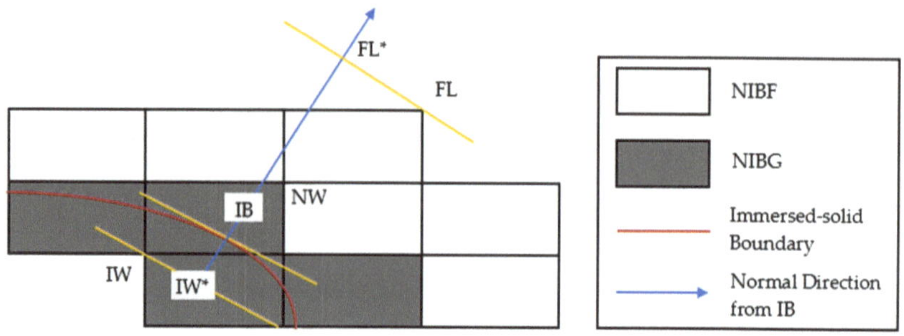

Figure 2. Schematic map of the mesh elements and nodes when using the immersed-solid method.

Table 1. The acronyms defined for the immersed-solid method.

Acronyms	Meaning
IW	In-wall node
IW*	In-wall point
NW	Near-wall node
IB	Point on the immersed-solid boundary
FL	Fluid node
FL*	Fluid point
NIBG	Elements between NW and IW nodes
NIBF	Elements between NW and FL nodes

In ANSYS CFX, there are some important settings for the immersed-solid method, as follows:

- Wall distance: 0 on the nodes inside the immersed solid; a function of the wall scales on the nodes near the inside of the immersed solid; the physical distance of the nodes near the immersed solid was achieved using the boundary-tracking method for wall functions.
- k and ω in SST turbulence-model cases: 0 on the nodes inside the immersed solid; scalable-wall treatment was used for ω in the logarithmic region by $\omega_{log} = u^{*2}/(C_\mu{}^{1/2}\kappa\nu y^*)$ where $u^* = C_\mu{}^{1/2}k^{1/4}$ and $y^* = u^*\Delta y/\nu$. The value of Δy is the distance between the node near the immersed solid and the immersed-solid boundary. The y^* value is lowered to 11.06 for dividing the logarithmic region and the linear viscous region.
- Friction velocity u_t: Based on the scalable-wall treatment, u_t in the logarithmic region is $u_t{}^{log} = \Delta U/[1/\kappa\ln(y^*) + C] = \Delta U/u^+$, where ΔU is the fluid tangential velocity relative to the immersed boundary.

According to the setup above, the velocity on the near-wall nodes can be forced as

$$\left(U_{NW}^F\right)_{tan} = \left(U_{IB}^{IS}\right)_{tan} + \frac{(u^*/u^+)_{FL}}{(u^*/u^+)_{NW}}\left[(U_{FL})_{tan} - \left(U_{IB}^{IS}\right)_{tan}\right] \quad (18)$$

where superscript F and IS denote the fluid and immersed solid, respectively. The momentum source can be calculated based on the tangential-force velocity $(U_{NW}{}^F)_{tan}$ and the wall-normal velocity $(U_{IB}{}^{IS})_{nm}$:

$$U_{NW}^F = \left(U_{NW}^F\right)_{tan} + \left(U_{IB}^{IS}\right)_{nm} \quad (19)$$

Hence, fluid can be prevented from getting through the immersed-solid domain.

3. Studied Object

3.1. NACA0015 Profile

The NACA0015 [35] profile belongs to the typical NACA four-digit-profile series. It is symmetric along the meanline. If we define the foil-meanline direction as m_f and the thickness direction (the origin from the meanline) as t_f, the NACA0015 m_f-t_f profile can be illustrated in Figure 3, where the total length along m_f is denoted as c and the maximum thickness is denoted as t_{fm}. The relationship between t_f and m_f can be expressed as:

$$\pm\frac{t_f}{c}=\frac{t_{mf}}{0.2c}\left[0.297\sqrt{\frac{m_f}{c}}-0.126\left(\frac{m_f}{c}\right)-0.352\left(\frac{m_f}{c}\right)^2+0.284\left(\frac{m_f}{c}\right)^3-0.102\left(\frac{m_f}{c}\right)^4\right] \quad (20)$$

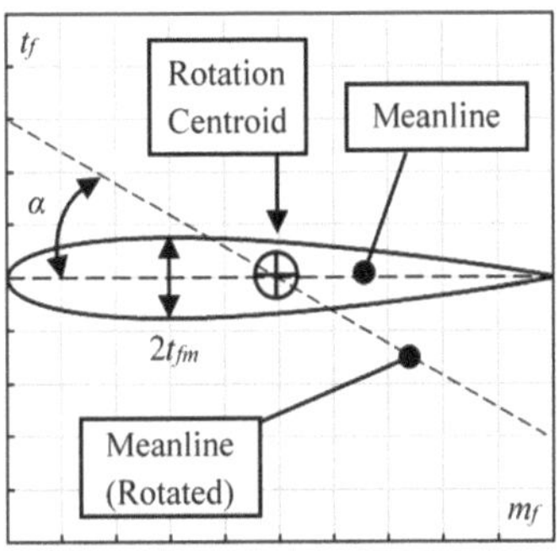

Figure 3. The NACA0015 hydrofoil profile.

In addition, the leading edge of the NACA0015 profile is an arc, whose radius is r_{LE}, as in the following:

$$\frac{r_{LE}}{c}=1.102\left(\frac{t_{mf}}{c}\right)^2 \quad (21)$$

Thus, the maximum thickness, t_{fm}, approximately locates at m_f/c = 0.3. The incidence angle (between the meanline and rotated meanline) can be denoted as α, the unit representing a degree. It represents the foil-rotation status, based on the rotation centroid.

3.2. Case Detail

In this study, the NACA0015 foil was studied at α = 8 degrees inside an $L_1 \times L_2 \times L_3$ cuboid domain, as shown in Figure 4. L_1, L_2 and L_3 were respectively equal to 4.35c, 1.04c and 0.70c. Coordinate components X, Y and Z denote the three orthogonal directions, where X is along the flow direction, Y is perpendicular to the flow direction and Z is along the foil span. Because the foil does not change shape and scale along Z, this 3D case can be treated as an X-Y quasi-2D case, which provides a good case simplification. Fluid flows into the domain at the inlet boundary and out of the domain at the outlet boundary. The four surrounding X-Y and X-Z planes are walls.

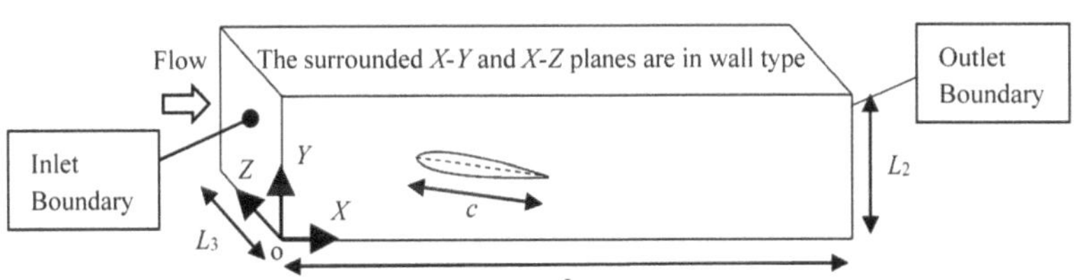

Figure 4. The 3D-flow domain with NACA0015 foil at incidence angle of 8 degrees.

3.3. Computational-Fluid-Dynamics Setups

In this case, the hydrodynamic characters were mainly studied. ANSYS CFX was used as the CFD solver by using the unsteady-RANS method with the SST turbulence model, as introduced in the above sections. The fluid medium was set as water at 20 °C under the 1 atm pressure environment. Compressibility, heat transfer and mass transfer were not considered. The advection scheme and the turbulence numeric were set as high resolution. The inlet boundary was set as velocity inlet with uniform V_{in} distribution, and the pressure at the inlet was at zero gradient. The outlet boundary was set as the static-pressure boundary of relatively 0 Pa, and the velocity at the outlet was at zero gradient. The surrounded walls were set as no-slip type. The foil surface was compared, between the no-slip type and the immersed-solid type. In this case, the Reynolds number, Re, was 5.0×10^5, according to the following equation:

$$Re = \frac{\rho c V_{in}}{\mu} \tag{22}$$

The unsteady simulation was based on the steady simulation of 600 iterations. The unsteady simulation covered a total of 0.1 s and spent 0.0001 s on each time step. The maximum iteration number for each time step was 10, with the convergence criteria of RMS 1.0×10^{-5}.

4. Numerical-Accuracy Check and Validation

4.1. No-Slip Wall Case

In the no-slip-wall case (the foil surface was set as the no-slip wall), the domain was meshed by structural hexahedral elements, using ICEMCFD. Six different mesh schemes, whose node number increased proportionally and gradually, as listed in Table 2, were comparatively checked for accuracy by comparing the numerically predicted pressure with the experimentally tested pressure [15] on the foil surface along the X direction. In this case, the dimensionless pressure coefficient, C_p, was used:

$$C_p = \frac{p - p_{in}}{0.5 \rho V_{in}^2} \tag{23}$$

where p is the pressure, and p_{in} is the pressure at the inlet boundary. The position along the X direction is also in a relative form, as X/c.

Table 2. Mesh schemes checked in the no-slip-wall case.

No.	Node Number	Description
1	30,884	Very Coarse
2	62,496	Coarse
3	107,652	Mid-Coarse
4	175,150	Mid-Fine
5	257,488	Fine
6	352,594	Very Fine

Figure 5 shows the pressure on the foil surface under different mesh schemes by treating the foil surface as no-slip wall boundary. In addition, the near-wall region where y^+ was 1.45~16.22 was predicted by automatic-wall functions. The Nos. one to six mesh schemes predicted the pressure well from $X/c \approx 0.4$ to $X/c \approx 1.0$ by comparing it with the experimental data [15]. However, the main difference occurred in the front 20% X/c region, especially on the foil-suction side. The minimum C_p value for the coarse mesh was larger than -3. Then, it decreased to $-3 \sim -3.25$ in the simulations for coarse, mid-coarse and mid-fine meshes. When the mesh nodes increased to 257488 (fine), the numerically predicted C_p curve overlapped well with the experimental data. This situation occurred when the mesh nodes increased to 352594 (very fine). Usually, we called this "mesh independence". Thus, the mesh scheme No. five with 257488 modes can be selected in this study, with good accuracy.

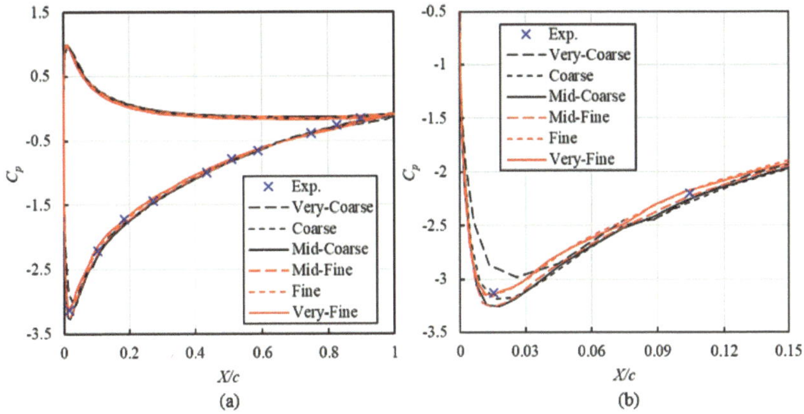

Figure 5. Pressure on foil surface under different mesh schemes. (**a**) C_p distribution on foil surface along X direction; (**b**) Enlarged view within X/c = 0~0.15 and C_p = −0.5~−3.5.

4.2. Immersed-Solid Case

The foil surface was also treated as the immersed-solid boundary. In this situation, four different mesh combinations were discussed, as listed in Table 3. The pressure distribution on the foil surface was also analyzed along the X direction. The dimensionless C_p and relative X/c were still used, as shown in Figure 6.

Table 3. Mesh combinations checked in the immersed-solid case.

No.	Node Number			Description
	Fluid	Solid	Total	
1	234,248	51,720	291,368	Coarse
2	753,984	378,840	1,132,824	Fine
3	234,248	51,720	291,368	Coarse with Boundary Tracking (Coarse-B)
4	753,984	378,840	1,132,824	Fine with Boundary Tracking (Fine-B)

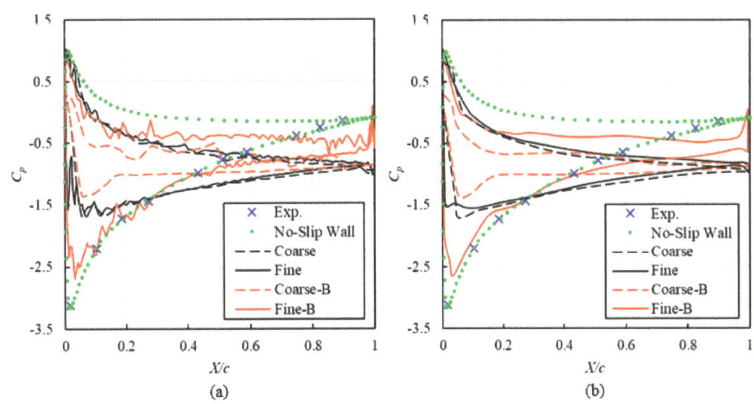

Figure 6. Pressure on foil surface under different mesh-scheme combinations. (**a**) C_p distribution on foil surface along X direction; (**b**) Smoothed C_p curves.

Figure 6a shows the pressure on the foil surface under different mesh-scheme combinations. The pulsations can be observed on the immersed-solid predicted C_p curves. These pulsations could be caused by the data interpolation on the fluid–solid interface. Thus, the smooth procedures [36] were conducted to create Figure 6b. The No. 1 combination had the same density as the selected No. 5 mesh in the no-slip-wall case. However, it produced a completely wrong C_p curve, compared with the no-slip-wall case and the experimental data. Therefore, the mesh density was refined to No. 2, but no improvement could be observed. Based on the two results, the boundary tracking was set to near-wall approximations on the immersed solid on the fluid side. The "Coarse-B" curve still produced wrong data. On the other hand, the "Fine-B" curve showed an obvious improvement, in that both the leading-edge pressure rise and drop were correctly captured in position. The amplitude of the maximum and minimum C_p values became closer to the experimental data. Therefore, the other flow characteristics will be analyzed in the next section.

5. Comparative-Flow Analysis

5.1. Forces and Torques

The lift force, F_Y, (along the Y direction) and the drag force, F_X, (along the X direction) are analyzed in Table 4. In the no-slip-wall case, the lift force was 113.01 N, which was larger than that in the immersed-solid case, of 90.82 N. The drag force was 4.95 N in the no-slip-wall case, which was smaller than that in the immersed-solid case, of 8.67 N. As a result, the immersed-solid method obtained a smaller lift/drag ratio compared with the no-slip-wall setup, with its lift/drag ratio being approximately 40% of the no-slip-wall case. The torques on the foil on the base of the rotation centroid, T_Z, were also predicted, as is also shown in Table 4. In the no-slip-wall case, the foil torque was 3.29 N·m. In the immersed-solid case, the foil torque was 2.31 N·m, which was approximately 70% of the no-slip-wall case. Under the two wall conditions, the difference between the force and torque obtained by numerical simulation is very obvious. In order to understand the reasons for the differences, we need to analyze the internal flow under different circumstances.

Table 4. Lift force, drag force and torque on hydrofoil.

	No-Slip-Wall Case	Immersed-Solid Case
Lift Force, F_Y	113.01 [N]	90.82 [N]
Drag Force, F_X	4.95 [N]	10.48 [N]
Lift/Drag Ratio, F_Y/F_X	22.83	8.67
Torque, T_Z	3.29 [N·m]	2.31 [N·m]

5.2. Pressure-Coefficient C_p Field

Figure 7 compares the C_p contour on the X-Y plane at middle Z position in the domain. The contours in the no-slip case and immersed-solid case were similar. Generally, the flow striking on the foil leading edge was correctly captured in both the two cases. The flow separation after the leading edge was also simulated. Therefore, the immersion-solid method can better simulate the pressure distribution in the flow. However, some differences can be found. Firstly, the very low C_p region ($C_p < -2.5$) on the foil upper side was wider in the no-slip-wall case than in the immersed-solid case. Secondly, C_p increased rapidly in the upper flow field in the no-slip-wall case. However, the C_p increase was relatively slow in the immersed-solid case. Thirdly, there was a high-pressure region in the lower flow field in the no-slip wall case. In the immersed-solid case, the pressure there was relatively low. In the case of the no-slip wall, the pressure difference between the upper and lower sides of the foil was large, which resulted in the lift of the foil being greater than in the case of the immersed solid. Fourthly, in the no-slip-wall case, except for the very low C_p region, the pressure coefficient in other regions was higher than that in the immersed-solid case.

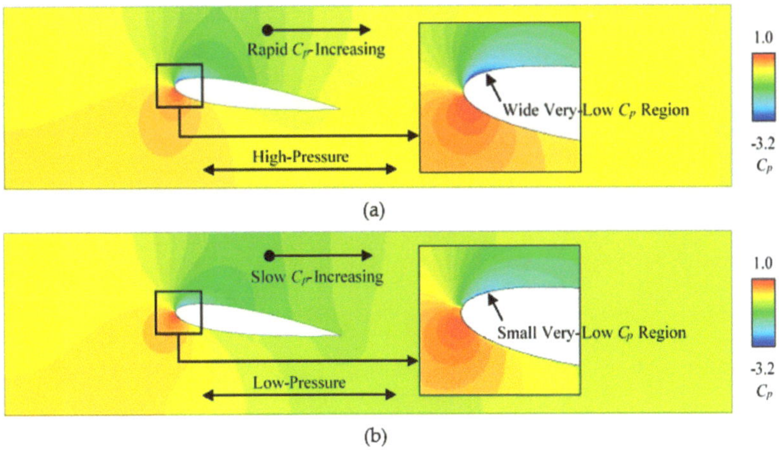

Figure 7. The C_p contour on the *X–Y* plane at middle *Z* position. (**a**) No-slip wall case; (**b**) Immersed-solid case.

5.3. Velocity-Coefficient C_v Field

Figure 8 compares the velocity field on the *X-Y* plane at middle *Z* position in the domain. The velocity was relative to the domain-inlet velocity as a velocity coefficient $C_v = V/V_{in}$. The contours in the no-slip case and immersed-solid case were qualitatively similar but quantitatively different. The leading-edge high-velocity region was captured in the two cases. However, the very high C_v region ($C_v > 1.5$) was wide in the no-slip-wall case but small in the immersed-solid case, and its distribution was basically consistent with the distribution of the very low C_p region. The low velocity after the trailing edge (the wake) was narrow in the no-slip-wall case, but very wide in the immersed-solid case. A high C_v region was in the lower flow field, as indicated in Figure 8b. This shows that the immersed-method underestimates the flow velocity at the leading and trailing edges of the foil, and overestimates the flow velocity of the upper and lower flow fields of the foil.

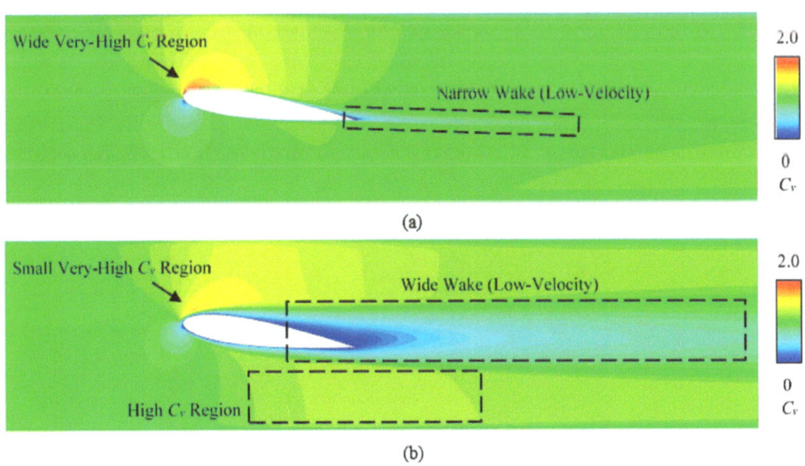

Figure 8. The C_v contour on the *X–Y* plane at middle *Z* position. (**a**) No-slip-wall case; (**b**) Immersed-solid case.

5.4. *Turbulence-Kinetic-Energy, k, Field*

Figure 9 shows the turbulence kinetic-energy, k, contour on the X-Y plane at middle Z position. The legend ranged from 0 to 3.9 m^2/s^2. It can be seen from the figure that the turbulence-kinetic-energy, k, contours in the two cases are quite different. In the no-slip-wall case, there is a relatively high k region from the foil trailing edge to downstream. It was narrow and relatively weaker than in the immersed-solid case. On the contrary, in the immersed-solid case, the high k region was very wide, covering the entire foil and extending to the downstream far field. The highest k (up to 3.7~3.9) occurred at two main locations, the upper surface near the leading edge and the lower surface, near the trailing edge. Compared with the no-slip-wall case, in the immersed-solid case, the turbulent-kinetic energy of the fluid was higher, resulting in a greater flow loss of the fluid and greater drag coefficient on the foil.

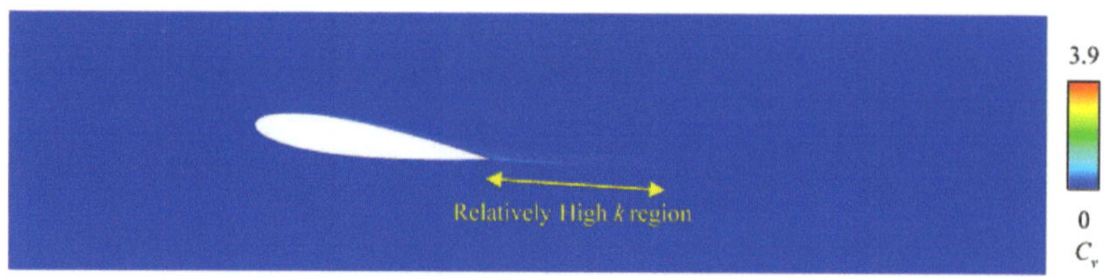

(**a**)

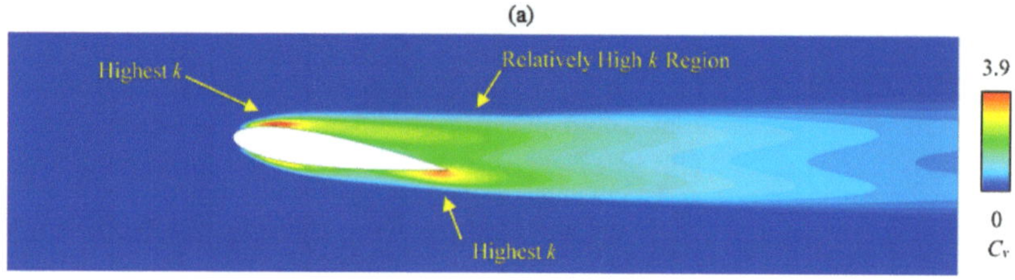

(**b**)

Figure 9. The k contour on the X–Y plane at middle Z position. (**a**) No-slip-wall case; (**b**) Immersed-solid case. Unit of k: [m^2/s^2].

6. Convergence-Ability Analysis

The convergence ability was analyzed based on the No. 5 mesh scheme in the no-slip-wall case and the No. 4 mesh combination in the immersed-solid case. The two computations were conducted on the same terminal, using 48-core 2.60 GHz CPU and 32.0 GB memory. The solver settings were also the same for the two cases. Table 5 compares the total mesh nodes, element number and timestep number. When the cases were converged to RMS 1.0×10^{-5}, the clock time until convergence, the clock time per timestep and the clock time per mesh node were also compared.

Table 5. Convergence Ability Based on the Steady-State Simulation.

	No-Slip-Wall Case	Immersed-Solid Case
Total Mesh-Node Number, N_{mn}	257,488	1,132,824
Total Mesh-Element Number	241,542	1,077,732
Timesteps Converged to RMS, 1.0×10^{-5} S_{cc}	236	81
Clock Time until Convergence, t_{cc}	1782.6 [s]	1440.3 [s]
Clock Time per Timestep, t_{cc}/S_{cc}	7.55 [s]	17.78 [s]
Clock Time per Mesh Node, t_{cc}/M_{mn}	0.0069 [s]	0.0013 [s]

According to the mesh check in Section 4, the immersed-solid method needed more mesh nodes than using the no-slip wall to obtain equal accuracy. However, the clock time per mesh node in the immersed-solid case was shorter than that in the no-slip-wall case. The immersed-solid case needed a lower timestep number to be converged than the no-slip-wall case. However, the clock time per timestep was longer in the immersed-solid case than that in the no-slip-wall case. Generally, the immersed-solid case converged to RMS 1.0×10^{-5} in 1440.3 s, which was shorter than the 1782.6 s in the no-slip-wall case. The time consumption of the immersed-solid setup in this study took 80.7% of the clock time of the traditional no-slip-wall setup. Because the memory consumption was proportional to the mesh number, the immersed-solid setup in this study consumed 4.39 times the memory of the traditional no-slip-wall setup.

7. Conclusions

As an available option in the commercial software ANSYS CFX, the immersed-solid method can flexibly add wall boundaries in the fluid domain. In this study, under the typical hydrodynamic condition of the flow around the hydrofoil, without considering the hydrofoil movement, the immersed-solid method is compared with the non-slip-wall boundary, and the effectiveness of the immersed-solid method to simulate extremely small leakage, changing leakage, and randomly moving/rotating solid and intermittent attaching/detaching solids is verified, and the following conclusions are drawn. (A) The results of the immersed-solid method were qualitatively correct but quantitatively inaccurate. (B) "Boundary tracking" was crucial for improving the near-wall solution. (C) More mesh nodes were required for the qualitatively correct result when using the immersed-solid method than when using the no-slip wall. (D) The curve-smoothing process was recommended to correct the "pulsation" around the fluid–solid interface, due to data interpolation.

In detail, the flow-striking and separation positions on the solid can be correctly captured by both the no-slip-wall and immersed-solid method. However, the pressure fields and velocity fields performed differently, especially at the downstream location. A much wider wake could be found using the immersed-solid method than in the no-slip-wall case. Because of the added momentum source, the immersed-solid method prevented fluid passing through the hydrofoil. At this time, high turbulence-kinetic energy was triggered around the foil surface, especially at the flow-striking, separation and wake positions.

Generally, the immersed-solid method can accurately simulate the flow field by setting reasonable mesh nodes, especially on the fluid–solid interface. The convergence ability of the immersed-solid method is better than the no-slip-wall setup. If we apply the immersed-solid method to special cases such as extremely small leakage, changing leakage, randomly moving/rotating solid and intermittent attaching/detaching solids, it will provide us with a fast hydrodynamic solution, by overcoming the problems caused by excessive mesh density, mesh deformation or domain change. The research results of this paper have great engineering value, especially in the engineering-flow cases.

Author Contributions: Conceptualization, R.X. and R.T.; writing—original draft preparation, Z.H. and D.Z.; writing—review and editing, R.X. and D.Z.; supervision, R.T. and D.Z. All authors have read and agreed to the published version of the manuscript.

Funding: This research is supported by the Open Research Subject of Key Laboratory of Fluid and Power Machinery (Xihua University), Ministry of Education (grant number LTDL-2022009).

Institutional Review Board Statement: Not applicable.

Informed Consent Statement: Not applicable.

Data Availability Statement: Data available on request from the corresponding author.

Acknowledgments: The authors would like to acknowledge the help from Zhengwei Wang and his research group at Tsinghua University.

Conflicts of Interest: The authors declare no conflict of interest.

References

1. Kan, K.; Zhang, Q.; Zheng, Y.; Xu, H.; Xu, Z.; Zhai, J.; Muhirwa, A. Investigation into Influence of Wall Roughness on the Hydraulic Characteristics of an Axial Flow Pump as Turbine. *Sustainability* **2022**, *14*, 8459. [CrossRef]
2. Tong, Z.; Xin, J.; Ling, C. Many-Objective Hybrid Optimization Method for Impeller Profile Design of Low Specific Speed Centrifugal Pump in District Energy Systems. *Sustainability* **2021**, *13*, 10537. [CrossRef]
3. Wendt, J.F.; Anderson, J.D. Basic Philosophy of CFD. In *Computational Fluid Dynamics: An Introduction*; Spinger: Berlin/Heidelberg, Germany, 2009; pp. 5–8.
4. Kajishima, T.; Taira, K. Numerical Simulation of Fluid Flows. In *Computational Fluid Dynamics*; Springer: Berlin/Heidelberg, Germany, 2017; pp. 1–22.
5. Grotjans, H.; Menter, F.R. Wall Functions for General Application CFD Codes. *Comput. Fluid Dyn.* **1998**, *1*, 1112–1117.
6. Kader, B.A. Temperature and Concentration Profiles in Fully Turbulent Boundary Layers. *Int. J. Heat Mass Transf.* **1981**, *24*, 1541–1544. [CrossRef]
7. Patel, V.C.; Rodi, W.; Scheuerer, G. Turbulence Models for Near-Wall and Low Reynolds Number Flows—A Review. *AIAA J.* **1985**, *23*, 1308–1319. [CrossRef]
8. Ha, T.W.; Choe, B.S. Numerical Simulation of Rotordynamic Coefficients for Eccentric Annular-Type-Plain-Pump Seal Using CFD Analysis. *J. Mech. Sci. Technol.* **2012**, *26*, 1043–1048. [CrossRef]
9. Blanco, A.M.; Oro, J.F. Unsteady Numerical Simulation of an Air-Operated Piston Pump for Lubricating Greases Using Dynamic Meshes. *Comput. Fluids* **2012**, *57*, 138–150. [CrossRef]
10. Akoz, M.S.; Kirkgoz, M.S.; Oner, A.A. Experimental and Numerical Modeling of a Sluice Gate Flow. *J. Hydraul. Res.* **2009**, *47*, 167–176. [CrossRef]
11. Qun-feng, Z.; Pan-pan, Y.A.N.; Jian-ping, S.; Wan-fa, H.E. Numerical Simulation of Inlet and Outlet Pressures Influencing Internal Flow of a Vane Pump. *Acta Armamentarii* **2014**, *35*, 1223.
12. Adkins, D.; Yan, Y.Y. CFD Simulation of Fish-like Body Moving in Viscous Liquid. *J. Bionic. Eng.* **2006**, *3*, 147–153. [CrossRef]
13. Jin, C.; Xu, K.; Chen, S. A Three Dimensional Gas-Kinetic Scheme with Moving Mesh for Low-Speed Viscous Flow Computations. *Adv. Appl. Math. Mech* **2010**, *2*, 746–762. [CrossRef]
14. Kajishima, T.; Takeuchi, S. Simulation of Fluid-Structure Interaction Based on an Immersed-Solid Method. *J. Mech. Eng. Sci.* **2013**, *5*, 555–561. [CrossRef]
15. Yoon, Y.; Park, B.-H.; Shim, J.; Han, Y.-O.; Hong, B.-J.; Yun, S.-H. Numerical Simulation of Three-Dimensional External Gear Pump Using Immersed Solid Method. *Appl. Therm. Eng.* **2017**, *118*, 539–550. [CrossRef]
16. Zhang, W.; Jiang, Q.; Bois, G.; Li, H.; Liu, X.; Yuan, S.; Heng, Y. Experimental and Numerical Analysis on Flow Characteristics in a Double Helix Screw Pump. *Energies* **2019**, *12*, 3420. [CrossRef]
17. Qiu, R.; Zhang, W.; Jiang, Q.; Shen, K.; Gread, B.; Yuan, S. Research on Full Flow Passage Characteristics of Double Screw Pump Based on Immersion Entity Method. In *Proceedings of the Journal of Physics: Conference Series*; IOP Publishing: Bristol, UK, 2022; Volume 2186, p. 012015.
18. Cervone, A.; Bramanti, C.; Rapposelli, E.; d'Agostino, L. Thermal Cavitation Experiments on a NACA 0015 Hydrofoil. *J. Fluids Eng.* **2006**, *128*, 326–331. [CrossRef]
19. Dreyer, M.; Decaix, J.; Münch-Alligné, C.; Farhat, M. Mind the Gap: A New Insight into the Tip Leakage Vortex Using Stereo-PIV. *Exp. Fluids* **2014**, *55*, 1849. [CrossRef]
20. Nowruzi, H.; Ghassemi, H.; Ghiasi, M. Performance Predicting of 2D and 3D Submerged Hydrofoils Using CFD and ANNs. *J. Mar. Sci. Technol.* **2017**, *22*, 710–733. [CrossRef]
21. Hu, C.; Chen, G.; Yang, L.; Wang, G. Large Eddy Simulation of Turbulent Attached Cavitating Flows around Different Twisted Hydrofoils. *Energies* **2018**, *11*, 2768. [CrossRef]
22. Day, M.A. The No-Slip Condition of Fluid Dynamics. *Erkenntnis* **1990**, *33*, 285–296. [CrossRef]
23. Schmitt, F.G. About Boussinesq's Turbulent Viscosity Hypothesis: Historical Remarks and a Direct Evaluation of Its Validity. *Comptes Rendus Mécanique* **2007**, *335*, 617–627. [CrossRef]
24. Menter, F.R.; Kuntz, M.; Langtry, R. Ten Years of Industrial Experience with the SST Turbulence Model. *Turbul. Heat Mass Transf.* **2003**, *4*, 625–632.
25. Launder, B.E.; Spalding, D.B. The Numerical Computation of Turbulent Flows. In *Numerical Prediction of Flow, Heat Transfer, Turbulence and Combustion*; Elsevier: Amsterdam, The Netherlands, 1983; pp. 96–116.
26. Wilcox, D.C. Formulation of the K-w Turbulence Model Revisited. *AIAA J.* **2008**, *46*, 2823–2838. [CrossRef]
27. Menter, F.R. Two-Equation Eddy-Viscosity Turbulence Models for Engineering Applications. *AIAA J.* **1994**, *32*, 1598–1605. [CrossRef]
28. Shu, J.-J.; Bin Melvin Teo, J.; Kong Chan, W. Fluid Velocity Slip and Temperature Jump at a Solid Surface. *Appl. Mech. Rev.* **2017**, *69*, 020801. [CrossRef]
29. Shu, J.-J.; Teo, J.B.M.; Chan, W.K. A New Model for Fluid Velocity Slip on a Solid Surface. *Soft Matter* **2016**, *12*, 8388–8397. [CrossRef]
30. Morris, D.L.; Hannon, L.; Garcia, A.L. Slip Length in a Dilute Gas. *Phys. Rev. A* **1992**, *46*, 5279–5281. [CrossRef]
31. Pope, S.B. The Scales of Turbulent Motion. *Turbul. Flows* **2000**, *1*, 195–197.

32. Pöschl, T. Mechanische Ähnlichkeit. In *Lehrbuch der Technischen Mechanik für Ingenieure und Physiker*; Springer: Berlin/Heidelberg, Germany, 1930; pp. 307–311.
33. Yang, X.I.A.; Xia, Z.-H.; Lee, J.; Lv, Y.; Yuan, J. Mean Flow Scaling in a Spanwise Rotating Channel. *Phys. Rev. Fluids* **2020**, *5*, 074603. [CrossRef]
34. Menter, F.; Esch, T. Elements of Industrial Heat Transfer Predictions. In Proceedings of the 16th Brazilian Congress of Mechanical Engineering (COBEM), Uberlândia, Brazil, 26–30 November 2001; Volume 109, p. 650.
35. Abbott, I.H.; Von Doenhoff, A.E. *Theory of Wing Sections*; Dover Publications: Mineola, NY, USA, 1959; pp. 112–115.
36. Arlinghaus, S. *Practical Handbook of Curve Fitting*; CRC Press: Boca Raton, FL, USA, 1994; ISBN 0-8493-0143-2.

sustainability

Article

Estimating Future Costs of Emerging Wave Energy Technologies

Pablo Ruiz-Minguela [1,2,*], Donald R. Noble [3], Vincenzo Nava [1], Shona Pennock [3], Jesus M. Blanco [1,2] and Henry Jeffrey [3]

1 TECNALIA, Basque Research and Technology Alliance (BRTA), Astondo Bidea, Edificio 700, 48160 Derio, Spain
2 School of Engineering, University of the Basque Country, Plaza Ingeniero Torres Quevedo, 1, 48013 Bilbao, Spain
3 Institute for Energy Systems, School of Engineering, The University of Edinburgh, Edinburgh EH9 3FB, UK
* Correspondence: jpablo.ruiz-minguela@tecnalia.com

Abstract: The development of new renewable energy technologies is generally perceived as a critical factor in the fight against climate change. However, significant difficulties arise when estimating the future performance and costs of nascent technologies such as wave energy. Robust methods to estimate the commercial costs that emerging technologies may reach in the future are needed to inform decision-making. The aim of this paper is to increase the clarity, consistency, and utility of future cost estimates for emerging wave energy technologies. It proposes a novel three-step method: (1) using a combination of existing bottom-up and top-down approaches to derive the current cost breakdown; (2) assigning uncertainty ranges, depending on the estimation reliability then used, to derive the first-of-a-kind cost of the commercial technology; and (3) applying component-based learning rates to produce the LCOE of a mature technology using the upper bound from (2) to account for optimism bias. This novel method counters the human propensity toward over-optimism. Compared with state-of-the-art direct estimation approaches, it provides a tool that can be used to explore uncertainties and focus attention on the accuracy of cost estimates and potential learning from the early stage of technology development. Moreover, this approach delivers useful information to identify remaining technology challenges, concentrate innovation efforts, and collect evidence through testing activities.

Keywords: emerging technology; future cost projection; learning rate; uncertainty propagation; wave energy

Citation: Ruiz-Minguela, P.; Noble, D.R.; Nava, V.; Pennock, S.; Blanco, J.M.; Jeffrey, H. Estimating Future Costs of Emerging Wave Energy Technologies. *Sustainability* **2023**, *15*, 215. https://doi.org/10.3390/su15010215

Academic Editors: Bin Huang, Ling Zhou, Yuquan Zhang and Jianhua Zhang

Received: 22 November 2022
Revised: 14 December 2022
Accepted: 15 December 2022
Published: 23 December 2022

1. Introduction

According to Rubin [1], a technology can be defined as emerging if is not yet deployed or ready for purchase on a commercial scale. The design details of an emerging technology are still preliminary or incomplete, performance has not been sufficiently validated, and it may require new components and subsystems that are not available off-the-shelf. Its current stage of development may range from concept to single device demonstration. On a technology readiness level (TRL) scale [2], emerging technologies encompass a TRL of 2 to 7, which is usually the main focus of research and development programmes. This is the case for wave energy technologies, which are still at the validation phase, or TRL 5 [3].

A common evaluation criterion to assess the feasibility and competitiveness of emerging renewable technologies is the future cost of the commercial-scale version once the technology is mature and widely deployed. The affordability metric typically used is the levelized cost of energy (LCOE) [4]. The LCOE provides a complete picture of technology development in the market by accounting for all lifetime costs and energy production. However, this is not a simple task for prototype technologies due to a lack of previous experience and various uncertainties and unknows. In fact, the direct quantification of the

LCOE for single prototypes yields unsuitable results, and therefore a future projection is needed. The LCOE estimate thus represents the future commercial cost that the emerging technology could achieve with sufficient replication provided its technical performance goals are met. The aim of such estimation is twofold: to allow comparison with other technologies currently exploited in the market and to benchmark different cost reduction targets or alternative technology concepts.

The estimation of future costs of wave energy has attracted great interest in order to demonstrate the potential of this renewable energy technology. Various studies have been published providing projections for the entire sector. The OES Technology Collaboration Programme by the IEA developed a study of the international levelized cost of energy for ocean energies in 2015 [5], which directly questioned developers on current costs and future targets. The estimations were updated in 2020 following the same methodology. Although the full report is not accessible, public results show that wave technologies will be able to reach the cost targets defined in the European Strategic Energy Technology Plan (SET-Plan) [6]. These targets are EUR 150/MWh by 2030 and EUR 100/MWh by 2035 for wave energy.

The Joint Research Centre (JRC) periodically publishes ocean energy status reports with cost estimations [3]. Cost are mainly based on the energy technology reference indicator (ETRI) projections for 2010–2050 [7] and the scenario-based cost trajectories to 2050 [8]. These in turn are derived from open literature (both primary and secondary sources), expert judgements, information from other similar technologies, and the application of learning curves with the cumulative capacity foreseen. The higher and lower cost estimates vary significantly due to a lack of a dominating technology and the associated uncertainties related to unproven technologies. Nonetheless, the limited data available from technologies currently under development suggest an LCOE in line or below the SET-Plan targets by 2030 in good resource sites after 1 GW installed capacity [3].

In the UK in 2018, ORE Catapult analysed the cost reduction pathway for wave energy [9]. Single devices reported a cost in excess of GBP 300/MWh together with key cost reductions to GBP 100/MWh after 1GW of deployment through learning by doing, process optimisation, engineering validation, and improved commercial terms. However, again, the lack of data, particularly energy generation, made it hard to accurately estimate the future cost of energy.

Although these studies provide a helpful tool to track the progress of the wave energy sector, they cannot be used to benchmark alternative technology options or assist in decision-making during the different stages of technology development. In this respect, various approaches have been recently proposed to assess wave energy technologies at early stages of development.

The detailed bottom-up techno-economic approach is the most common costing method used for energy technologies. Some future LCOE projections found for wave energy technologies are Oscilla Power's Triton [10], M3 Wave [11], UGEN [12], or Sandia's Reference Model Project [13]. In this approach, the design of the wave energy device and balance of plant and array layout, together with the corresponding technical and operational performance parameters, are specified. Based on this information, the capital expenditure (CAPEX) and operational expenditure (OPEX) costs are estimated for a particular deployment site. This cost is then aggregated with other costs such as project development, insurance, and decommissioning costs, and, as a result, an LCOE is obtained. However, this direct method of estimating the future cost of a commercial-scale technology is only suitable for technologies that are close to commercialisation and whose design is well defined. Experience demonstrates that cost estimates for emerging technologies tend to be rather optimistic and significantly lower than the actual cost of the first commercial plant deployed [14]. As the design is further detailed and unforeseen technological issues uncovered in the development process, the estimated costs of these technologies tend to escalate. Subsequently, the relatively high cost of early deployments declines as the technology is replicated and learning is capitalised through more efficient designs and processes.

Alternatively, Têtu [15] proposed a top-down approach based on the target LCOE for the specific market and a technology-agnostic breakdown of costs to derive thresholds for the different cost centres. Various ranges of uncertainties are suggested per development stage as described in [16]. Developers can use this approach internally to inform their technical decisions, but the method lacks transparency where the cost estimation and the allocation of different levels of uncertainty to the detailed breakdown are concerned. A similar approach based on a reverse cost calculation was developed by Pennock [17] for emerging technologies. In this case, the current cost thresholds for early-stage devices are calculated in order to reach the 2030 SET-Plan levelised cost targets [6]. Component-based learning curves are applied, and the resultant breakdown of costs is compared with cost estimates for current devices to identify areas where further cost reduction is still needed. This method provides more clarity but still requires the external assumption of a standard breakdown of costs for the particular class of device (e.g., point absorber), which might differ for the wave energy concept considered. Moreover, uncertainties in the initial estimations are not embedded, but only a sensitivity to the learning rates applied.

To overcome the limitations of the previous methods, this paper proposes a combination of both the bottom-up and top-down approaches. Starting from the current breakdown of costs, uncertainty ranges are assigned depending on the reliability of the estimation and used to derive the first-of-a-kind cost of the commercial technology. Component-based learning rates are then applied to produce the LCOE of a mature technology after achieving installation of a certain capacity through various commercial projects. Learning rates are also subjected to varying uncertainty.

This paper provides a transparent method for estimating of the future costs of wave energy technologies at early stages of development that counters the propensity toward over-optimism. It supplements the IEA-OES international evaluation framework [18], which prescribes the LCOE as the highest-level affordability metric but fails to provide the specific procedures to perform such an estimation. This novel approach is illustrated with the assistance of a case study. The Reference Model Project [13] provides the underlying information to implement this approach. Future LCOE estimations will be compared with direct methods and overall estimations based on expert judgements.

The proposed method is mainly intended for wave energy technology developers. The ultimate goal is to assist them in reducing the high development cost, time, and risk of wave energy technologies. This is quite relevant since there are 87 active wave energy developers, 60% of which are still in the early phases of technology development [19].

2. Methodology

As explained before, the direct quantification of the LCOE is highly unsuitable for prototype technologies. The affordability assessment of an emerging technology requires the future projection of costs with regard to the commercial technology and a first-of-a-kind commercial deployment. To be precise, this farm project should be the smallest size of a wave energy array for the LCOE to yield a meaningful value.

The proposed approach for estimating future costs of emerging wave energy technologies is an indirect method which consists of three main steps as shown in Figure 1:

- Step 1: Estimation of current cost and performance based on a standardised cost and performance breakdown. The emerging technology is assessed with reference to a first-of-a-kind commercial deployment.
- Step 2: Cost escalation to account for uncertainties in the estimations. Uncertainty ranges (lower and upper bounds) are assigned based on the reliability of the input data at each stage of development. Incorporation of standardised contingencies allows for the estimation of costs for the evolving technology with regard to the same first-of-a-kind commercial deployment.

- Step 3: Projection of the future cost based on technology replication. Component-based learning rates are then applied to the upper bound obtained in the previous step. The upper bound is used to counterbalance the inherent optimism bias in early-stage estimates. The technology is assessed in its mature format and when it has been widely deployed.

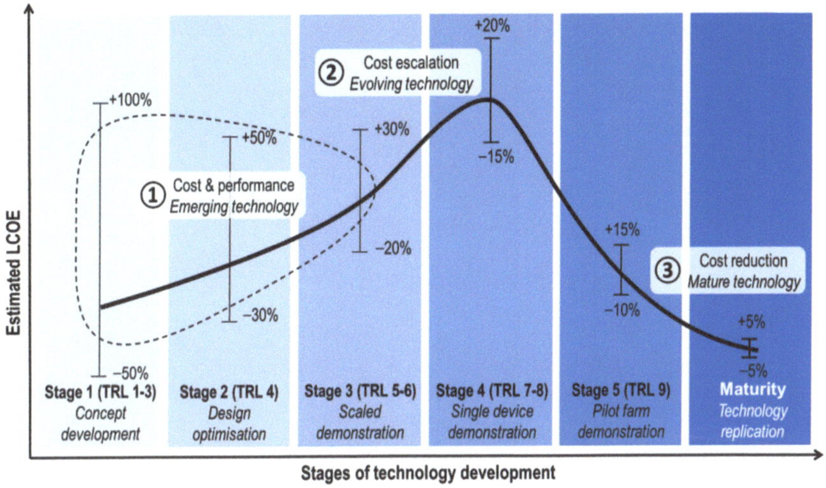

Figure 1. Proposed 3-step approach for estimating the future cost of an emerging wave energy technology at different stages of technology development, with an illustrative LCOE estimate and uncertainty at each stage.

The reader should note that the stages of technology development are not drawn in a time scale in Figure 1. In fact, time is not evenly distributed through the development stages. More time and effort should be allocated to the initial stages, and the overall development time depends on the selected development trajectory [20].

2.1. Step 1: Estimation of Current Cost and Performance

The first step of this approach involves the bottom-up estimation of the LCOE for the emerging technology at its current state of development. Wave energy technology is decomposed into major cost centres. For emerging technologies which are at lower TRLs, this can include a simplified list of subsystems and cost centres. Further granularity (more breakdown levels) can be added as the technology moves up the TRL scale. Parametric modelling is used to identify functional relationships between physical and performance characteristics of an item and its costs, derived from experience and engineering judgement [21].

The standardised cost and performance breakdown used in this work is shown in Figure 2 to the fourth level of detail. It builds upon several published guidance documents and tools such as the US Department of Energy reporting guidance [22], BVGA ocean energy value chain study [23], the COE Calculation Tool commissioned by the Danish Transmission System Operator [24], or the DTOceanPlus design tools [25]. These guidance documents are useful to avoid omitting any relevant cost centre.

For the estimation of future costs, a wave energy farm model is created representing an illustrative first commercial project of 50 units. Considering that rated capacity for utility-scale wave energy technologies usually ranges 200–1500 kW [26], this means between 10 and 75 MW in total. The array size lies in the range of the capacity used for cost estimation of commercial farms [27]. The wave farm model should describe deployment site characteristics, water depth, distance to the shore, and other design parameters used.

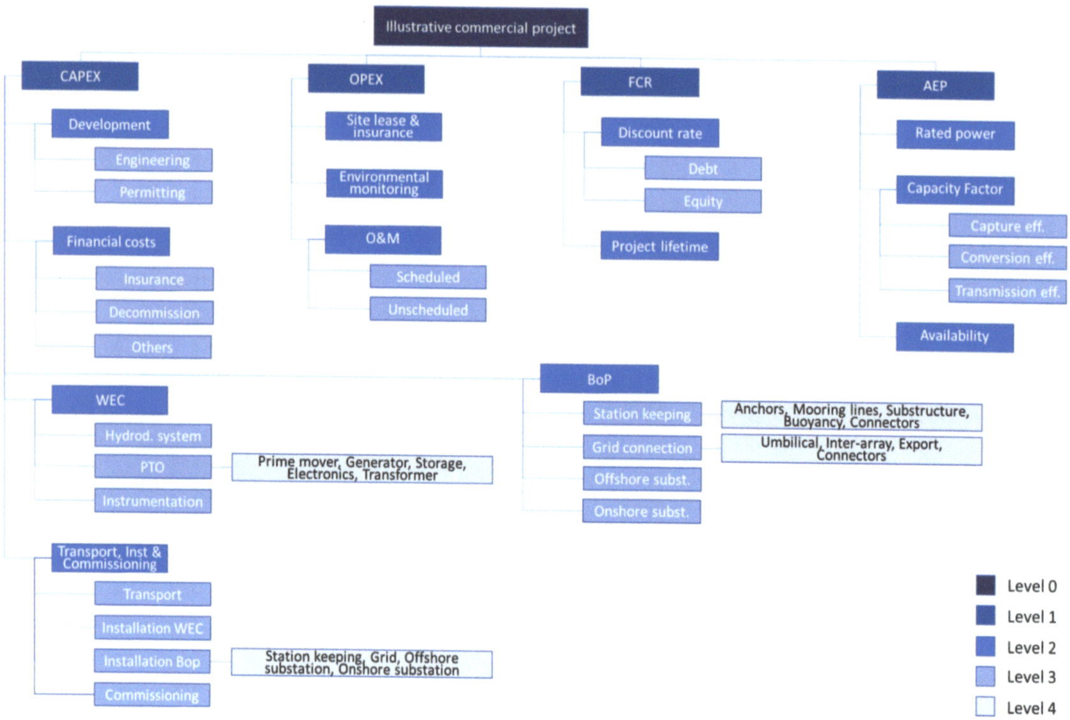

Figure 2. Standard cost and performance breakdown for an illustrative commercial project (adapted from [22–25]).

The first breakdown level fully aligns with the general *LCOE* equation. Due to the emerging nature of this technology, it is assumed that the annual O&M costs and annual energy production will remain constant during its lifetime. This is a common hypothesis in most techno-economic models and is reasonable provided the long-term average system uptime and site resource are used for the calculation of energy. In this case, the simplified *LCOE* can be represented using the following expression [28]:

$$LCOE = \frac{CAPEX \times FCR + OPEX}{AEP} \tag{1}$$

where

- *CAPEX*, capital expenditure, represents all capital costs associated with the farm development, manufacturing, installation, and decommissioning at the end of the project's life.
- *FCR*, fixed charge rate, is the annual return, that is the fraction of the *CAPEX* needed to meet investor revenue requirements,
- *OPEX*, annual operating expenditures, include all routine maintenance, operations, and monitoring activity,
- *AEP*, annual energy production, represents the average net annual energy generated (after accounting for availability) and delivered to the grid.

A brief description of this breakdown is provided in the sections below.

2.1.1. CAPEX

CAPEX can be broken down into farm development costs, financial costs, and all the expenditures associated with the manufacture, installation, and commissioning of both the wave energy converters (WEC) and the balance of plant (BoP).

Development costs comprise engineering (e.g., project management, design engineering, planning, and certification) and permitting services (e.g., environmental studies, consenting, and licenses). Financial costs include insurance during construction and decommissioning bonds.

The generic WEC system breakdown [29] has been used to structure the costs of WEC and BoP manufacture. The WEC contains:

- A hydrodynamic system, comprising structural elements, ballast, and ancillary systems (e.g., navigation lights, bollards, and deck crane).
- Power take-off (PTO), including the prime mover (either mechanical, pneumatic, hydraulic, or direct drive), electrical generator, short-term storage, and power electronics.
- Instrumentation, control, and safety systems, ranging from sensors, comms, control software, cooling, lubrication, firefighting, and back-up power.

On the other hand, the BoP includes all the supporting infrastructure and auxiliary systems of the wave farm needed to deliver the energy other than the WEC itself [30], i.e.:

- Station-keeping, including the foundation (e.g., anchors and piles), mooring lines for compliant systems, or substructure for rigid systems.
- Grid connection, comprising the umbilical, intra-array, and export power cables.
- Offshore substation and switchgear.

The installation and commissioning cost of the WEC and the different subsystems comprising the BoP are considered.

A basic estimate of some of these costs, such as development and financial costs, can be expressed as a percentage of total *CAPEX* costs. Guidance can be found in [31], where Têtu and Fernandez Chozas performed a comprehensive literature review in order to build a cost database for wave energy projects. However, as we will see in Section 2.2, whenever feasible, it is much better to use more sophisticated techno-economic methods to increase the accuracy of the cost estimations.

2.1.2. OPEX

OPEX is usually measured on an annual basis. These costs can be broken down into operation and maintenance (O&M) costs, as well as site leases and insurance during operation. Costs related to site leases and insurance are self-evident. Insurance transfers the risks associated with the replacement of faulty components during the underwritten period of time (usually 5 years).

O&M costs include servicing of the WECs and BoP. Depending on the ability to plan the activities, these costs can be split between:

- Scheduled maintenance, which includes periodic inspections and preventive actions.
- Unscheduled maintenance, which comprises all corrective actions to restore the operational capabilities of the farm and the logistical cost of waiting for a suitable weather window.

Again, when data is scarce, *OPEX* can be estimated as a percentage of *CAPEX* [31]. This is a basic estimate with high uncertainty. As the technology developer starts designing operational plans, techno-economic estimations based on the failure rate of components and subsystems, vessel cost, operation time, and the cost of spares should be a more appropriate tool to improve the accuracy.

2.1.3. Financial Assumptions

A key consideration for utility-scale renewable energy technologies is the impact of the availability and cost of capital on LCOE values. The discount rate (a proxy of the cost of capital) and the project lifetime are the two main parameters.

Assumptions of discount rates are crucial for the assessment of wave energy technology and investment decisions. However, they are subject to a significant degree of uncertainty since the expectations and risk perceptions of investors and project sponsors differ significantly. Discount rates are often estimated based on the weighted average cost of capital (WACC) [32]. The WACC gives an estimate of the cost of raising capital, which is equivalent to the approximate return required by potential creditors (debt) and investors (equity).

The simplified LCOE expression uses the fixed charge rate (*FCR*) [33]:

$$FCR = \frac{d}{\left[1 - (1+d)^{-n}\right]} \tag{2}$$

where d is the discount rate and n is the project lifetime in years.

2.1.4. AEP

Calculating net *AEP* should closely follow the IEC's technical specification 62600-100, "Electricity producing wave energy converters—Power performance assessment" [34]. Assumptions regarding the wave energy resource at the intended deployment site and the numerical method for estimating performance should be documented and justified. Particularly, the estimations should account for losses due to directionality, shallow water, and array interaction effects, together with WEC ancillary energy consumption needs.

The *AEP* is the product of the rated power of the array, the capacity factor, and the availability

$$AEP = 8766 \times P \times CF \times AF \tag{3}$$

where:

- 8766 is the average total hours in a year.
- P is the rated power of the farm.
- CF is the capacity factor.
- AF is the availability factor.

CF represents the ratio of the energy produced by the technology continuously operating over a year compared to the energy that could have been produced at the rated power during the same period. In turn, *CF* can be computed as the product of the device capture efficiency (i.e., the ratio of absorbed and rated power), the conversion efficiency (i.e., the ratio of converted and absorbed power), and the transmission-to-grid efficiency (i.e., the ratio of grid and device output power).

AF is the fraction of time in a year that the wave energy technology is capable of producing energy [35]. By convention, the zero production periods (i.e., wave resources lie below or above certain limits) are counted against the *CF* but not against the *AF*.

2.2. Step 2: Cost Escalation to Account for Uncertainties

For commercial technologies, the costs of a farm project are commonly calculated based on quotes or published data, and when costs are not readily available, they can be estimated using engineering handbooks and numerical models. However, for emerging technologies that have not yet been built at a commercial scale, the direct estimation method might be misleading due to the associated uncertainty in making the cost appraisals. The importance of estimating initial costs is paramount since it will determine the total additional spending required for an emerging technology to be cost competitive in the market.

Actually, LCOE estimates of wave energy technologies can vary widely across studies depending on the external properties and the complexity of the analysis methods utilized [21]. Both aspects were highlighted in the previous step. For a correct interpretation of results, it is essential to undertake a careful examination of the underlying assumptions of farm size, deployment site characteristics, cost of capital, materials, and service vessels.

The current step of the method deals with a third source of variability, namely the uncertainty of the input data for the wave energy farm model. Assigning a range with a nominal confidence band is a good practice that provides much more useful information for decision-making. However, emerging technologies imply that little experience is available to assign ranges of uncertainty to costs.

Several strategies can be used to allocate expected ranges of accuracy to the estimations based on expert judgement. Previsic [16] assigns uncertainty ranges as a double function of the stage of technology development and the source of input data for the estimation for wave energy technologies. Hence, estimation accuracy may vary from −30% to +80% for simplified estimations and technologies at the concept stage and from −5% to +5% for detailed estimates of mature technologies. Fernandez-Chozas [24] applies Previsic's uncertainty ranges to the *AEP* data for each development stage and source of performance estimates (i.e., power matrix and standard sea states). Likewise, organisations including EPRI [36], the DOE [37], and the Association for the Advancement of Cost Engineering International (AACE) [38] have defined several cost estimate classes ranging from "simplified" to "finalised". Parsons performed an exhaustive review and comparison of cost contingency practices and standards to conclude that AACE represents best industry practices [39]. Cost estimation should require increasing levels of effort (and expense) as the technology moves from concept and preliminary design to demonstration and replication.

The ability to properly combine uncertainties from different cost factors is crucial. The individual estimates and their uncertainties can be combined statistically provided they can be calculated with statistical techniques. Rothwell [40] shows that the current engineering guidelines are consistent with contingencies equal to the standard deviation of the cost estimate. He derives the standard deviation from an 80% confidence level using a lognormal probability distribution, since most cost estimate accuracy ranges are non-symmetric. This is because final costs are usually higher than those estimated, and there is no probability that the final cost will ever be less than zero (which is a possibility with the normal distribution).

Table 1 presents the suggested contingencies and expected accuracy ranges used by current engineering guidelines for the different types of cost estimates as well as the corresponding lognormal property fit of the uncertainty ranges. Statistical properties have been normalised by the mode, the most likely estimate. The median represents the 50% probability, which is an indication of the basic uncertainty factor. The standard deviation (Std) has been adjusted with reference to the upper bound in AACE guidelines for an 80% confidence level interval. It can be noted that the statistical fit results in a Std within the range of the expected accuracy values except for the final estimate, in which it is slightly lower.

Table 1. Suggested contingencies and lognormal properties of uncertainty ranges normalised by mode (adapted from [40]).

Type of Estimate	AACE			Statistical Properties			
	Class	**Contingency**	**Accuracy Range**	**Median**	**Mean**	**Std**	**80% Confidence**
Concept	Class 5	50%	−50% to +100%	1.159	1.249	43%	−33% to +101%
Simplified	Class 4	30%	−30% to +50%	1.068	1.104	27%	−24% to +51%
Preliminary	Class 3	20%	−20% to +30%	1.031	1.047	18%	−18% to +30%
Detailed	Class 2	15%	−15% to +20%	1.017	1.025	13%	−14% to +20%
Final	Class 1	5%	−10% to +15%	1.005	1.007	7%	−8% to +10%

Assuming independence of each factor, the probability distributions can now be combined. This is particularly simple if each distribution can be treated as lognormal. In such instances, the final distribution is also lognormal, with the logarithmic standard deviation given by the square root of the sum of squares of the individual geometric standard deviations. Moreover, the error propagation technique can be used to combine uncertainties from multiple variables in the techno-economic expressions of the wave energy LCOE model.

Propagation of error (or uncertainty) is a calculus-derived statistical calculation designed to combine uncertainties from multiple quantities to another quantity. It is based on a set of simple mathematical rules. The standard deviations are used to calculate the resulting uncertainty. Furthermore, provided variables are independent, covariances will be avoided. The general formula for error propagation is given by:

$$\delta q = \sqrt{\left(\frac{\partial q}{\partial x}\delta x\right)^2 + \ldots + \left(\frac{\partial q}{\partial z}\delta z\right)^2} \quad (4)$$

where q is a function that depends on the estimated quantities, $x, \ldots, z$ and their associated uncertainties, $\delta x, \ldots, \delta z$.

The process for error propagation involves:

1. Identifying the uncertain variables in the techno-economic expression for cost estimation.
2. Taking partial derivatives with respect to each of the variables identified in the previous step.
3. Multiplying the partial derivatives by the associated uncertainty to calculate the error contribution from each variable.
4. Adding the contributions in quadrature.

The uncertainty estimation in the LCOE is not direct, but it is calculated by means of its formula involving *CAPEX*, *OPEX*, *FCR*, and *AEP*. In turn, each of these factors were derived in Step 1 using basic parametric relationships. Error propagation is used to calculate the aggregated uncertainty in a cascading manner from the lowest level of the standard cost and performance breakdown. For instance, the structural cost of the hydrodynamic system can be calculated from three techno-economic variables: the unit cost of the main raw material (EUR/kg), a coefficient to account for the manufacturing complexity (-), and the structural weight (kg). Ranges of uncertainty in the material unit cost (exogenous factor), maturity of manufacturing processes (suppliers' capability), and estimation of the structural weight (design accuracy) will determine the aggregated uncertainty in the estimation of the hydrodynamic system cost, in this case, the geometric mean of the standard deviations. This estimate in turn will be combined with other capital expenditures to derive the uncertainty in the WEC, farm *CAPEX*, and finally the LCOE.

2.3. Step 3: Projection of the Future Cost with Technology Replication

The third and final step of the methodology involves the application of learning curves to project the future costs of wave energy technology once it has been sufficiently replicated and the estimation of uncertainties in the forecast due to learning. Different learning mechanisms have been described in the literature, such as in [41–43]. The most important mechanism is technological learning. Other learning factors may include:

- Economic learning, i.e., shifting production to low-wage countries.
- Social learning; as stakeholders become more familiar, they increase trust in one another.
- Financial learning; as banks and investors gain confidence in a new technology, they reduce the expected interest rates.

These exogenous factors have a significant impact on the LCOE estimation, but unfortunately, they can only be accounted for within the initial assumptions or through sensitivity analysis.

Technological learning is an endogenous factor that encompasses learning by research in the early stages due to R&D investments, learning by doing during the production stage due to higher efficiency of manufacturing processes, learning by using in the initial stage of introduction of the technology into the market, and learning by interaction in the diffusion of the technology incidentally reinforcing the previous factors. Scale effects are also part of the technological learning mechanism, both upsizing (i.e., the increase in rate power) through technology redesign leading to lower unit costs, and economies of scale (i.e., mass production) through standardisation allowing upscaling of production facilities.

The analysis presented in this paper uses the learning curve method as the most applied approach. The commonly used formulation originates from empirical observations across diverse energy technologies that often evidence a log-linear relationship between cost reductions driven by manufacturing, standardisation, scale of production and use, and cumulative installed capacity or production [44]. In the simplest form, it can be expressed as:

$$Y = ax^b \tag{5}$$

where Y is the cost of the technology and x represents the cumulative experience (often characterised by the installed capacity in MW). The constants a and b denote the cost of the first commercial deployment and the rate of cost reduction, respectively. Note that b represents the slope in a log-log scale in Equation (5). The cost reduction associated with duplication of experience is referred to as the learning rate (LR).

$$LR = 1 - 2^b \tag{6}$$

The independent variable x in Equation (5) reflects all the factors that influence the cost trajectory of the technology. Often, combinations of technological learning occur at each stage, and their contributions may change during the development of a technology over time. Furthermore, single-factor learning curves do not necessarily describe the underlying factors of cost reduction [44]. Some components and subsystems in wave energy farms, such as electrical infrastructure and offshore operations, are not entirely new to the market. They build on the experience gained from more mature sectors, a disaggregated approach that can account for individual learning effects at the component level leading to improved cost reduction estimations for emerging technologies which lack historical data. This can take advantage of past learning rates for direct comparable technologies in order to build a composite learning rate. In addition, it can break apart the impact of raw material spending (an exogenous parameter) from other cost reductions due to cumulative experience.

Learning rates found in the literature for wave energy technologies mainly rely on expert judgements, expectations, and assumptions. They tend to differ widely even at the subsystem level [8]. Overall LRs range from a low 9% [45] to an optimistic 30% [5]. Component-based learning rates range 1% to 12% [27]. Similarly, SI Ocean [46] included a learning rate of 3% for the capacity factor in its LCOE projections.

Since there is little empirical evidence to establish the learning rates for WEC technologies, the component-based learning approach used in this work allocates them depending on the stage of development of the individual components. Three main categories are defined:

- Mature components. These are technologies already established in the market that have well-known characteristics and limited potential for cost reduction due to learning. Low learning rates of 0–5%. E.g., export power cables.
- Evolving components. These have niche market commercialisation and have the potential for significant cost reductions due to learning. Medium learning rates of 5–10%. E.g., prime mover.
- Emerging components. These have not been commercialised yet, but their potential learning cost reductions are high. High learning rates of 10–20%. E.g., maintenance operations.

The upper bound of learning rates is consistent with analyses such as the PelaStar cost of energy [47] and WaveBoost [48]. In these studies, the technological maturity of each major cost item is categorised as "mature", "emerging", or "nascent/emerging 2" with 5%, 10%, and 15–20% learning rates, respectively. The lower bound refer to more conservative analyses such as NEMS [49]. Technologies classified as "conventional", "evolutionary", and "revolutionary" are assigned 1%, 5%, and 10% learning rates, correlatively.

Assigning error margins to *LR*s is recommended to avoid overrepresentation in cost reduction estimates [50]. Forecasts are highly sensitive to uncertainties in the progress ratio. As in the previous step, the error can be calculated from the error propagation theory [43]:

$$\delta LR = \frac{\partial\left(1-2^{-b}\right)}{\partial b}\delta b = ln2{\cdot}2^{-b}{\cdot}\delta b = ln2{\cdot}LR{\cdot}\delta b \tag{7}$$

where δb is the uncertainty in the experience parameter and δLR is the resulting uncertainty.

Moreover, the cost reduction for a technology cannot be realised continually. There will be a bare minimum or baseline cost necessary to build a technology. As suggested in Section 2.2, segregating the price of raw materials from the estimation of manufactured component costs is a recommended strategy to prevent this situation.

3. Case Study and Results

The application of the proposed cost estimation methodology is illustrated with the help of one of the reference models (RMs) for wave energy technologies [13]. The RM project team, led by Sandia National Laboratories (SNL), included a partnership between the US Department of Energy (DOE), the National Renewable Energy Laboratory (NREL), and other US laboratories. The RMs provide a non-proprietary, open-source instrument for technical and economic assessment and validation of design tools, as well as the identification of cost reduction pathways and research priorities to meet the affordability targets. The wave energy models [51] reproduce three common archetypes, namely, a point absorber (RM3), an oscillating wave surge converter (RM5), and an oscillating water column (RM6).

The present case study is based on the RM5, a floating oscillating wave surge converter (OWSC) designed for a wave site near Eureka in Humboldt County, California. The OWSC is one of the most promising wave energy technologies in terms of its energy absorption capabilities [52]. It basically consists of a vertical flap that faces the waves and articulates in its lower part for rotation. The surge motion of waves creates a back-and-forth movement from which energy is extracted [53]. Several OWSC designs have been proposed, including Aquamarine Power's Oyster [54], AW-Energy's WaveRoller [55], Resolute Marine's Wave$_2$O [56], and Langlee's Robusto [57]. The floating version of OWSCs tackle the potential environmental restrictions of nearshore shallow waters, at the same time opening the way to harness the higher wave energy resource in deep-water sites [58].

Figure 3 shows a schematic of the floating OWSC device. The flap rotates against the supporting frame to convert wave energy into electrical power from the motion induced by incoming waves. An oleo-hydraulic PTO with two rams, high pressure accumulators, electrical generator, and corresponding switchgear is used to transform the oscillation in electrical power. The device is tension-moored to the seabed in deep waters (50 to 100 m) through four tendons.

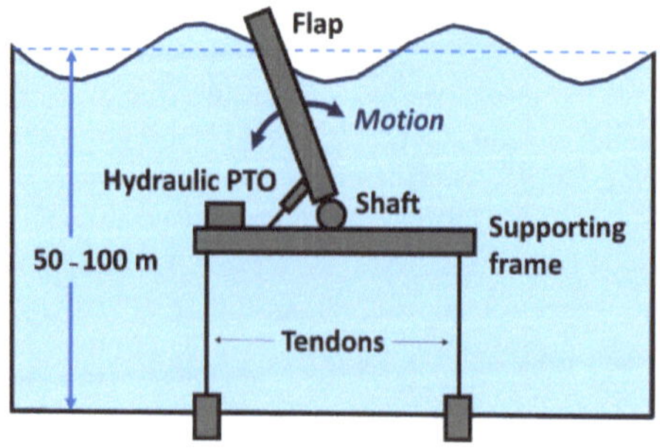

Figure 3. Schematic of the RM5 floating OWSC.

NREL created a techno-economic model for the assessment of the LCOE with multiple scenarios ranging from a single RM5 device to arrays of 10, 50, and 100 units [59]. For the estimation of future costs, this case study uses the cost breakdown of the 50-unit farm model, which can be considered representative of the first commercial project. The RM5 has a rated capacity of 360 kW, which results in an 18 MW wave energy farm.

The array configuration is depicted in Figure 4. A staggered configuration with 600 m spacing between the devices to accommodate moorings is considered to avoid collisions with vessels and produce negligible hydrodynamic losses. Groups of 10 devices are interconnected by umbilical cables as shown in the figure. Electricity is then transmitted to a junction box. Intra-array cables connect the five junction boxes. Lastly, a three-phase AC export cable delivers energy to the shore. Cable landing is accomplished using directional drilling. Close to the deployment site, there is a port with facilities well-suited for installation and maintenance activities and a 60 kV onshore substation.

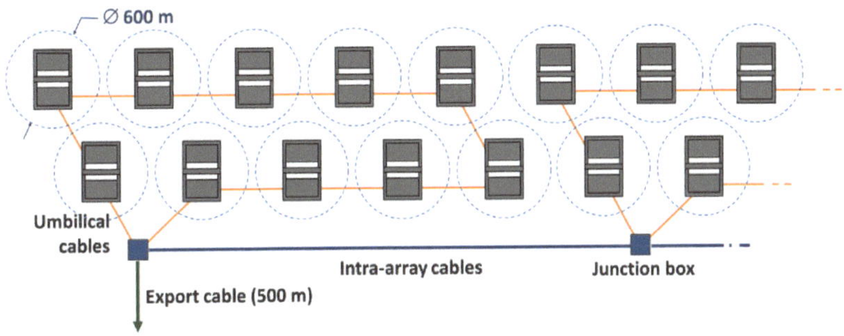

Figure 4. 50-unit farm array layout (not drawn to scale).

The key design parameters and main assumptions are included in Table 2. Further details of RM5 design can be found in [58].

Table 2. Case study specifications.

Category	Parameter	Specification
Site	Water depth	70 m
	Seabed	Soft sediments (sand and clay)
	Wave resource	30 kW/m, unidirectional
	Distance to shore	500 m
Device	Rated power	360 kW
	Hydrodynamic system	Flap (25 m × 19 m), shaft (∅3 m); fiberglass and steel
	PTO	Oleo-hydraulic (2 rams, HP accumulators, hydraulic motor, generator)
	Control	Optimal velocity-dependent damping per sea state
Balance of Plant	Station keeping	Steel frame (45 m × 29 m), four polyester lines and suction anchors
	Grid connection	Umbilical, inter-array, and export (30 kV); terminators and connectors
Array	Device spacing	600 m
Performance	Capture efficiency	37%
	Conversion efficiency	82%
	Transmission efficiency	95%
	Availability	98%
Financial	Discount rate	8.8%
	Project lifetime	20 years

3.1. Step 1: Cost and Performance of the 50-Unit Farm

NREL's model for the 50-unit farm results in an estimated LCOE of USD 0.78/kWh [59]. The proposed method yields a slightly lower estimate (USD 0.72/kWh) due to the 10%

contingency in *CAPEX* costs included in NREL's model. Contingency is a consequence of the propagation of uncertainties, and consequently it is accounted for in Step 2.

The detailed breakdown of *CAPEX* and *OPEX* costs, financial assumptions, and annual energy production taken directly from the RM5 model are presented in Table 3. The last column outlines the modelling basis directly extracted from [59]. The resulting percentage contribution to the lifetime costs of the main cost centres is shown in Figure 5.

Table 3. Detailed breakdown of cost and performance (adapted from [59]).

ID	Breakdown	50-Unit Farm	Basis (Equations Refer to Subsequent Row IDs)
1	*CAPEX* (USD)	240,016,908	= 1.1 + 1.2 + 1.3 + 1.4 + 1.5 + 1.6
1.1	Development	10,558,725	= 1.1.1 + 1.1.2
1.1.1	Engineering	4,589,164	Percentage of CAPEX (2%)
1.1.2	Permitting	5,969,561	Average of PNNL estimates
1.2	Financial costs	0	= 1.2.1 + 1.2.2 + 1.2.3
1.2.1	Insurance (during construction)	0	Not considered
1.2.2	Decommission	0	Percentage of installation cost (70%), depreciation
1.2.3	Other	0	Percentage of CAPEX (0%)
1.3	WEC	109,478,032	= 1.3.1 + 1.3.2 + 1.3.3
1.3.1	Hydrodynamic system	86,670,989	Weight (499 t), unit cost (UDS 3161/t), subsystem integration (10%)
1.3.2	PTO	22,561,677	= 1.3.2.1 + 1.3.2.2 + 1.3.2–3 + 1.3.2.3 + 1.3.2.4 + 1.3.2.5
1.3.2.1	Prime mover	19,208,071	Mass (32,920 kg), unit cost (USD 10.61/kg), subsystem integration (10%)
1.3.2.2	Generator	1,467,120	Mass (908 kg), unit cost (USD 29.38/kg), subsystem integration (10%)
1.3.2.3	Storage	0	Included in the hydraulic prime mover
1.3.2.4	Power electronics	1,143,890	Mass (1200 kg), unit cost (USD 17.32/kg), subsystem integration (10%)
1.3.2.5	Transformer	742,597	Mass (1590 kg), unit cost (USD 8.49/kg), subsystem integration (10%)
1.3.3	Instrumentation and control	245,366	Unit cost (USD 4461), subsystem integration (10%)
1.4	BoP	91,009,936	= 1.4.1 + 1.4.2 + 1.4.3 + 1.4.4
1.4.1	Station-keeping	81,681,936	= 1.4.1.1 + 1.4.1.2 + 1.4.1.3 + 1.4.1.4 + 1.4.1.5
1.4.1.1	Anchors and piles	14,500,828	No./device (8), weight (13 t), unit cost (USD 2789/t)
1.4.1.2	Mooring lines	15,789,988	No./device (4), length (80 m), unit cost (USD 987/m)
1.4.1.3	Substructure	44,087,621	Weight (301 t), unit cost (USD 2663/t), subsystem integration (10%)
1.4.1.4	Buoyancy	2,700,000	Bulk discount factor (0.9), unit cost (USD 60,000)
1.4.1.5	Connecting hardware	4,603,500	Bulk discount factor (0.9), unit cost (USD 102,300)
1.4.2	Grid connection	9,328,000	= 1.4.2.1 + 1.4.2.2 + 1.4.2.3 + 1.4.2.4
1.4.2.1	Umbilical	4,400,000	Length (40,000 m) and unit cost (USD 110/m)
1.4.2.2	Inter-array	2,880,000	Length (14,400 m) and unit cost (USD 200/m)
1.4.2.3	Export	1,200,000	Length (6000 m) and unit cost (USD 200/m)
1.4.2.4	Connectors	848,000	Percentage of cable cost (10%)
1.4.3	Offshore substation	0	Not considered
1.4.4	Onshore infrastructure	0	Not considered
1.5	Transp, install, and commission	23,320,215	= 1.5.1 + 1.5.2 + 1.5.3 + 1.5.4
1.5.1	Transport	1,487,500	Unit cost (USD 29,750)
1.5.2	Installation WEC	3,854,375	Days (55 days), rate (USD 70,080/day)
1.5.3	Installation BoP	14,123,965	= 1.5.3.1 + 1.5.3.2 + 1.5.3.3 + 1.5.3.4
1.5.3.1	Station-keeping	8,852,950	Days (127 day), rate (USD 69,483/day)
1.5.3.2	Grid connection	4,503,815	Days (50 day), rate (USD 90,949/day)
1.5.3.3	Offshore substation	0	Not considered
1.5.3.4	Onshore infrastructure	767,200	Cable landing distance (500 m), unit cost (USD 1534/m)
1.5.4	Commissioning	3,854,375	Percentage of WEC installation (100%)
1.6	Dedicated O&M vessels	5,650,000	Number (1), vessel cost (USD 5.65 mo)

Table 3. *Cont.*

ID	Breakdown	50-Unit Farm	Basis (Equations Refer to Subsequent Row IDs)
2	Annual *OPEX* (USD)	5,870,427	= 2.1 + 2.2 + 2.3
2.1	Site lease and insurance	2,414,582	Lease cost (USD 120,000), percentage of CAPEX (1%)
2.2	Environmental monitoring	1,785,000	Data taken from PNNL study
2.3	O&M	1,670,845	= 2.3.1 + 2.3.2
2.3.1	Scheduled	1,009,692	Staff (6.5), salary (USD 51,491/year), consumables (USD 13,500)
2.3.2	Unscheduled	661,153	Days (109 days), rate (USD 5680/day), cost spares (USD 24,830), no. (1.75)
3	Financial assumptions (*FCR*, %)	0.11	= 3.1/(1 − 1/(1 + 3.1) ^ 3.2)
3.1	Discount rate (%)	0.09	= 3.1.1 + 3.1.2
3.1.1	Debt (%)	0.05	Return on debt (9.5%), percentage (50%)
3.1.2	Equity (%)	0.04	Return on equity (8.1%), percentage (50%)
3.2	Project lifetime (years)	20	n/a
4	*AEP* (kWh)	44,101,201	= 8766 × N × 4.1 × 4.2 × 4.3
4.1	Rated power (kW)	360	n/a
4.2	Capacity factor (%)	0.29	= 4.2.1 × 4.2.2 × 4.2.3
4.2.1	Capture efficiency (%)	0.37	Average extracted power (132 kW)
4.2.2	Conversion efficiency (%)	0.82	NREL's assumption
4.2.3	Transmission efficiency (%)	0.95	NREL's assumption
4.3	Availability (%)	0.98	NREL's assumption
5	LCOE (USD/kWh)	0.72	= (1 × 3 + 2)/4

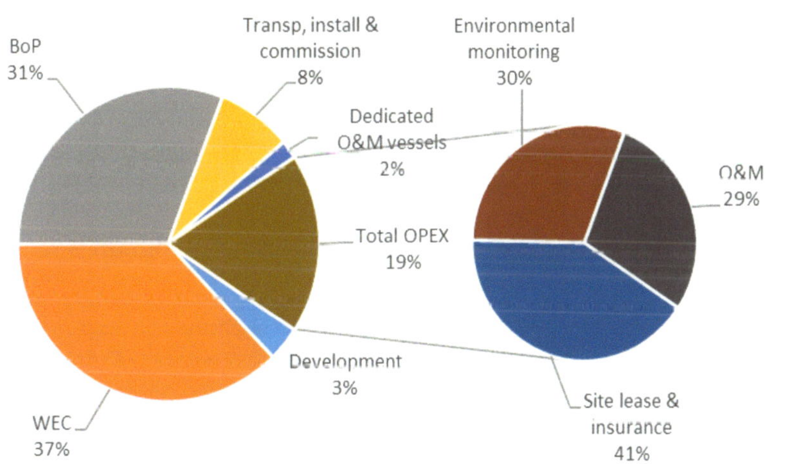

Figure 5. Breakdown of costs for the RM5 farm. **Left**: percentage of total lifetime costs; **Right**: distribution of *OPEX* costs.

3.2. Step 2: Cost Escalation to Account for Uncertainties

The RM5 model has inherent uncertainties regarding performance, design, and economics. NREL carried out a qualitative uncertainty assessment of both design and performance [58]. Levels of uncertainty, from low to very high, were assigned to various components of the model depending on whether this facet was assessed using test/field data (low), modelled data (medium), or engineering judgment (high). Aspects that were not addressed were assigned a "very high" level of uncertainty.

The qualitative assessment has been mapped to the AACE's uncertainty classes and corresponding quantitative standard deviation (Std). Sometimes "low to medium" and "medium to high" levels of uncertainty were used. In these two cases, an average value between the two adjacent classes is assumed as shown in Table 4. None (0%) is only used whenever the parameter has no implicit uncertainty.

Table 4. Uncertainty categories, associated standard deviation, and 80% confidence intervals.

Uncertainty	AACE	Std	80% Confidence
Very high	Class 5	43.0%	−33% to +101%
High	Class 4	27.0%	−24% to +51%
Med/High	-	22.5%	−21% to +40%
Medium	Class 3	18.0%	−18% to +30%
Low/Med	-	15.5%	−16% to +25%
Low	Class 2	13.0%	−14% to +20%
Very low	Class 1	7.0%	−8% to +10%
None	-	0.0%	–

Uncertainty is propagated upwards in the breakdown structure using the generic Equation (4) until a final LCOE is was obtained. The method comprises four specific categories of functions:

- Addition of several components (applicable to *CAPEX* and *OPEX* cost centres). The absolute uncertainty is the geometric mean of individual absolute uncertainties.

$$\delta q = \sqrt{(\delta x)^2 + \ldots + (\delta z)^2} \tag{8}$$

- Multiplication or division of several components (applicable to *AEP*). The relative uncertainty is the geometric mean of the individual relative uncertainties.

$$\frac{\delta q}{|q|} = \sqrt{\left(\frac{\delta x}{|x|}\right)^2 + \ldots + \left(\frac{\delta z}{|z|}\right)^2} \tag{9}$$

- Financial uncertainty with a variable discount rate (d) and constant lifetime (n) and differentiation of the *FCR* with respect to the discount rate.

$$\delta q = \frac{(1+d)^{n-1}((1+d)^n + d((1+d)^n - n - 1) - 1)}{((1+d)^n - 1)^2} \delta d \tag{10}$$

- Uncertainty in LCOE. A sequential combination of multiplication ($CAPEX \times FCR$), addition (*OPEX*), and division (*AEP*) computed with the help of Equations (8) and (9).

The detailed results are presented in Table 5. Following this procedure, the LCOE results in an upper and lower bound of USD 1.33/kWh and USD 0.50/kWh, respectively. The Std of the LCOE uncertainty is 38.2%, which gives an indication of the contingency to be considered. It can be noticed that the *AEP* is the greatest contributor to the global uncertainty. The rest of the components in the LCOE (Equation (1)) are slightly above 10%, which fairly matches the aforementioned assumption of contingency in the NREL's model. Figure 6 displays the resulting uncertainties for the high-level components in the LCOE equation. It is also worth mentioning that the NREL's model estimates USD 1.44/kWh for a small array of 10 units [59].

Table 5. Propagation of uncertainties and corresponding costs. For assigned uncertainties, the associated Std and 80% confidence interval in Table 4 is used. Propagated uncertainties use Equations (8)–(10) as appropriate. Units of upper/lower bound as per breakdown.

ID	Breakdown	Uncertainty	Std	80% Conf Interval		Lower Bound	Upper Bound
1	*CAPEX* (USD)	Propagated	11.4%	−12%	17%	210,033,907	281,374,854
1.1	Development	Propagated	17.3%	−18%	29%	8,704,876	13,563,431
1.1.1	Engineering	High	27.0%	−24%	51%	3,468,629	6,928,178
1.1.2	Permitting	Med/High	22.5%	−21%	40%	4,690,419	8,348,819
1.2	Financial costs	Propagated	0.0%	0%	0%	0	0
1.2.1	Insurance (during construction)	None	0.0%	0%	0%	0	0
1.2.2	Decommission	None	0.0%	0%	0%	0	0
1.2.3	Other	None	0.0%	0%	0%	0	0
1.3	WEC	Propagated	21.6%	−21%	38%	86,709,892	150,851,161
1.3.1	Hydrodynamic system	High	27.0%	−24%	51%	65,508,563	130,845,643
1.3.2	PTO	Propagated	15.4%	−16%	25%	18,948,422	28,125,143
1.3.2.1	Prime mover	Medium	18.0%	−18%	30%	15,730,180	24,951,644
1.3.2.2	Generator	Medium	18.0%	−18%	30%	1,201,478	1,905,817
1.3.2.3	Storage	Medium	18.0%	−18%	0%	0	0
1.3.2.4	Power electronics	Medium	18.0%	−18%	30%	936,773	1,485,934
1.3.2.5	Transformer	Medium	18.0%	−18%	30%	608,139	964,647
1.3.3	Instrumentation and control	Medium	18.0%	−18%	30%	200,939	318,735
1.4	BoP	Propagated	14.7%	−15%	23%	77,004,901	112,141,127
1.4.1	Station-keeping	Propagated	16.3%	−17%	26%	68,002,855	103,257,710
1.4.1.1	Anchors and piles	High	27.0%	−24%	51%	10,960,166	21,891,640
1.4.1.2	Mooring lines	High	27.0%	−24%	51%	11,934,552	23,837,862
1.4.1.3	Substructure	High	27.0%	−24%	51%	33,322,761	66,558,293
1.4.1.4	Buoyancy	High	27.0%	−24%	51%	2,040,742	4,076,142
1.4.1.5	Connecting hardware	High	27.0%	−24%	51%	3,479,465	6,949,822
1.4.2	Grid connection	Propagated	10.5%	−12%	16%	8,239,279	10,793,785
1.4.2.1	Umbilical	Medium	18.0%	−18%	30%	3,603,318	5,715,683
1.4.2.2	Inter-array	Medium	18.0%	−18%	30%	2,358,536	3,741,174
1.4.2.3	Export	Medium	18.0%	−18%	30%	982,723	1,558,823
1.4.2.4	Connectors	Medium	18.0%	−18%	30%	694,458	1,101,568
1.4.3	Offshore substation	None	0.0%	0%	0%	0	0
1.4.4	Onshore infrastructure	None	0.0%	0%	0%	0	0
1.5	Transp, install, and commission		11.0%	−12%	17%	20,485,546	27,191,364
1.5.1	Transport	Med/High	22.5%	−21%	40%	1,168,762	2,080,365
1.5.2	Installation WEC	Med/High	22.5%	−21%	40%	3,028,470	5,390,594
1.5.3	Installation BoP	Propagated	15.9%	−16%	26%	11,808,255	17,734,310
1.5.3.1	Station-keeping	Med/High	22.5%	−21%	40%	6,955,964	12,381,426
1.5.3.2	Grid connection	Med/High	22.5%	−21%	40%	3,538,749	6,298,877
1.5.3.3	Offshore substation	Med/High	22.5%	−21%	0%	0	0
1.5.3.4	Onshore infrastructure	Med/High	22.5%	−21%	40%	602,806	1,072,979
1.5.4	Commissioning	Med/High	22.5%	−21%	40%	3,028,470	5,390,594
1.6	Dedicated O&M vessels	High	27.0%	−24%	51%	4,270,441	8,529,704
2	Annual *OPEX* (USD)	Propagated	9.0%	−10%	13%	5,271,888	6,642,826
2.1	Site lease and insurance	Low	13.0%	−14%	20%	2,079,003	2,897,679
2.2	Environmental monitoring	Low/Med	15.5%	−16%	25%	1,498,118	2,227,544
2.3	O&M	Propagated	19.5%	−19%	33%	1,349,152	2,223,782
2.3.1	Scheduled	High	27.0%	−24%	51%	763,156	1,524,314
2.3.2	Unscheduled	High	27.0%	−24%	51%	499,720	998,132
3	Financial assumptions (*FCR*, %)	Propagated	12.1%	−13%	18%	0.09	0.13
3.1	Discount rate	Propagated	19.1%	−19%	32%	0.07	0.12
3.1.1	Debt	High	27.0%	−24%	51%	0.04	0.07
3.1.2	Equity	High	27.0%	−24%	51%	0.03	0.06
3.2	Project lifetime	None	0.0%	0%	0%	20	20

Table 5. *Cont.*

ID	Breakdown	Uncertainty	Std	80% Conf Interval		Lower Bound	Upper Bound
4	*AEP* (kWh)	Propagated	35.6%	−29%	76%	31,118,603	77,555,818
4.1	Rated power (kWh)	None	0.0%	0%	0%	360	360
4.2	Capacity factor (%)	Propagated	23.3%	−22%	42%	0.22	0.40
4.2.1	Capture efficiency (%)	Medium	18.0%	−18%	30%	0.30	0.48
4.2.2	Conversion efficiency (%)	Low	13.0%	−14%	20%	0.71	0.98
4.2.3	Transmission efficiency (%)	Very low	7.0%	−8%	10%	0.87	1.00
4.3	Availability (%)	High	27.0%	−24%	51%	0.74	1.00
5	LCOE (USD/kWh)		38.2%	−31%	84%	0.50	1.33

Figure 6. Uncertainties of the high-level components in the LCOE equation. Note LCOE uncertainty is propagated and not simply added.

3.3. Step 3: Projecting the Future Cost of the Mature Technology

The last step of the methodology involves the optimisation of the current version of the technology through learning by doing and economies of scale (endogenous factors) leading to cost reduction. Learning is proportional to the installed capacity, having impact on the *CAPEX*, *OPEX*, and to a certain extent on the *AEP*. Component-based learning rates are applied to the upper bound obtained in the previous step. In this case study, LCOE results are projected once 1 GW of the emerging technology has been deployed. Selection of 1 GW installed capacity allows comparison with JRC forecasts [3]. The NREL's model provides component-based learning rates for the PTO. For other cost centres, they only provide a qualitative indication depending the predicted innovation potential [58]. A baseline cost has also been included marking a hard threshold beyond which no more learning would be possible. This baseline is based on the 100-unit model, which corresponds to a fully commercial project.

The component-based learning rates are classified in three main categories according to the technology type as shown in Table 6. Learning rates of mature technologies are matched with low uncertainty, whereas evolving and emerging technologies are assumed to have medium and high uncertainties, respectively. The same standard deviations as in Table 4 are used.

Detailed results are shown in Table 7. Component-based projections are combined using the same basis as in Table 3 to derive the corresponding LRs at the immediate upper level. This process is repeated until the aggregated LR, 10.6%, is finally obtained. Figure 7 displays the resulting LR for the high-level components in the LCOE equation. The proposed method estimates the future cost of energy at USD 0.69/kWh. The baseline cost suggested is USD 0.62/kWh, which is higher than the lower bound of USD 0.50/kWh identified in Step 2. Finally, the NREL's 100-unit model results in exactly the same estimate of USD 0.69/kWh [59].

Table 6. Component-based LR, uncertainty and standard deviation.

Technology Type	Learning Rate (LR)		Uncertainty	Std
	From	Up to		
Mature	0.0%	5.0%	Low	13%
Evolving	5.0%	10.0%	Medium	18%
Emerging	10.0%	20.0%	High	27%

Table 7. Component-based learning and future cost projections. Upper bound from Table 5. Units of projection and baseline as per breakdown.

ID	Breakdown	Upper Bound	Learning Rate		Category	Projection	Baseline
1	*CAPEX* (USD)	281,374,854	Aggregated	3.9%	Mature	223,023,237	210,412,696
1.1	Development	13,563,431	Aggregated	8.0%	Evolving	8,389,156	5,399,407
1.1.1	Engineering	6,928,178	Assigned	7.5%	Evolving	4,409,418	2,414,626
1.1.2	Permitting	8,348,819	Assigned	12.0%	Emerging	3,979,737	2,984,781
1.2	Financial costs	0	Aggregated	0.0%	Mature	0	0
1.2.1	Insurance (during construction)	0	Assigned	5.0%	Mature	0	0
1.2.2	Decommission	0	Assigned	6.0%	Evolving	0	0
1.2.3	Other	0	Assigned	5.0%	Mature	0	0
1.3	WEC	150,851,161	Aggregated	6.7%	Evolving	100,817,663	96,237,347
1.3.1	Hydrodynamic system	130,845,643	Assigned	8.0%	Evolving	80,700,943	76,546,193
1.3.2	PTO	28,125,143	Assigned	5.9%	Evolving	19,833,205	19,472,556
1.3.2.1	Prime mover	24,951,644	Assigned	8.0%	Evolving	16,708,154	16,708,154
1.3.2.2	Generator	1,905,817	Assigned	3.7%	Mature	1,531,727	1,284,397
1.3.2.3	Storage	0	Assigned	5.0%	Mature	0	0
1.3.2.4	Power electronics	1,485,934	Assigned	20.0%	Emerging	831,920	831,920
1.3.2.5	Transformer	964,647	Assigned	4.0%	Mature	761,404	648,084
1.3.3	Instrumentation and control	318,735	Assigned	2.0%	Mature	283,516	218,599
1.4	BoP	112,141,127	Aggregated	4.1%	Mature	88,247,013	84,640,133
1.4.1	Station-keeping	103,257,710	Assigned	4.5%	Mature	78,919,013	75,312,133
1.4.1.1	Anchors and piles	21,891,640	Assigned	7.0%	Evolving	14,375,068	13,281,286
1.4.1.2	Mooring lines	23,837,862	Assigned	7.0%	Evolving	15,789,988	15,789,988
1.4.1.3	Substructure	66,558,293	Assigned	8.0%	Evolving	41,050,790	38,937,360
1.4.1.4	Buoyancy	4,076,142	Assigned	6.0%	Evolving	2,847,752	2,700,000
1.4.1.5	Connecting hardware	6,949,822	Assigned	6.0%	Evolving	4,855,416	4,603,500
1.4.2	Grid connection	10,793,785	Assigned	2.5%	Mature	9,328,000	9,328,000
1.4.2.1	Umbilical	5,715,683	Assigned	6.0%	Evolving	4,400,000	4,400,000
1.4.2.2	Inter-array	3,741,174	Assigned	5.0%	Mature	2,880,000	2,880,000
1.4.2.3	Export	1,558,823	Assigned	5.0%	Mature	1,200,000	1,200,000
1.4.2.4	Connectors	1,101,568	Assigned	6.0%	Evolving	848,000	848,000
1.4.3	Offshore substation	0	Assigned	6.0%	Evolving	0	0
1.4.4	Onshore infrastructure	0	Assigned	2.0%	Mature	0	0
1.5	Transp, install, and commission	27,191,364	Aggregated	4.4%	Mature	20,937,816	20,045,808
1.5.1	Transport	2,080,365	Assigned	6.0%	Evolving	1,487,500	1,487,500
1.5.2	Installation WEC	5,390,594	Assigned	7.0%	Evolving	3,763,500	3,763,500
1.5.3	Installation BoP	17,734,310	Assigned	6.6%	Evolving	11,920,737	11,031,308
1.5.3.1	Station-keeping	12,381,426	Assigned	10.0%	Evolving	6,723,055	6,622,419
1.5.3.2	Grid connection	6,298,877	Assigned	6.0%	Evolving	4,400,641	3,641,689
1.5.3.3	Offshore substation	0	Assigned	6.0%	Evolving	0	0
1.5.3.4	Onshore infrastructure	1,072,979	Assigned	5.0%	Mature	797,040	767,200
1.5.4	Commissioning	5,390,594	Assigned	6.0%	Evolving	3,766,079	3,763,500
1.6	Dedicated O&M vessels	8,529,704	Assigned	10.0%	Evolving	4,631,589	4,090,000

Table 7. *Cont.*

ID	Breakdown	Upper Bound	Learning Rate		Category	Projection	Baseline
2	Annual *OPEX* (USD)	6,642,826	Aggregated	3.9%	Mature	5,270,454	4,478,928
2.1	Site lease and insurance	2,897,679	Assigned	8.0%	Evolving	1,787,185	1,102,959
2.2	Environmental monitoring	2,227,544	Assigned	4.0%	Mature	1,785,000	1,785,000
2.3	O&M	2,223,782	Aggregated	4.5%	Mature	1,698,269	1,590,969
2.3.1	Scheduled	1,524,314	Assigned	7.0%	Evolving	1,000,935	952,317
2.3.2	Unscheduled	998,132	Assigned	6.0%	Evolving	697,334	638,652
3	Financial assumptions (*FCR*)	0.13	Aggregated	2.0%	Mature	0.11	0.11
3.1	Discount rate	0.12	Aggregated	3.5%	Mature	0.10	0.09
3.1.1	Debt	0.07	Assigned	10.0%	Evolving	0.05	0.05
3.1.2	Equity	0.06	Assigned	10.0%	Evolving	0.04	0.04
3.2	Project lifetime	20	Aggregated	0.0%	Mature	20	20
4	AEP (kWh)	31,118,603	Aggregated	−6.2%	Evolving	44,071,015	44,071,015
4.1	Rated power	360	Assigned	0.0%	Mature	360	360
4.2	Capacity factor	0.22	Aggregated	−4.4%	Mature	0.29	0.29
4.2.1	Capture efficiency	0.30	Assigned	−10.0%	Evolving	0.37	0.37
4.2.2	Conversion efficiency	0.71	Assigned	−6.0%	Evolving	0.82	0.82
4.2.3	Transmission efficiency	0.87	Assigned	−2.0%	Mature	0.95	0.95
4.3	Availability	0.74	Assigned	−5.0%	Mature	0.98	0.98
5	LCOE (USD/kWh)	1.33	Aggregated	10.6%	Emerging	0.69	0.62

Figure 7. Learning Rates (LR) of the high-level components in the LCOE equation. Note resulting LR for the LCOE is propagated and not simply added.

Based on Table 6, a 10.6% LR implies a medium uncertainty of 18.5% in the cost reduction exponent (δb). Using Equation (7), the LR uncertainty is rescaled to 12.8% (low). Now taking an 80% confidence interval as per Table 4 would result in an LCOE within USD 0.60/kWh and USD 0.83/kWh.

4. Discussion

The implementation of this novel method for estimating costs of the RM5 WEC leads to an initial LCOE (Step 1) for this emerging technology of USD 0.72/kWh. The application of uncertainties (Step 2) shows that the LCOE could be as high as USD 1.33/kWh in its first commercial deployment. Finally, the projection of future costs using component-based learning rates (Step 3) forecasts that LCOE could be reduced to USD 0.69/kWh after 1 GW of cumulative capacity has been deployed (USD 0.60–0.83/kWh accounting for uncertainties in the learning rate). These results follow the cost progress pattern of a wave energy technology along the development stages depicted in Figure 1. Moreover, results align with the NREL's direct estimation method for 50-unit, 10-unit, and 100-unit farms, respectively. The reader should note that while the NREL's model envisages a progressive cost reduction with the increase in farm size, the proposed method follows the development cost pattern peaking at an intermediate stage.

Despite the significant cost reduction that can be achieved through learning, the projection of future commercial costs for the RM5 technology is still far from the SET Plan EUR 0.15/kWh target for 2030, since the starting cost for this emerging technology is well above this target. A closer look at the case study results unveils two main factors for the discouraging result leading to a very high projection of costs.

On the one hand, the *AEP* is subjected to large uncertainty (35.7%) penalising the LCOE from which learning can start to happen. In fact, the lowest bound in Step 2, USD 0.50/kWh, remains far distant from the SET Plan target for wave technologies. On the other hand, the baseline costs are established for the 100-unit farm which limit the ability to capitalise cost reductions through component-based learning beyond a certain deployment level. This outcome reinforces the recommendation to technology developers of deploying R&D activities aimed at collecting evidence that can reduce uncertainty with regard to the availability factor, capture efficiency, and baseline costs, since they will significantly lower the overall uncertainty in the LCOE and open the way to a starker cost reduction.

The case study results only illustrate one of the possible trajectories that an emerging technology can experience in connection with the estimation of future costs. The methodology described before can be repeated with several other wave energy archetypes such as reference models RM3 and RM6 [13] leading to potentially dissimilar results. Actually, three scenarios, depicted in Figure 8, can be envisaged through combining different levels of uncertainty (U) and learning capacity (L).

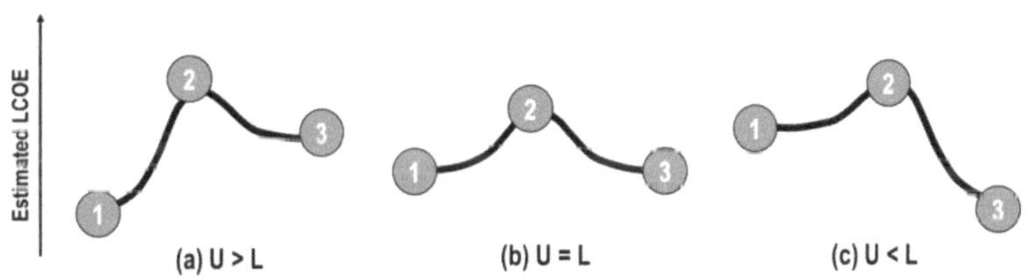

Figure 8. Three possible cost trajectories for emerging technologies, with different uncertainty (U) and learning capacity (L). Numbers ①②③ reference methodology steps, illustrated in Figure 1.

(a) Uncertainty overshadows potential learning (U > L). This trajectory leads to a long-term projection of cost in Step 3 higher than the initial LCOE. The LCOE calculated in Step 1 should be much below the energy price in the addressed market. Otherwise, radical changes must be implemented in the emerging technology. Provided technology development is continued, efforts should be driven to collecting evidence that lowers the cost estimation uncertainty in Step 2. If successful, the LCOE reassessment should lie in either scenario (b) or (c) at the next development stage.

(b) Uncertainties in the same range of learning capacity (U $\approx$ L). This scenario leads to a similar future projection of costs as the initial LCOE estimation in Step 1. For emerging technologies with a high uncertainty level, such as the RM5, there is still room to fill up the critical cost and performance knowledge gaps. Again, if efforts are successful, the LCOE reassessment at the next development stage should lie in scenario (c). Nevertheless, technologies that combine relatively low levels of uncertainty and learning capacity should exhibit an LCOE in Step 1 below the energy price in the addressed market, or else technology breakthroughs should be implemented to meet the commercial goals.

(c) Learning potential dominates uncertainty (U < L). A higher learning capacity provides a more favourable scenario for the emerging technology. The future projection of costs will be lower than the initial estimation in Step 1. If the long-term estimation in Step 3 is below the energy price in the addressed market, the technology can pass to the next development stage without major changes. However, care must be taken that the emerging technology is not stuck in scenarios (a) or (b) above when a new LCOE assessment is performed.

The affordability of emerging wave energy technologies could be improved thanks to a combined exploitation with other marine-space activities. Shared infrastructure will effectively reduce future cost estimations. The cost centres involved could be either structural elements for fixed devices integrated in breakwaters and existing platforms, or electrical components, if connected to the same onshore grid point. Although this strategy can offset the LCOE, it is worth noting that the global uncertainty of the cost estimation will not be significantly altered since the *AEP* is the greatest contributor.

The error propagation method proved to be a useful tool to identify the greatest contributors to uncertainty in the standardised breakdown of *CAPEX*, *OPEX*, *FCR*, and *AEP*. The addition of several components as per Equation (8) decreases the relative uncertainty, which suggests that expanding the breakdown levels in the CAPEX and *OPEX* is a useful strategy to improve the quality of future LCOE projections. However, the product of components as per Equation (9), such as *AEP*, will always enlarge the relative uncertainty, which indicates that the emerging technology should strive to enhance the accuracy of the performance estimations and keep to a minimum the number of energy transformation stages in the PTO design.

The statistical fit of lognormal properties with an 80% confidence interval, borrowed from previous research in other engineering applications, led to quantitative results in line with the NREL's cost model assumptions for the *CAPEX*. The propagated uncertainty obtained with the proposed method (11.4%) is close to the 10% contingency for the 50-unit RM5 farm assumed by the NREL's cost model. It matches the AACE's Class 2 estimate (i.e., detailed estimate, project definition between 30% and 75%). Additionally, the proposed cost estimation method points out other sources of uncertainty in the *OPEX* and in the financial assumptions, which can carry out similar contingencies (i.e., 9% for *OPEX* and 12.1% for the *FCR*), but not considered by the NREL. Furthermore, the overall uncertainty in *AEP* matches the equivalent of a simplified estimate (Class 4).

Component-based learning rates and baseline costs are also useful to avoid over-optimism. LRs between 2% (mature) and 20% (emerging) were assigned to the standardised breakdown resulting in an aggregated *LR* of 10.6%. Although the method of quantification is highly qualitative, this indirect estimation helps identify inherent limitations in cost reduction that could be hidden if considered in the *LR* of the emerging WEC as a whole.

The main merit of this cost estimation method is to provide a transparent and traceable way to assess the future affordability of wave energy technologies under development. It focuses the attention on the accuracy of the cost estimates at an early stage. Instead of oversimplifying the LCOE quantification when there are still many knowledge gaps, the approach includes these unknowns in the form of uncertainties to counter any optimism bias for the first commercial deployment. The method delivers useful information to identify remaining technology challenges, concentrate innovation efforts, and collect evidence through testing activities. The importance of estimating first commercial farm costs is paramount since it impacts the total additional spending required for an emerging technology to be cost competitive in the market and achieve long-term LCOE projection. Starting a cost reduction from an over-optimistic point will ultimately yield highly unrealistic figures of LCOE for mature technology.

The consideration of a first commercial deployment is useful for tracking the evolution of costs in the development cycle of the emerging technology, but it cannot be used to estimate the learning investment or the timescales to achieve future LCOE. The wave energy sector needs to achieve certain deployment level before consistent cost reduction occurs, as the wind industry has shown. This will offset the forecasts by some years and increase the amount of learning investment required to converge to this cost.

This method contains some limitations:

- The statistical treatment of cost centres, and particularly the assumption of independence, will tend to underestimate the overall uncertainty and therefore the resulting LCOE. For instance, the failure rate correlates both the availability and the unscheduled maintenance cost. To counterbalance, this method takes the conservative upper bound of the 80% confidence interval. Alternatively, Monte Carlo methods could be implemented to combine the individual uncertainties, provided the technology developer can build a fully parametric model of the emerging technology.
- Some costs do not scale linearly with the installed capacity, such as grid connection. As the array gets bigger in subsequent projects, the share of grid connection in the LCOE will be reduced. This method considers a constant size of the farm, but economies of scale (larger farms) can be included in the learning rates to account for these situations.
- The uncertainty in learning rates follows the same engineering guidelines as costs due to a lack of previous experience.
- Learning by research, innovation, and upscaling leading to performance and reliability increase is not considered in the future LCOE projection. The cost estimation method can also be used in the next iteration of the technology and the results compared. Note, however, that the innovations introduced in the emerging technology should bring greater benefits than the corresponding uncertainty increase due to lower maturity in order to lead to a more attractive cost projection.

5. Conclusions

This paper presents a novel method to estimate future costs of emerging wave energy technologies that counters the human propensity to over-optimism. Compared with state-of-the-art direct estimation methods, it provides a tool that can be used to explore uncertainties and focus attention on the accuracy of the cost estimates and potential learning from the early stage of technology development. Moreover, this approach delivers useful information to identify remaining technology challenges, concentrate innovation efforts, and collect evidence through testing activities.

A case study was used to illustrate this method. Results show that the uncertainties are in the same range of potential future learning, leading to a future projection of costs similar to the initial LCOE estimation. Technology development efforts should be driven to fill the critical cost and performance knowledge gaps.

The quantitative results are very specific to this case study. Actually, three possible cost trajectories have been discussed in this paper, depending on how the learning potential of the emerging technology weights against the inherent uncertainties. The most favourable scenario is when the learning potential of the emerging technology dominates the inherent uncertainties, and hence the technology can pass to the next development stage without major changes.

While this method has been demonstrated for wave energy technologies, the approach is fully transferrable to any nascent electricity generation technology without any loss of generality.

Future work could apply this approach to a state-of-the-art wave device which is currently undergoing full-scale demonstration, as well as improve the quantification of component-based learning rates and the propagation of corresponding uncertainties. Complementary to this research stream, further work could also pinpoint promising innovation strategies to overcome the challenges that have been identified through this approach.

Author Contributions: Conceptualization and writing—original draft, P.R.-M.; methodology, P.R.-M., D.R.N. and S.P.; review and editing, D.R.N., V.N., S.P. and J.M.B.; visualization, P.R.-M. and D.R.N.; funding acquisition, J.M.B.; resources and supervision, H.J. All authors have read and agreed to the published version of the manuscript.

Funding: This research received no external funding.

Institutional Review Board Statement: Not applicable.

Informed Consent Statement: Not applicable.

Data Availability Statement: Not applicable.

Acknowledgments: The authors would like to thank the Basque Government through the support of the research group IT-1514-22.

Conflicts of Interest: The authors declare no conflict of interest. The funders had no role in the design of the study; in the collection, analyses, or interpretation of data; in the writing of the manuscript; or in the decision to publish the results.

Abbreviations

The following abbreviations are used in this manuscript:

AEP	Annual Energy Production
AF	Availability Factor
BoP	Balance of Plant
CAPEX	CAPital EXpenditure
CF	Capacity Factor
D	Discount rate
FCR	Fixed Charge Rate
LCOE	Levelized Cost of Energy
LR	Learning Rate
N	Project lifetime
O&M	Operation and Maintenance
OPEX	OPerational EXpenditure
OWSC	Oscillating Wave Surge Converter
P	Rated power
PR	Progress Ratio
PTO	Power Take-Off
RM	Reference Model
TRL	Technology Readiness Level
WACC	Weighted Average Cost of Capital
WEC	Wave Energy Converter

References

1. Rubin, E.S. *Evaluating the Cost of Emerging Technologies*; Carnegie Mellon: Oslo, Norway, 2016.
2. De Rose, A.; Buna, M.; Strazza, C.; Olivieri, N.; Stevens, T.; Peeters, L.; Tawil-Jamault, D. *Technology Readiness Level: Guidance Principles for Renewable Energy Technologies*; European Commission: Petten, The Netherlands, 2017; pp. 17–27.
3. Magagna, D.; European Commission; Joint Research Centre. *Ocean Energy: Technology Development Report*; Publication Office of the European Union: Luxembourg, 2020; ISBN 978-92-76-12428-3.
4. Mai, T.; Mowers, M.; Eurek, K. *Competitiveness Metrics for Electricity System Technologies*; National Renewable Energy Laboratory: Golder, CO, USA, 2021; p. 1765599.
5. OES | Cost of Energy. Available online: https://www.ocean-energy-systems.org/oes-projects/levelised-cost-of-energy-assessment-for-wave-tidal-and-otec-at-an-international-level/ (accessed on 29 August 2022).
6. Implementation Working Group Ocean Energy. *SET-Plan—Ocean Energy Implementation Plan*; European Commision: Brussels, Belgium, 2021; p. 51.
7. Lacal Arantegui, R.; Jaeger-Waldau, A.; Vellei, M.; Sigfusson, B.; Magagna, D.; Jakubcionis, M.; Perez Fortes, M.; Lazarou, S.; Giuntoli, J.; Weidner Ronnefeld, E.; et al. *ETRI 2014—Energy Technology Reference Indicator projections for 2010-2050*; Publications Office of the European Union: Luxembourg, 2014.
8. Tsiropoulos, I.; Tarvydas, D.; Zucker, A. *Cost Development of Low Carbon Energy Technologies: Scenario-Based Cost Trajectories to 2050, 2017 Edition*; Publications Office of the European Union: Luxembourg, 2018.
9. Smart, G.; Noonan, M. *Tidal Stream and Wave Energy Cost Reduction and Industrial Benefit*; Offshore Renewabel Energy Catapult: Glasgow, UK, 2018; p. 21.
10. Mundon, T. *Oscilla Power Triton 1310 System Overview and Baseline LCOE Calculations*; Marine and Hydrokinetic Data Repository; Oscilla Power, Inc.: Washington, DC, USA, 2022.
11. Morrow, M. M3 Wave DMP/APEX WEC Projected LCOE. Available online: https://mhkdr.openei.org/submissions/300 (accessed on 29 August 2022).
12. Moliner, E.G. *Cost Analysis of the UGEN*; MSc Dissertation, Instituto Superior Tecnico: Lisboa, Portugal, 2016.

13. Sandia National Laboratories Reference Model Project (RMP). Available online: https://energy.sandia.gov/programs/renewable-energy/water-power/projects/reference-model-project-rmp/ (accessed on 29 August 2022).
14. Simon, R.; Rubin, E.S.; van der Spek, M.; Booras, G.; Berghout, N.; Fout, T.; Garcia, M.; Gardarsdottir, S.; Kuncheekanna, V.N.; Matuszewski, M.; et al. Towards Improved Guidelines for Cost Evaluation of Carbon Capture and Storage; Zenodo. 2021. Available online: https://zenodo.org/record/4643649 (accessed on 14 December 2022).
15. Têtu, A.; Fernandez Chozas, J.A. Proposed Guidance for the Economic Assessment of Wave Energy Converters at Early Development Stages. *Energies* **2021**, *14*, 4699. [CrossRef]
16. Previsic, M.; Siddiqui, O.; Bedard, R. *EPRI Global E2I Guideline Economic Assessment Methodology for Offshore Wave Power Plants*; Electric Power Research Institute: Palo Alto, CA, USA, 2004.
17. Pennock, S.; Garcia-Teruel, A.; Noble, D.R.; Roberts, O.; de Andres, A.; Cochrane, C.; Jeffrey, H. Deriving Current Cost Requirements from Future Targets: Case Studies for Emerging Offshore Renewable Energy Technologies. *Energies* **2022**, *15*, 1732. [CrossRef]
18. Hodges, J.; Henderson, J.; Ruedy, L.; Soede, M.; Weber, J.; Ruiz-Minguela, P.; Jeffrey, H.; Bannon, E.; Holland, M.; MacIver, R.; et al. *An International Evaluation and Guidance Framework for Ocean Energy Technology*; IEA-OES: Lisbon, Portugal, 2021.
19. ELBE European Strategic Cluster Partnership in Blue Energy. Available online: http://www.elbealliance.eu/home (accessed on 14 December 2022).
20. Weber, J.; Costello, R.; Nielsen, K.; Roberts, J. Requirements for Realistic and Effective Wave Energy Technology Performance Assessment Criteria and Metrics. In Proceedings of the 13th European Wave and Tidal Energy Conference, Naples, Italy, 1–6 September 2019; p. 10.
21. Mukora, A.; Winksel, M.; Jeffrey, H.F.; Mueller, M. Learning Curves for Emerging Energy Technologies. *Proc. Inst. Civ. Eng. Energy* **2009**, *162*, 151–159. [CrossRef]
22. LaBonte, A.; O'Connor, P.; Fitzpatrick, C.; Hallett, K.; Li, Y. Standardized Cost and Performance Reporting for Marine and Hydrokinetic Technologies. In Proceedings of the 1st Marine Energy Technology Symposium (METS13), Washington, DC, USA, 10–11 April 2013; pp. 10–11.
23. BVG Associates. *Ocean Power Innovation Network Value Chain Study: Summary Report*; BVG Associates: Swindon, UK, 2019; p. 26.
24. Chozas, J.F.; Kofoed, J.P.; Jensen, N.E.H. *User Guide—COE Calculation Tool for Wave Energy Converters: Ver. 1.6—April 2014*; Department of Civil Engineering, Aalborg University: Aalborg, Denmark, 2014.
25. DTOceanPlus DTOceanPlus—Advanced Design Tools for Ocean Energy Systems Innovation, Development and Deployment. Available online: https://www.dtoceanplus.eu/ (accessed on 29 August 2022).
26. OceanSET. *OceanSET Third Annual Report*; European Commission: Brussels, Belgium, 2022; p. 89.
27. Carbon Trust. *Accelerating Marine Energy: The Potential for Cost Reduction—Insights from the Carbon Trust Marine Energy Accelerator*; Carbon Trust: London, UK, 2011; p. 64.
28. NREL LCOE Calculator. Available online: https://sam.nrel.gov/financial-models/lcoe-calculator.html (accessed on 14 December 2022).
29. Hamedni, B.; Mathieu, C.; Bittencourt Ferreira, C. SDWED Deliverable D5.1 Generic WEC System Breakdown; The Danish Council for Strategic Research: 2014; p. 33. Available online: https://www.sdwed.civil.aau.dk/digitalAssets/97/97538_d5.1.pdf (accessed on 14 December 2022).
30. Marques, M.I. *Deliverable D8.2 Analysis of the Supply Chain*; DTOceanPlus; European Commission: Brussels, Belgium, 2020; p. 78.
31. Têtu, A.; Fernandez, C. *Deliverable D8.1—Cost Database*; LiftWEC; European Commision: Brussels, Belgium, 2020.
32. Innovation Fund. *Methodology for Relevant Costs Calculation*; InnovFund-LSC-2021; European Commision: Brussels, Belgium, 2022; p. 38.
33. Previsic, M. *Economic Methodology for the Evaluation of Emerging Renewable Technologies*; RE Vision Consulting, LLC: Sacramento, CA, USA, 2011; p. 14.
34. *IEC TS 62600-100:2012*; Marine Energy—Wave, Tidal and Other Water Current Converters—Part 100: Electricity Producing Wave Energy Converters—Power Performance Assessment. International Electrotechnical Commission: Geneva, Switzerland, 2012; p. 35.
35. *ISO 14224:2016*; Petroleum, Petrochemical and Natural Gas Industries—Collection and Exchange of Reliability and Maintenance Data for Equipment. ISO: Geneva, Switzerland, 2016; p. 178.
36. EPRI. *Technical Assessment Guide (TAG)—Power Generation and Storage Technology Options*; Electric Power Research Institute: Palo ALto, CA, USA, 2013.
37. DOE. *Cost Estimating Guide*; U.S. Department of Energy: Washington, DC, USA, 1997; p. 298.
38. AACE International. *Cost Estimate Classification System—As Applied In Engineering, Procurement, and Construction for the Hydropower Industry Tcm Framework: 7.3—Cost Estimating and Budgeting*; AACE: Morgantown, WV, USA, 2013; p. 15.
39. Parsons, E.L. *Waste Management Project Contingency Analysis*; U.S. Department of Energy, Federal Energy Technology Center: Pittsburgh, PA, USA, 1999; p. 23.
40. Rothwell, G. Cost Contingency as the Standard Deviation of the Cost Estimate. 2005. Available online: https://www.researchgate.net/profile/Geoffrey-Rothwell/publication/237635336_Cost_Contingency_as_the_Standard_Deviation_of_the_Cost_Estimate_for_Cost_Engineering/links/5cc8084092851c8d220e7b6e/Cost-Contingency-as-the-Standard-Deviation-of-the-Cost-Estimate-for-Cost-Engineering.pdf (accessed on 14 December 2022).

41. Carbon Trust. *Future Marine Energy—Results of the Marine Energy Challenge: Cost Competitiveness and Growth of Wave and Tidal Stream Energy*; Carbon Trust: London, UK, 2006; p. 40.
42. Weiss, M.; Junginger, M.; Patel, M.K.; Blok, K. A Review of Experience Curve Analyses for Energy Demand Technologies. *Technol. Forecast. Soc. Chang.* **2010**, *77*, 411–428. [CrossRef]
43. Junginger, M.; Lako, P.; Lensink, S.; van Sark, W.; Weiss, M. *Technological Learning in the Energy Sector*; Netherlands Environmental Assessment Agency: Bilthoven, The Netherlands, 2010; p. 190.
44. Rubin, E.S.; Azevedo, I.M.L.; Jaramillo, P.; Yeh, S. A Review of Learning Rates for Electricity Supply Technologies. *Energy Policy* **2015**, *86*, 198–218. [CrossRef]
45. Dalton, G.J.; Alcorn, R.; Lewis, T. A 10 Year Installation Program for Wave Energy in Ireland: A Case Study Sensitivity Analysis on Financial Returns. *Renew. Energy* **2012**, *40*, 80–89. [CrossRef]
46. European Commision. *Ocean Energy: Cost of Energy and Cost Reduction Opportunities*; European Commision: Brussels, Belgium, 2013; p. 29.
47. Hurley, W.L.; Nordstrom, C.J. *PelaStar Cost of Energy: A Cost Study of the PelaStar Floating Foundation System in UK Waters*; Enery Technologies Institute: Seattle, DC, USA, 2014; p. 61.
48. Pennock, S. *Deliverable D7.2—Techno-Economic Analyses*; WaveBoost; European Commision: Brussels, Belgium, 2019; p. 57.
49. Gumerman, E.; Marnay, C. *Learning and Cost Reductions for Generating Technologies in the National Energy Modeling System (NEMS)*; Ernest Orlando Lawrence Berkeley National Laboratory: Berkeley, CA, USA, 2004; p. 36.
50. Santhakumar, S.; Meerman, H.; Faaij, A. Improving the Analytical Framework for Quantifying Technological Progress in Energy Technologies. *Renew. Sustain. Energy Rev.* **2021**, *145*, 111084. [CrossRef]
51. Jenne, D.S.; Yu, Y.-H.; Neary, V. Levelized Cost of Energy Analysis of Marine and Hydrokinetic Reference Models. In Proceedings of the 3rd Marine Energy Technology Symposium, Washington, DC, USA, 27–29 April 2015; p. 8.
52. Babarit, A. A Database of Capture Width Ratio of Wave Energy Converters. *Renew. Energy* **2015**, *80*, 610–628. [CrossRef]
53. Babarit, A. *Ocean Wave Energy Conversion: Resource, Technologies and Performance*; ISTE Press, Elsevier Science: London, UK, 2017; ISBN 978-0-08-102390-7.
54. European Marine Energy Centre Aquamarine Power's Oyster. Available online: https://www.emec.org.uk/about-us/wave-clients/aquamarine-power/ (accessed on 14 December 2022).
55. AW Energy Oy WaveRoller. Available online: https://aw-energy.com/waveroller/ (accessed on 14 December 2022).
56. Resolute Marine Wave2O Technology. Available online: https://www.resolutemarine.com/technology/ (accessed on 14 December 2022).
57. Langlee Wave Power Langlee's Robusto Technology. Available online: http://www.langleewp.com/?q=langlee-technology (accessed on 29 August 2022).
58. Yu, Y.H.; Jenne, D.S.; Thresher, R.; Copping, A.; Geerlofs, S.; Hanna, L.A. *Reference Model 5 (RM5): Oscillating Surge Wave Energy Converter*; National Renewable Energy Lab. (NREL): Golden, CO, USA, 2015.
59. NREL Reference Model 5 Cost Breakdown—Marine and Hydrokinetic Data Repository (MHKDR). Available online: https://mhkdr.openei.org/ (accessed on 14 December 2022).

sustainability

MDPI

Article

Numerical Analysis and Experimental Study of Unsteady Flow Characteristics in an Ultra-Low Specific Speed Centrifugal Pump

Yangyang Wei [1], Yuhui Shi [1], Weidong Shi [2,*] and Bo Pan [3]

[1] National Research Center of Pumps, Jiangsu University, Zhenjiang 212013, China
[2] School of Mechanical Engineering, Nantong University, Nantong 226019, China
[3] Shandong Xinchuan Mine Electromechanical Equipment Co., Ltd., Jining 272300, China
* Correspondence: wdshi@ntu.edu.cn; Tel.: +86-0511-88799918

Abstract: Low specific speed centrifugal pumps are widely used in urban water supply, agricultural irrigation, petrochemical and other fields due to their small flow rate and high head. Therefore, the study of unsteady flow characteristics plays a vital role in its safe and stable operation. In this paper, numerical simulation and experimental methods are used to explore the unsteady performance of the pump. The results show that the fluctuations of the external characteristic such as head, shaft power and energy loss are due to the periodic disturbance of the flow field of pump. But the transient performance of shaft power and head shows different changing trends due to different influencing factors. In this paper, the transient process of hydraulic performance is divided into three stages according to the causes and characteristics of hydraulic fluctuations. Most of the hydraulic losses occur inside the impeller, so the impeller flow field determines the level of time average hydraulic performance. Although the hydraulic loss of the spiral case is small, it is greatly affected by the rotor-stator interaction, which affects the strength of the hydraulic fluctuation. This study is of great significance to the mechanism of rotor-stator interaction and the stable operation of low specific speed centrifugal pumps.

Keywords: low specific speed; transient pulsation; hydraulic performance; rotor-stator interaction

Citation: Wei, Y.; Shi, Y.; Shi, W.; Pan, B. Numerical Analysis and Experimental Study of Unsteady Flow Characteristics in an Ultra-Low Specific Speed Centrifugal Pump. *Sustainability* **2022**, *14*, 16909. https://doi.org/10.3390/su142416909

Academic Editor: Domenico Mazzeo

Received: 27 October 2022
Accepted: 9 December 2022
Published: 16 December 2022

1. Introduction

From the perspective of energy saving, ultra-low specific speed centrifugal pumps are widely used in various fields of social development, and their safe and efficient operation has attracted much attention [1,2]. Their specific speed is not greater than 30, and these pumps has the characteristics of small flow and high head. Due to the relatively small geometric size of the ultra-low specific speed centrifugal pump, its internal flow is complex when pumping viscous fluid, often accompanied by turbulent flow such as cavitation, secondary flow, axial eddy current, and complex pressure pulsation signals [3,4]. This also results in a low hydraulic efficiency. At present, the research on ultra-low specific speed centrifugal pump mainly focuses on two aspects: design optimization and internal flow characteristics research [5–7].

Zhao et al. [8] adopted a genetic algorithm based on NSGA-II to optimize the design of a centrifugal pump with a specific speed of 30. Based on the Plackett-Burman test, the parameters were screened to obtain the optimal geometric parameters of the impeller of the centrifugal pump. After optimization, the hydraulic efficiency was increased by about 5.82%. The quadratic regression orthogonal composite method was used to optimize low specific speed centrifugal pumps in the study of Zhong et al. [9]. They took 7 main geometric parameters of the impeller as research factors and discovered that the blade number played a dominant role in improving the hydraulic performance of the centrifugal pump. Zhang et al. [10] optimized the design of the impeller blade wrap angle of an ultra-low specific speed centrifugal pump based on the Bezier curve. It was found that the centrifugal pump efficiency reached a maximum when the blade wrap angle was 120°. Sato et al. [11] and Daqiqshirazi et al. [12] optimized the shape of the pump casing in order to reduce the disc friction loss of the ultra-low specific speed centrifugal pump.

Scholars have also done a lot of research on the influence of different pump structures on the pump performance. Siddique et al. [13,14] studied the influence of splitter blades on the hydraulic performance of an ultra-low specific speed centrifugal pump, and the results showed that splitter blades can reduce the amplitude of pressure pulsation as well as hydraulic loss, and improve the hydraulic performance of the pump. Gao et al. [15] studied the influence of blade trailing edge profile on the performance and unsteady pressure fluctuation of a centrifugal pump. It was found that, when the pressure surface of the blade was elliptical or both sides of the blade were elliptical, the pump efficiency was significantly improved, and the pressure fluctuation was also reduced. Chabannes et al. [16] analyzed the effect of volute throat area on the performance of ultra-low specific speed centrifugal pumps. Within a certain range, the pump head and efficiency increased by increasing the throat area.

The external characteristics of centrifugal pumps are the external manifestations of internal flow features, such as rotational stall and rotor-stator interaction, which often lead to the hump and unsteady fluctuations of the external characteristic curve. Based on numerical simulation, Zeng et al. [17] and Cui et al. [18] studied the fluid excitation phenomenon in an ultra-low specific speed centrifugal pump and its influence on the dynamic characteristics of the rotor. The results indicated that the main frequency of the pressure pulsation was the passing frequency of the impeller blades, and the pressure pulsation amplitude of the main frequency was much larger than that of the secondary main frequency. Zhang et al. [19] conducted research on the unsteady flow field of an ultra-low specific speed centrifugal pump by PIV test. At low flow rate, there was a typical jet wake flow pattern at the blade outlet. However, this pattern was not obvious at large flow rate because the high-momentum fluid was concentrated on the suction side of the blade. Ye et al. [20–22] conducted a research on the ultra-low specific speed centrifugal pump and discovered that the liquid flow angle at the inlet of the pump impeller channel was relatively larger due to the strong separation flow and the evolution of the vortex. Besides, many researchers had studied the internal flow of ultra-low specific speed centrifugal pumps through experiments and numerical simulations, such as the pressure pulsation, radial force, and vortex dynamics [23–26].

According to the above discussion, there are relatively few studies on the internal velocity fluctuation, the velocity-pressure coupling characteristics under the influence of rotor-stator interaction, and its effects on the internal flow field. In order to study this problem, this paper mainly discusses the influence mechanism of rotor-stator interference on the pump energy characteristics and the unsteady variation of the pressure on the impeller-volute interface. And Through numerical simulation and experimental analysis, it is proposed that the SST k-w turbulence model is more suitable for ultra-low specific speed centrifugal pumps. In addition, the research in this paper also enriches the coupling mechanism of the velocity field and the pressure field in the ultra-low specific speed centrifugal pump, and provides a theoretical basis and reference for its stable operation.

2. Numerical Methods

2.1. Centrifugal Pump Parameters

The basic external characteristic parameters of the centrifugal pump include head and efficiency. The dimensionless head coefficient ϕ and efficiency η are defined as:

$$\phi = \frac{gH}{0.5u_2^2} \tag{1}$$

$$\eta = \frac{(P_2 - P_1)Q}{W_s} \tag{2}$$

where H is the head, m; u_2 is the circumferential velocity at the impeller outlet, m/s; P_1 and P_2 are the inlet and outlet total pressures, Pa; W_s is the shaft power, W; Q is the flow rate, m^3/s; g is the acceleration due to the gravity, m/s^2.

To qualitatively analyze the change in the internal pressure of the pump, it is necessary to analyze the time-averaged value and pulsation pattern of pressure. The time-averaged value of the pressure pulsation is defined as:

$$\overline{p} = \frac{1}{N}\sum_{i=1}^{N} p(i) \tag{3}$$

The equation for the coefficient of RMS pressure fluctuations (root mean square) can be expressed as follows:

$$C_{Psdv} = \frac{\sqrt{\frac{1}{N}\sum\limits_{i=1}^{N}(p(i)-\overline{p})^2}}{0.5\rho u_2^2} \tag{4}$$

where $p(i)$ is the instantaneous pressure of the measuring point, Pa; $\overline{p}$ is the average pressure, Pa. When the number of samples is high enough, the root mean square change of the pressure fluctuation coefficient is very small, which equals to a constant. And the dimensionless pressure fluctuation amplitude is defined as follows:

$$C_P = \frac{p(i)-\overline{p}}{0.5\rho u_2^2} \tag{5}$$

where ρ is the medium density, kg/m^3.

2.2. Total Pressure Loss Coefficient

The internal flow field of the centrifugal pump can be considered as periodic flow. The total pressure loss of flow passage components, namely hydraulic loss, can be calculated by using the total pressure difference. The total pressure drop coefficient of the volute is defined as follows:

$$\phi_s = \frac{P_{w,in,tp} - P_{w,out,tp}}{0.5\rho {u_2}^2} \tag{6}$$

The total pressure rise coefficient of impeller is defined as follows:

$$\phi_r = \frac{P_{r,out,tp} - P_{r,in,tp}}{0.5\rho {u_2}^2} \tag{7}$$

where *out* is the outlet of the component; *in* is the inlet of component; *tp* represents the total pressure, Pa; The subscripts w and r represent the volute and impeller, respectively.

The dimensionless coefficient of impeller work is defined as follows:

$$W_{sft} = \frac{2\pi Mn}{60u_2A_2 0.5\rho u_2^2} \tag{8}$$

where A_2 is the outlet area of the impeller.

3. Numerical Model and Setting

3.1. Experimental System

Figure 1 shows the test rig of the model pump. The motor and pump are fixed on the metal mounting plate, and several pressure monitoring points are arranged at the circumferential position of the volute to measure the internal pressure pulsation. The sensor parameters are given in Table 1. The accuracy of the instrument refers to the percentage of its allowable error in the dial scale value.

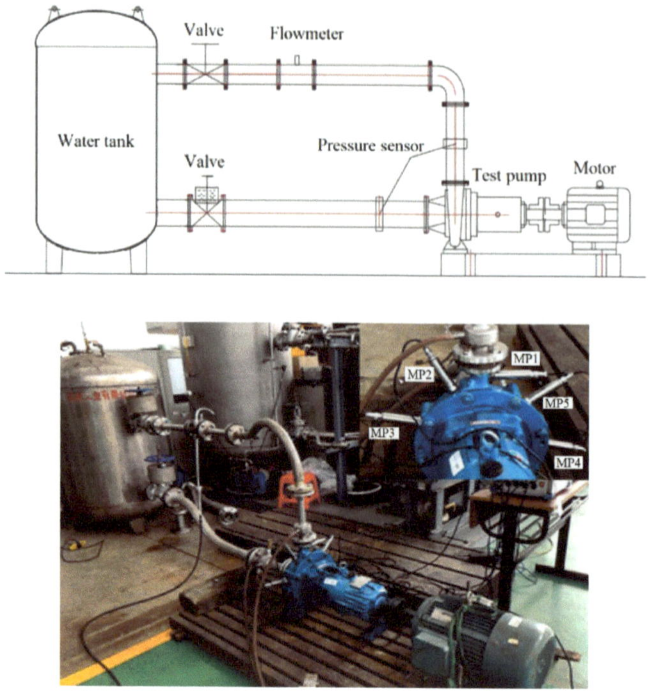

Figure 1. The test rig of the centrifugal pump.

Table 1. Parameters of the sensor.

Sensor	Model	Range	Precision
Flowmeter	LW-40	0~25 m^3/h	±0.5%
Inlet pressure sensor	WT2000	−0.1~0.1 MPa	±0.1%
Outlet pressure sensor	WT2000	0~1.6 MPa	±0.1%
High-frequency dynamic pressure sensor	CY200	0~2.0 MPa	±0.1%
Temperature sensor	PT100	0~100 °C	±0.5%

3.2. Model Parameter Settings

The key parameters of the model pump used in this study are shown in Table 2. The specific speed ns is defined as the following equation:

$$n_s = \frac{3.65 \times n_d \times \sqrt{Q_{des}}}{H_{des}^{0.75}} \tag{9}$$

where the units of all variables are consistent with those in Table 2. The mesh for numerical simulations is generated by the ANSYS ICEM. The computational domains of the pump mainly include the inlet pipe, the impeller, the volute, and the outlet pipe. In order to obtain a relatively stable flow field, the length of the inlet and outlet pipes is set to be 10 times the pipe diameter. Hexahedral mesh is used in the computational domain and mesh refinement is performed on key parts such as blade surface and volute tongue. In order to ensure the accuracy of data transmission and capture the flow characteristics at the interface of different computational domains, mesh refinement is performed at the interface of different computational domains. The mesh division is shown in Figure 2.

Table 2. Parameters of the pump model.

Parameter	Value
Rotating speed n_d (r/min)	2900
Design flow rate Q_{des} (m^3/h)	10
Design head H_{des} (m)	80
Pump inlet diameter D_1 (mm)	50
Pump outlet diameter D_2 (mm)	20
Outlet width of impeller (mm)	6.5
Exit angle of blade (°)	25
Volute tongue inner diameter d_3 (mm)	290
Outlet diameter of impeller D_i (mm)	259
Blade thickness at trailing edge	6
Number of blades Z	3

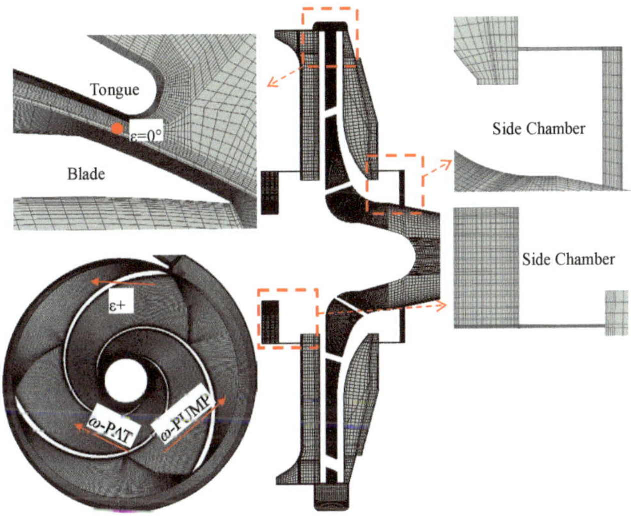

Figure 2. Grid of computational domain.

Figure 3 shows the comparison between the calculated and experimental values at the design condition for different turbulence models [27]. The calculation results of the SST k-ω turbulence model agrees well with the experimental data compared to other turbulent models [28–30]. Thus, it is chosen for the turbulence model of the numerical simulation in this study. The boundary conditions are set as the total pressure at the inlet of the pump, the mass flow at the outlet of the pump. The outlet pressure is set to 1 atm. No-slip wall is applied to all the walls. The interface between the rotor and the stator selects the frozen rotor condition when in the steady calculation and selects the transient rotor-stator condition when in the unsteady calculation [31]. The initial field of transient simulation is set to the result of the steady calculation. The unsteady calculation time step is 5.75×10^{-5} which equals to 1° rotation of the impeller. A transient simulation is conducted for 6 revolutions of the impeller. And the calculation data of the last revolution is chosen for the analysis of the pressure and power fluctuation.

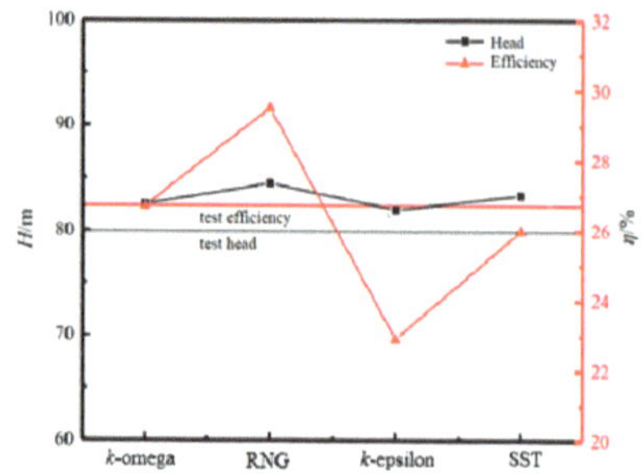

Figure 3. Comparison of external characteristics under different turbulence models.

3.3. Grid Independence Verification

The validation of grid independence is made to ensure the accuracy of simulation results. Three different grids are used to calculate it numerically, and the differences between the experimental and numerical simulation results are compared. The number of cells corresponding to each grid scheme is shown in Table 3.

Table 3. Mesh schemes.

Scheme	Mesh Number
G1	2.01×10^6
G2	5.12×10^6
G3	7.48×10^6

The comparison of the performance between the numerical calculation and experiment is shown in Figure 4a. The head of the pump decreases with the increase of the flow rate. As the number of grids increases, the difference between the experimental head and the numerical head becomes smaller. Considering the computational cost, scheme G2 was selected for numerical simulation in this study. Figure 4b presents the pressure amplitude at the dominant frequency of the monitoring points in Figure 1. The dominant frequency of pressure pulsation obtained by numerical simulation and experiment both occurs at the blade frequency, and the amplitude is similar.

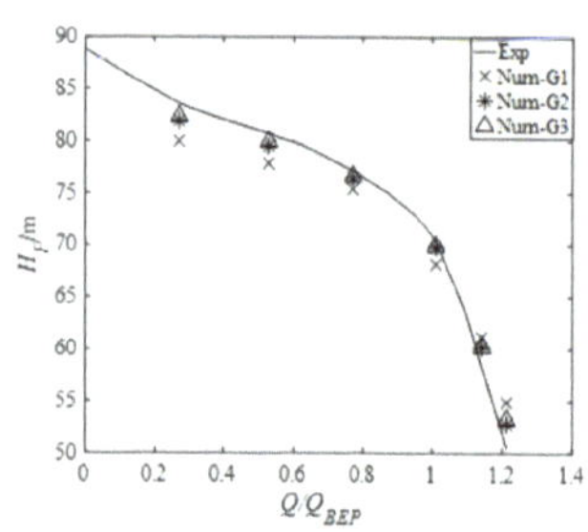

(**a**) External performance

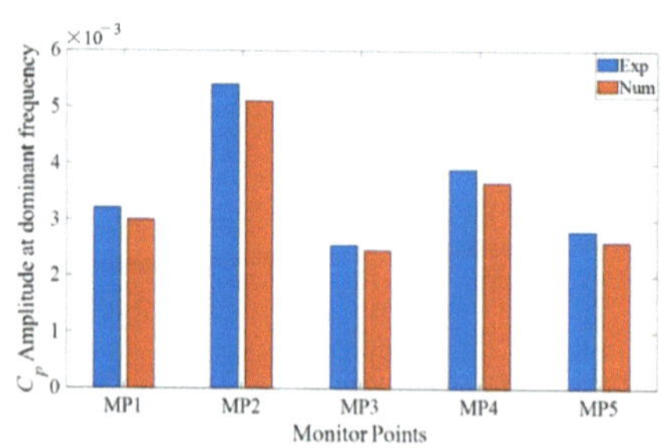

(**b**) Pressure amplitude at the dominant frequency

Figure 4. Validation of grid independence.

4. Results and Discussions

Figure 5 shows the unsteady head of the pump in one rotation cycle. Here, ε is the angle between the trailing edge of the blade and the volute tongue. The ε is zero when the blade trailing edge is passing the tongue. The head gradually increases with the decrease in flow rate. The head coefficient shows obvious periodic fluctuation, and the fluctuation law is similar under different discharge. It is well understood that the number of cycles is the same as the number of blades of the impeller. Therefore, the key to the problem is the reason that affects the transient change trend. For the minimum value of pulsation, it is obviously the interaction between the fluid at the tail of the impeller and the tongue. In particular, there is a quadratic fluctuation at the valley of the curve shown in Figure 5. This may be related to the rotor-stator interaction between the impeller and the tongue. When the pressure side of the blade sweeps across the tongue, its unsteady head reaches the minimum value. With the rotation of the blade, a secondary fluctuation occurs when the suction side of the blade sweeps across the tongue. As the blade continues to rotate, the unsteady head gradually rises to the maximum value. Therefore, the blade thickness may be the cause of secondary fluctuation. The comprehensive fluctuation of the unsteady head coefficient is closely related to the unsteady flow field.

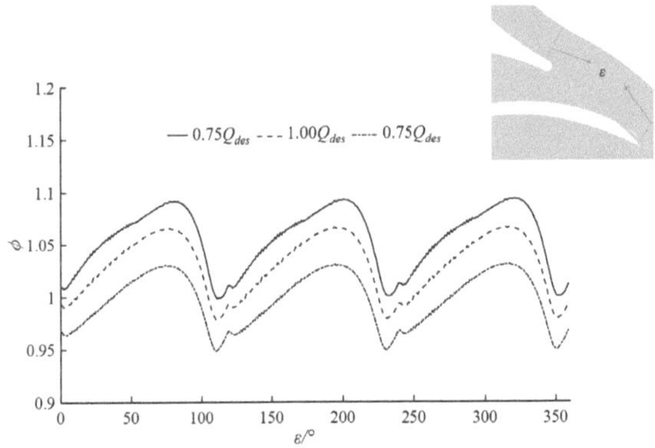

Figure 5. Transient change of head coefficient for three flow rates.

Figure 6 shows the unsteady fluctuation curve of the impeller shaft power during one impeller rotation. The shaft power increases with the increase of the flow rate, and the oscillation trend of the shaft power coefficient curve is similar under different flow rates, which shows obvious periodic fluctuations. With the blades rotating from the tongue, the shaft power rapidly increases to a maximum value. The reduction of shaft power will go through three stages. First, the shaft power will slowly decrease, then increase in a gentler trend, and finally, rapidly reduce to the minimum value. Compared with Figure 5, when the blade rotates 120°, there is no secondary fluctuation like the head coefficient, which indicates that the influence of this transient effect on the impeller is very weak.

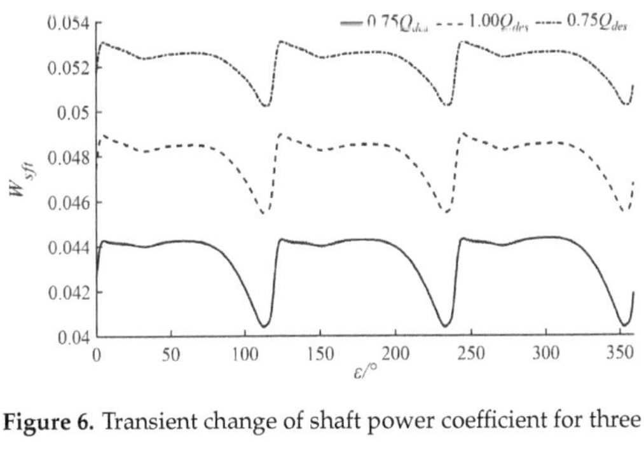

Figure 6. Transient change of shaft power coefficient for three flow rates.

Within one cycle of the impeller, the pressure rises of impeller and pressure drop of volute are shown in Figure 7. With the increase in the flow rate, the pressure rise coefficient in the impeller presents a decreasing trend while it has the opposite trend in the volute. Moreover, the pressure rise coefficient of the impeller is close to the head coefficient, which means that most of the water losses occur in the impeller. Therefore, the flow field in the impeller determines the hydraulic performance. However, the fluctuation of the impeller pressure rise coefficient is 0.01 times of the head coefficient and mainly occurs near the tongue, which indicates that the loss of the impeller is not the main reason for determining the hydraulic fluctuation. The pressure drop coefficient of volute is very small, but the fluctuation is more obvious, and the fluctuation amplitude can be close to 0.1. Therefore, although the hydraulic loss of volute is small, it is the main factor affecting the fluctuation range of hydraulic performance. This also indirectly shows that the rotor-stator interaction has a greater impact on the flow field of the volute.

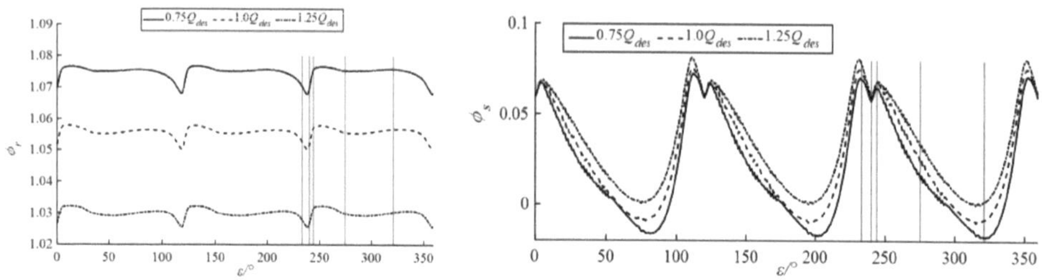

(**a**) Impeller (**b**) Volute

Figure 7. Transient change of pressure coefficient at three flow rates.

Before studying the large hydraulic fluctuation of the volute, it is necessary to explore the hydraulic loss and flow characteristics of the impeller. Figure 8 shows the average value and variance distribution of pressure in the mid-span section of the pump. The time-averaged values of turbulent kinetic energy and dissipation rate under design flow rate are presented at Figure 9. The pressure increases gradually from the impeller inlet to the outlet. Due to the pressure loss caused by the leakage of the balance hole in the pump casings, the pressure at the impeller inlet is very small. In general, the maximum pressure appears in the impeller outlet. With the increasing flow rate, pressure of inlet gradually increases. As can be seen from the oval marked area, the low pressure is mainly distributed in the region around the impeller inlet. The maximal variance value of the pressure occurs on the interface of the impeller and volute, especially in the regions near the blade trailing edge. As the flow rate increases, the pressure at the inlet of the impeller also increases, causing the head to drop. As shown in Figure 9, the dissipation of turbulent kinetic energy mainly occurs at the balance hole at the impeller inlet. This is the main reason for inducing hydraulic loss. Since it is far from the interface between impeller and volute and located in the upstream area, it is less subject to rotor-stator interaction. Therefore, the pressure rise coefficient of the impeller changes slightly, and only changes greatly before and after passing through the tongue.

One cycle of the above four external characteristic curves is shown in Figure 10a, and five typical blade rotation angles are shown in Figure 10b. The positions are named T1, T2, T3, T4, and T5 in chronological order. The three flow channels are named C1, C2 and C3 in anticlockwise direction. For convenience, the interface between impeller and volute is hereinafter referred to as interface. Figure 10a more clearly proves that the hydraulic performance can be affected by changing the hydraulic loss and flow field characteristics of the impeller under different flow rates. However, the hydraulic fluctuation characteristics under different flow rates are similar, which is mainly controlled by the flow field in the volute. The rotor-stator interaction mainly affects the flow field of the volute, and the mechanism of this influence will be explained later. There are very similar fluctuation characteristics under different flows, however, the following article pays more attention to the transient changes of hydraulic characteristics in a cycle. Therefore, only the flow field characteristics of design flow are studied to obtain the mechanism of rotor-stator interaction on hydraulic performance.

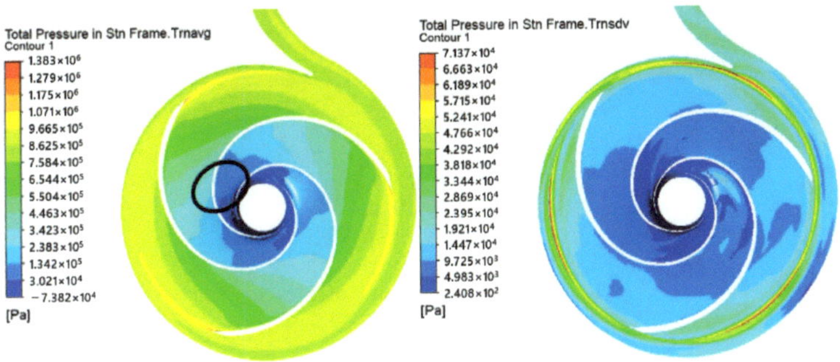

(**a**) $0.75Q_{des}$

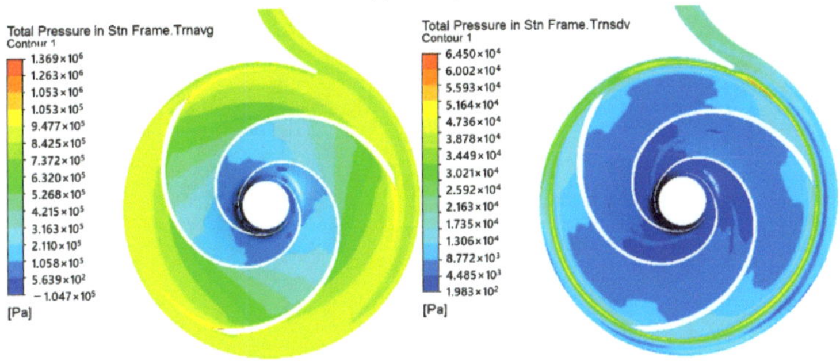

(**b**) $1.0Q_{des}$

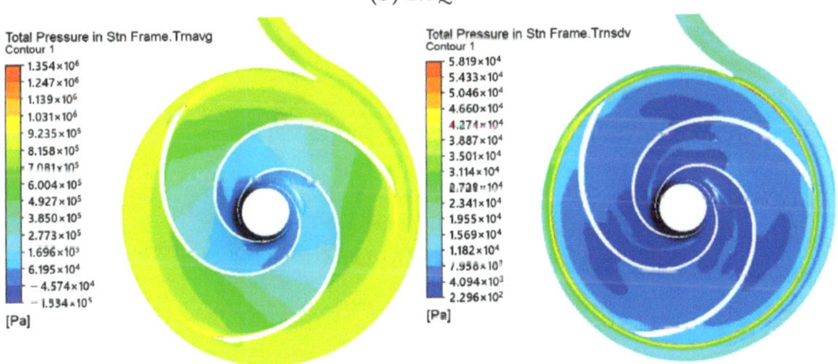

(**c**) $1.25Q_{des}$

Figure 8. Mean value and variance distribution of total pressure at mid-span surface under three flow rates.

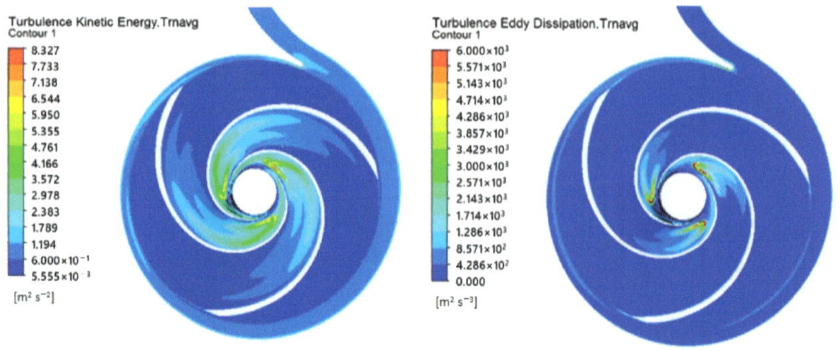

Figure 9. Mean distribution of turbulent kinetic energy and turbulent eddy dissipation rate at Q_{des}.

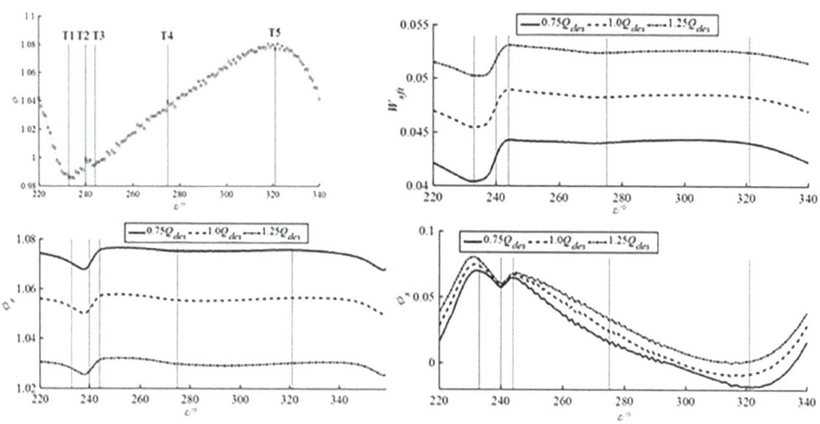

(**a**) Performance change in 1/3 impeller rotation cycle

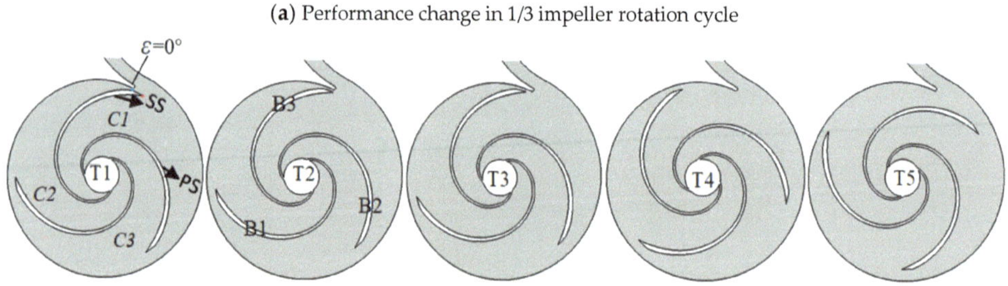

(**b**) position of impeller corresponding to typical time

Figure 10. Performance change characteristics of 1/3 impeller in rotating cycle.

According to Figure 10a, the hydraulic performance change process can be divided into three stages according to the shaft power. In the first stage (T1~T3): the power rise period, the shaft power increases, and the head first increases and then decreases. In the second stage: power stability period (T3~T5), shaft power is stable and head increases. In the third stage (T5~T1): power reduction period, shaft power and head decrease. The hydraulic loss of the volute basically presents the opposite trend to the head. The analysis of the transient characteristics of the flow field at the interface is of great significance to the study of the rotor-stator interaction of centrifugal pumps. Figure 11 shows the distribution of radial velocity and circumferential velocity at the interface, in which HS represents the hub surface and SS represents the shroud surface. And the contours from top to button corresponds to the position of T1, T2, T3, T4, T5 in turn. The velocity distribution at the inlet will directly affect the flow field and energy loss in the volute. The radial velocity difference at different times is small, but the inflow at time T5 is better than that at time T1. This indicates that the volute has lower energy loss and the head will increase at T5. The high value of circumferential velocity is distributed in the region of y/b = 0.25~0.75. The width of the volute inlet is twice that of the impeller outlet, so the area where y/b = 0.0~0.25 is the front pump casing, and the area where y/b = 0.75~1 is the rear pump casing. The area near the vanes and pump casing has a slightly slower speed. Similar to the radial velocity, the flow field from T1 to T3 has no obvious change, but the flow field at T4 and T5 has been significantly improved, which will also improve the hydraulic performance. In addition, for the flow passage near the tongue, the radial velocity and circumferential velocity have large gradients, which makes the fluid have a certain impact on the impeller. Therefore, the impeller shaft power will also change between T1 and T3 (the blade goes across the tongue).

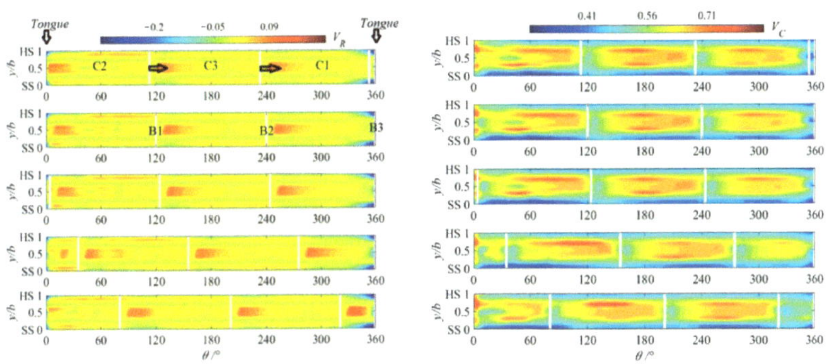

(**a**) Radial velocity (**b**) Circumferential velocity

Figure 11. Velocity Field at Interface.

Internal pressure distribution of pump determines the variation of fluid mechanical energy and the transmission of shaft power. Figure 12 shows velocity vector and pressure distribution at interface. Obviously, the pressure fluctuation gradient at T2 is smaller than T1 and T3. This resulted in a small fluctuation of loss coefficient and head at T2. T4 and T5 are similar to the velocity field, and the inlet pressure field of the volute has been further improved. High pressure area and low-pressure area are produced on the left and right sides of the tongue, which affect the flow field near the tongue. To further study the flow field characteristics that cause this phenomenon, the pressure and velocity vectors at the mid-span position in the 180° area near the tongue arc selected to obtain Figure 12b. The red dotted line represents the tongue, and the red solid line represents the suction side of the trailing edge of the blade. They are respectively the blue point and the red point in Figure 10b. The solid black line represents the pressure. At time T3, the blade has started to move away from the tongue. With the increase of time, the flow field on the left side of the tongue has been improved. The distance between the tongue and the right blade is not enough to cause greater interaction. Therefore, at T5, the hydraulic loss of the volute is the smallest and the head coefficient is the largest. However, after T5, as the blade approaches the tongue, the tail jet will cause great impact near the tongue, which leads to rapid increase of hydraulic loss and reduction of head. The velocity vector distribution of T1, T2 and T3 has little difference. However, according to the conclusion in Figure 12a and the relative position of the blade and tongue, it can be inferred that the small fluctuations at T2 are related to the thickness of the blade. Assuming that the blade has no thickness, the changes of various hydraulic properties will be monotonic during T1–T5. However, due to the thickness and tangential trailing edge structure of the blade, the flow field near the trailing edge is distorted greatly, so a small secondary wave is finally generated.

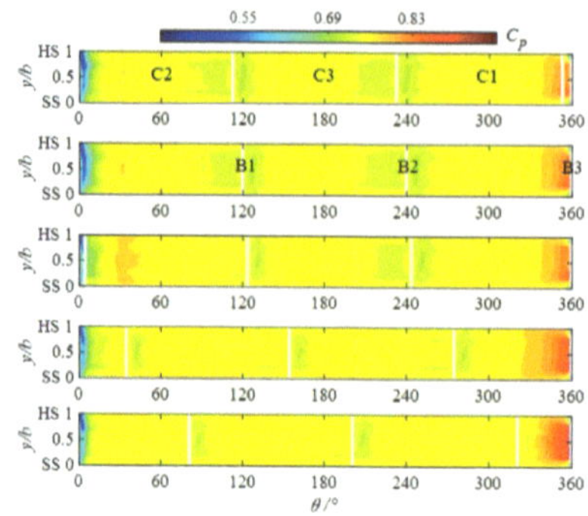

(**a**) Pressure distribution of interface

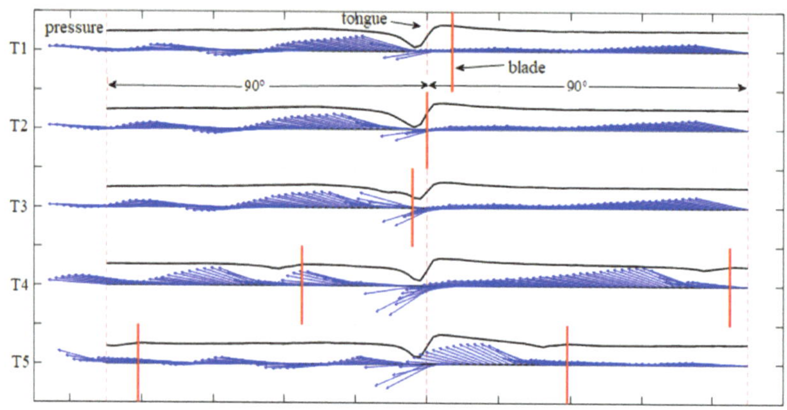

(**b**) Velocity vector and pressure at mid-span position

Figure 12. Velocity vector and pressure distribution at interface.

5. Conclusions

To study the transient fluctuation characteristics of ultra-low specific speed centrifugal pump, numerical simulation and experimental methods were applied. The unsteady flow pattern is studied by analyzing the hydraulic performance and pressure loss coefficient under different flow rates. According to the characteristics of the internal flow field, the mechanism of the rotor-stator interaction that causes the transient changes is discussed.

(1) The head of the numerical simulation presents a good agreement with that of the experimental data. And the maximum error of hydraulic head is calculated to be lower than 5% at the selected grid scheme. Thus, the numerical results in this study are fairly reliable.

(2) The number of cycles of all hydraulic characteristics is equal to the number of blades, and the hydraulic performance shows a similar change trend under different flow rates. But the transient performance of shaft power and head shows different changing trends due to different influencing factors. Most of the hydraulic losses occur inside the impeller, so it is the impeller flow field that determines the level of time averaged hydraulic performance. The hydraulic loss of volute is small, but it determines the fluctuation range of hydraulic performance.

(3) The hydraulic loss and turbulent dissipation in the impeller mainly occur near the balance hole, so it is less affected by rotor-stator interaction, about 10% of the total fluctuation range. The circumferential velocity and pressure distribution at the volute inlet show obvious changes with time. The effect of rotor-stator interaction on the flow field in the volute is more significant.

(4) According to the characteristics and causes, the transient hydraulic performance is divided into three stages. In the period of power increase, the rotor-stator interaction is the strongest, causing the increase of shaft power. However, due to the thickness of the blade and the structural form of the trailing edge, the loss in volute and head of pump have non-linear changes. During the power stability period, the rotor-stator interaction has little influence on the impeller, and the continuously improved flow field improves the hydraulic performance. During the power reduction period, the distance between the blade and the tongue is shortened, which makes the blade wake flow have a greater impact on the tongue, resulting in increased hydraulic loss.

Author Contributions: Conceptualization, Y.W. and W.S.; methodology, Y.W. software, Y.S.; validation, W.S. and B.P.; formal analysis, Y.S.; investigation, Y.W.; resources, W.S.; data curation, Y.S.; writing—original draft preparation, Y.W.; visualization, W.S. and B.P.; supervision, Y.W. and Y.S.; project administration, W.S. All authors have read and agreed to the published version of the manuscript.

Funding: This work was supported by the Jiangsu Province Universities Natural Sciences Foundation (Grant No. 20KJB570002), Project of self-made experimental instruments and equipment of Jiangsu University (Grant No. ZZYQSB202206), National Natural Science Foundation of China (Grant No. 51979138), and National key research and development program (Grant No. 2019YFB2005300).

Institutional Review Board Statement: Not applicable.

Informed Consent Statement: Not applicable.

Data Availability Statement: Not applicable.

Conflicts of Interest: The authors declare no conflict of interest.

Nomenclature

A_2	Area of the impeller outlet [m^2]
C_{Psdv}	Coefficient of RMS pressure fluctuation [-]
C_p	Pressure coefficient [-]
D_1	Pump inlet diameter [mm]
D_2	Pump outlet diameter [mm]
D_i	Outlet diameter of impeller [mm]
H_{des}	Design head [m]
f_b	Blade passing frequency [Hz]
g	Acceleration due to the gravity [m/s^2]
k_{AVE}	Time-average of turbulent kinetic energy [m^2/s^2]
ke_{AVE}	Time-average of turbulent eddy dissipation [m^2/s^3]
k_{STD}	Standard deviation of turbulent kinetic energy [-]
ke_{STD}	Standard deviation of turbulent eddy dissipation [-]
M_n	Torque [N·m]
M_P	Torque of the fluid acting on the pressure side of the flow channel [N·m]
M_S	Torque of the fluid acting on the suction side of the flow channel [N·m]
n_d	Rotating speed [r/min]
n_s	Specific speed [-]
N	Number of samples [-]
P_1	Total pressure at the pump inlet [Pa]
P_2	Total pressure at the pump outlet [Pa]
$p(i)$	Pressure at the monitor point [Pa]

$\overline{p}$	Average pressure [Pa]
$P_{w,out,tp}$	Total pressure at the volute outlet [Pa]
$P_{w,in,tp}$	Total pressure at the volute inlet [Pa]
$P_{r,out,tp}$	Total pressure at the impeller outlet [Pa]
$P_{r,in,tp}$	Total pressure at the impeller inlet [Pa]
Q	Flow rate [m^3/s]
Q_{des}	Flow rate at design point [m^3/s]
u_2	Circumferential velocity [m/s]
W_s	Shaft power [W]
W_{sft}	Dimensionless coefficient of impeller work [-]
Z	Number of impeller blades [-]
ϕ	Dimensionless head coefficient [-]
ϕ_S	Total pressure loss coefficient of the volute [-]
ϕ_r	Total pressure loss coefficient of the impeller [-]
η	Efficiency [%]
ρ	Density [kg/m^3]

References

1. Wang, C.; Zhang, Y.; Hou, H.; Yuan, Z.; Liu, M. Optimization Design of an Ultra-Low Specific-Speed Centrifugal Pump Using Entropy Production Minimization and Taguchi Method. *Int. J. Fluid Mach. Syst.* **2020**, *13*, 55–67. [CrossRef]
2. Jia, X.; Cui, B.; Zhu, Z.; Yu, X. Numerical Investigation of Pressure Distribution in a Low Specific Speed Centrifugal Pump. *J. Therm. Sci.* **2018**, *27*, 9. [CrossRef]
3. Wang, C.; Zhang, Y.; Zhu, J.; Yuan, Z.; Lu, B. Effect of cavitation and free-gas entrainment on the hydraulic performance of a centrifugal pump. *Proc. Inst. Mech. Eng. Part A J. Power Energy* **2020**, *235*, 095765092093934. [CrossRef]
4. Wei, Y.; Yang, Y.; Zhou, L.; Jiang, L.; Shi, W.; Huang, G. Influence of Impeller Gap Drainage Width on the Performance of Low Specific Speed Centrifugal Pump. *J. Mar. Sci. Eng.* **2021**, *9*, 106. [CrossRef]
5. Zhang, N.; Gao, B.; Li, Z.; Jiang, Q. Cavitating flow-induced unsteady pressure pulsations in a low specific speed centrifugal pump. *R. Soc. Open Sci.* **2018**, *5*, 180408. [CrossRef]
6. Alemi, H.; Nourbakhsh, S.A.; Raisee, M.; Najafi, A.F. Effect of the volute tongue profile on the performance of a low specific speed centrifugal pump. *Proc. Inst. Mech. Eng. Part A J. Power Energy* **2015**, *229*, 210–220. [CrossRef]
7. Del Rio, A.; Casartelli, E.; Mangani, L.; Roos, D. Assessment of advanced RANS turbulence models for the stability analysis of low specific speed pump-turbines. *IOP Conf. Ser. Earth Environ. Sci.* **2021**, *774*, 205–218. [CrossRef]
8. Zhou, L.; Hang, J.; Bai, L.; Krzemianowski, Z.; El-Emam, M.A.; Yasser, E.; Agarwal, R. Application of entropy production theory for energy losses and other investigation in pumps and turbines: A review. *Appl. Energy* **2022**, *318*, 119211. [CrossRef]
9. Zhong, W.; Zhou, J.; Zhou, S. The Effects of Different Splitter Blades Number on Characteristics of Miniature Super-Low Specific Speed Centrifugal Pump. *IOP Conf. Ser. Earth Environ. Sci.* **2021**, *772*, 012065. [CrossRef]
10. Zhang, R.H.; Guo, R.; Yang, J.H.; Luo, J.Q. Inverse Method of Centrifugal Pump Impeller Based on Proper Orthogonal Decomposition (POD) Method. *Chin. J. Mech. Eng.* **2017**, *30*, 1025–1031. [CrossRef]
11. Sato, K.; Shigemitsu, T. Research on Performance Improvement of Low Specific Speed Centrifugal Pump. *Proc. Mech. Eng. Congr. Jpn.* **2017**, *2017*, J0550101. [CrossRef]
12. DaqiqShirazi, M.; Torabi, R.; Riasi, A.; Nourbakhsh, S.A. The effect of wear ring clearance on flow field in the impeller sidewall gap and efficiency of a low specific speed centrifugal pump. *Arch. Proc. Inst. Mech. Eng. Part C J. Mech. Eng. Sci. 1989–1996* **2017**, *232*, 095440621772942. [CrossRef]
13. Siddique, M.H.; Samad, A.; Hossain, S. Centrifugal pump performance enhancement: Effect of splitter blade and optimization. *Proc. Inst. Mech. Eng. Part A J. Power Energy* **2022**, *236*, 391–402. [CrossRef]
14. Chabannes, L.; Tefan, D.; Rudolf, P. Effect of Splitter Blades on Performances of a Very Low Specific Speed Pump. *Energies* **2021**, *14*, 3785. [CrossRef]
15. Gao, B.; Zhang, N.; Li, Z.; Ni, D.; Yang, M. Influence of the Blade Trailing Edge Profile on the Performance and Unsteady Pressure Pulsations in a Low Specific Speed Centrifugal Pump. *J. Fluids Eng. Trans. ASME* **2019**, *138*, 051106. [CrossRef]
16. Chabannes, L.; Tefan, D.; Rudolf, P. Volute throat area and wall modelling influence on the numerical performances of a very low specific speed pump. *IOP Conf. Ser. Earth Environ. Sci.* **2021**, *774*, 012007. [CrossRef]
17. Zeng, G.; Li, Q.; Wu, P.; Qian, B.; Huang, B.; Li, S.; Wu, D. Investigation of the impact of splitter blades on a low specific speed pump for fluid-induced vibration. *J. Mech. Sci. Technol.* **2020**, *34*, 2883–2893. [CrossRef]
18. Cui, B.; Zhang, Y.; Huang, Y.; Zhu, Z. Analysis of unsteady flow and fluid exciting forces of multistage centrifugal pump based on actual size. *Proc. Inst. Mech. Eng. Part A J. Power Energy* **2022**, *236*, 21–32. [CrossRef]
19. Zhang, N.; Gao, B.; Li, Z.; Ni, D.; Jiang, Q. Unsteady flow structure and its evolution in a low specific speed centrifugal pump measured by PIV. *Exp. Therm. Fluid Sci.* **2018**, *97*, 133–144. [CrossRef]

20. Ye, W.; Luo, X.; Huang, R.; Jiang, Z.; Li, X.; Zhu, Z. Investigation of flow instability characteristics in a low specific speed centrifugal pump using a modified partially averaged Navier–Stokes model. *Proc. Inst. Mech. Eng. Part A J. Power Energy* **2019**, *233*, 834–848. [CrossRef]
21. Gu, Y.; Li, J.; Wang, P.; Cheng, L.; Qiu, Y.; Wang, C.; Si, Q. An Improved One-Dimensional Flow Model for Side Chambers of Centrifugal Pumps Considering the Blade Slip Factor. *J. Fluids Eng.* **2022**, *144*, 091207. [CrossRef]
22. Gu, Y.; Cheng, J.; Wang, P.; Cheng, L.; Si, Q.; Wang, C.; Shouqi, Y. A Flow Model for Side Chambers of Centrifugal Pumps Considering Radial Wall Shear Stress. *Proc. Inst. Mech. Eng. Part C J. Mech. Eng. Sci.* **2022**, *236*, 09544062211073023. [CrossRef]
23. Lu, G.; Zuo, Z.; Sun, Y.; Liu, D.; Tsujimoto, Y.; Liu, S. Experimental evidence of cavitation influences on the positive slope on the pump performance curve of a low specific speed model pump-turbine. *Renew. Energy* **2017**, *113*, 1539–1550. [CrossRef]
24. El-Emam, M.; Zhou, L.; Yasser, E.; Bai, L.; Shi, W. Computational methods of erosion wear in centrifugal pump: A state-of-the-art review. *Arch. Comput. Methods Eng.* **2022**, *29*, 3789–3814. [CrossRef]
25. Yang, Y.; Zhou, L.; Zhou, H.; Lv, W.; Wang, J.; Shi, W.; He, Z. Optimal Design of Slit Impeller for Low Specific Speed Centrifugal Pump Based on Orthogonal Test. *J. Mar. Sci. Eng.* **2021**, *9*, 121. [CrossRef]
26. Zhang, N.; Jiang, J.; Gao, B.; Liu, X.; Ni, D. Numerical analysis of the vortical structure and its unsteady evolution of a centrifugal pump. *Renew. Energy* **2020**, *155*, 748–760. [CrossRef]
27. Han, Y.; Zhou, L.; Bai, L.; Shi, W.; Agarwal, R. Comparison and validation of various turbulence models for U-bend flow with a magnetic resonance velocimetry experiment. *Phys. Fluids* **2021**, *33*, 125117. [CrossRef]
28. Al-Obaidi, A.R. Effects of different turbulence models on three-dimensional unsteady cavitating flows in the centrifugal pump and performance prediction. *Int. J. Nonlinear Sci. Numer. Simul.* **2019**, *20*, 487–509. [CrossRef]
29. Zhang, J.; Li, G.; Mao, J.; Yuan, S.; Qu, Y.; Jia, J. Numerical investigation of the effects of splitter blade deflection on the pressure pulsation in a low specific speed centrifugal pump. *Proc. Inst. Mech. Eng. Part A J. Power Energy* **2020**, *234*, 420–432. [CrossRef]
30. Gu, Y.; Yuan, S.; Pei, J.; Zhang, J.; Zhang, F.; Huang, X. Effects of the impeller–volute tongue interaction on the internal flow in a low-specific-speed centrifugal pump with splitter blades. *Proc. Inst. Mech. Eng. Part A J. Power Energy* **2018**, *232*, 170–180. [CrossRef]
31. Zhang, Y.; Zang, W.; Zheng, J.; Cappietti, L.; Zhang, J.; Zheng, Y.; Fernandez-Rodriguez, E. The influence of waves propagating with the current on the wake of a tidal stream turbine. *Appl. Energy* **2021**, *290*, 116729. [CrossRef]

sustainability

MDPI

Technical Note

Variation Characteristics of Hydrodynamic Coefficients of Two-Dimensional Rectangular Moonpool Resonance under Sway Motion

Zhouhao Zhang [1], Hongsheng Zhang [1], Yuxin Wang [1] and Yu Zhang [2,*]

[1] College of Ocean Science and Engineering, Shanghai Maritime University, Shanghai 201306, China
[2] College of Harbour, Coastal and Offshore Engineering, Hohai University, Nanjing 210024, China
* Correspondence: jessicazhang@hhu.edu.cn

Abstract: The phenomenon of sudden changes in hydrodynamic coefficients cannot be neglected when the resonance of a two-dimensional rectangular moonpool occurs; however, it may not be meaningful to discuss the detailed values of hydrodynamic coefficients when the resonance occurs. The analytical solution of the hydrodynamic coefficients of a two-dimensional rectangular moonpool under sway motion derived based on the linear potential flow theory is first improved. The improved analytical results and the results with the far-field method are then compared and verified. The results show that the hydrodynamic coefficients of a two-dimensional rectangular moonpool under sway motion change suddenly at the second resonant frequency, which was not found in the original analytical solution of damping coefficient.

Keywords: moonpool; resonance; hydrodynamic coefficient; analytical solution

Citation: Zhang, Z.; Zhang, H.; Wang, Y.; Zhang, Y. Variation Characteristics of Hydrodynamic Coefficients of Two-Dimensional Rectangular Moonpool Resonance under Sway Motion. *Sustainability* **2022**, *14*, 15952. https://doi.org/10.3390/su142315952

Academic Editors: Yuquan Zhang, Jianhua Zhang, Ling Zhou and Bin Huang

Received: 13 October 2022
Accepted: 25 November 2022
Published: 30 November 2022

1. Introduction

Most engineering ships and offshore platforms are equipped with a moonpool structure with openings at the bottom, and the water in the moonpool is connected to the outside. The fluid movement in the moonpool can be divided into vertical piston mode and horizontal sloshing mode [1]. On one hand, the moonpool can provide a stable marine operating environment for the drilling equipment and riser, and can improve the efficiency of wave energy conversion devices. On the other hand, when the external excitation frequency is the same as the natural frequency of the moonpool, fluid resonance occurs in the moonpool, which significantly affects the motion of the ship and the stability of navigation. Therefore, it is of great significance to study the resonant phenomenon of the fluid inside the moonpool as well as its hydrodynamic characteristics of the moonpool.

The natural modes of oscillation of the inner free surfaces of the moonpool are determined under the assumption of infinite water depth via the linearized potential flow theory, and the problem is addressed in two and three dimensions by Molin (2001) [2]. A numerical method for the study of sloshing in tanks with a two-dimensional flow was presented by Faltinsen (1978) [3]. The hydrodynamic interactions of water wave diffraction on a super large floating structure composed of numerous box-shaped modules were studied; a new resonant phenomenon, which is related to the draft of the structure, was revealed, and the relationship between the draft of the structure and the gap width and resonance frequencies was discussed by Miao et al. (2000) [4]. Combined theoretical and experimental studies of the two-dimensional piston-like steady-state motions of the fluid in a moonpool are presented, and a high-precision, analytically oriented, linear-potential-flow method was developed by Faltinsen et al. (2007) [5]. The equations for solving the radiation and diffraction problems of a two-dimensional rectangular moonpool were established via the linearized potential flow theory, the expressions of the hydrodynamic coefficients were derived, and the effect of the opening on the excitation force and hydrodynamic coefficients was evaluated, by Zhou et al. (2013) [6]. Based on Zhou et al. (2013) [6], the radiation and diffraction problems of a two-dimensional rectangular moonpool in front of a straight wall at a finite water depth were studied by Zhang and Zhou (2013) [7]. Based on the CFD theory, a method that includes both numerical simulation and analysis of the ship added mass and damping coefficient was outlined for ships oscillating in waves by Zhu et al. (2009) [8]. Potential and N.-S. equation-based flows were numerically simulated around a two-dimension floating rectangular body with a moonpool using Fluent software, and the influences of the vorticity on the vertical motion of the moonpool under a forced heave motion were evaluated by Heo et al. (2014) [9]. The differences in the moonpool resonant frequencies between the fixed and free-floating conditions were studied using the radiation-diffraction code WAMIT, and the effect of the vessel configuration on the offset of the moonpool resonant frequencies was examined by Huang et al. (2019) [10]. Theoretical models for calculating the natural frequencies and modal shapes of two-dimensional asymmetric and symmetric moonpools in finite water depth were proposed by Zhang et al. (2019) [11]. In the case of a symmetric moonpool, single-mode approximations (SMA) have been derived and can be used to quickly estimate the natural frequencies of the piston and sloshing modes.

In this study, the analytical solution of a two-dimensional rectangular moonpool under the sway motion deduced in Zhou et al. (2013) [6] was improved, and the variation characteristics of the hydrodynamic coefficients at the second resonance frequency were supplemented. The far-field method results of the hydrodynamic coefficients were compared with the improved analytical results.

2. Analysis of the Original Analytical Solution

Suppose that a rectangular moonpool floats on the free surface of water, and the structure oscillates with a small amplitude under external excitation. The Cartesian coordinate system $Oxyz$ is obtained, where the Oxy plane coincides with the still water level, and the origin of the coordinate axis is located at the intersection of the symmetry axis of the moonpool and still water level. The water depth is h. Assuming that the rectangular moonpool is infinite in the y direction, the influence in the length direction can then be neglected. The moonpool and flow field can thus be simplified into a two-dimensional form, as shown in Figure 1, where the width of the moonpool is $2b$, the width of the opening is $2a$, the draft depth of the moonpool is h_1, and the height of the moonpool is h_2.

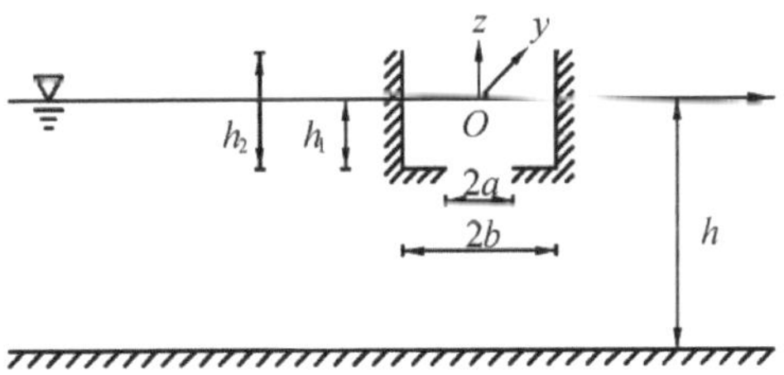

Figure 1. Sketch of two-dimensional flow field and rectangular moonpool.

The hydrodynamic force on a body can be obtained from the integration of the pressure based on the linear Bernoulli equation over the wetted body surface. For wave radiation due to body oscillation, Zhou et al. (2013) [6] provided the expression of the excitation force, τ_{pq}, in mode p due to the motion of unit amplitude in mode q:

$$\tau_{pq} = i\omega\rho \int_{S_b} \left[-i\omega\phi^{(q)}(x,z)\right] n_p dS = \rho\omega^2 \int_{S_b} \phi^{(q)} n_p dS \tag{1}$$

where, ω is the circular frequency, ρ is the water density, S_b represents the wetted body surface, $\phi^{(q)}$ is the normalized velocity potential of motion in mode q, and n_p is the normal direction of the wetted body surface.

The force due to wave radiation can be written in terms of the added mass μ_{pq} and damping coefficient λ_{pq}. The non-dimensional added mass μ_{pq} and damping coefficient λ_{pq} derived by Zhou et al. (2013) [6] are expressed as follows:

$$\mu_{pq} = Re(\tau_{pq}) / \left[2\rho h_1 \omega^2 b b^{(\delta_{3p}+\delta_{3q})}\right] \tag{2}$$

$$\frac{\lambda_{pq}}{\omega} = Im(\tau_{pq}) / \left[2\rho h_1 \omega^2 b b^{(\delta_{3p}+\delta_{3q})}\right] \tag{3}$$

where Re indicates that the real part of the complex function is taken, Im indicates that the imaginary part of the complex function is taken, and δ_{pq} is the Kronecker delta function given by

$$\delta_{pq} = \begin{cases} 0 & p \neq q \\ 1 & p = q \end{cases}. \tag{4}$$

After the authors' rederivation, it is inferred that the analytical solution derived by Zhou et al. (2013) [6] is accurate. However, we found that the numerical variation of the hydrodynamic coefficients, especially the damping coefficient under the sway motion at the second resonant frequency, cannot be accurately captured because the calculation accuracy is insufficient.

Assume $h = 3.0$ m, $h_1 = 1.0$ m, $2b = 1.0$ m, and $a/b = 0.8$. For this case, $2kb$ is the same as the value of k. The hydrodynamic coefficients of the moonpool were calculated according to the code provided by Dr. Zhou. It can be observed from Figure 2a that, under the heave motion, significant numerical changes in the added mass appear in the frequency range of $2kb = 0 \sim 1$, and two extrema, one large and one small, appear in this range. It can be observed from Figure 2b that, under the heave motion, significant numerical changes in the damping coefficient also appear in the same frequency range, and the extremum variation of λ_{22} corresponds to the value change of the added mass.

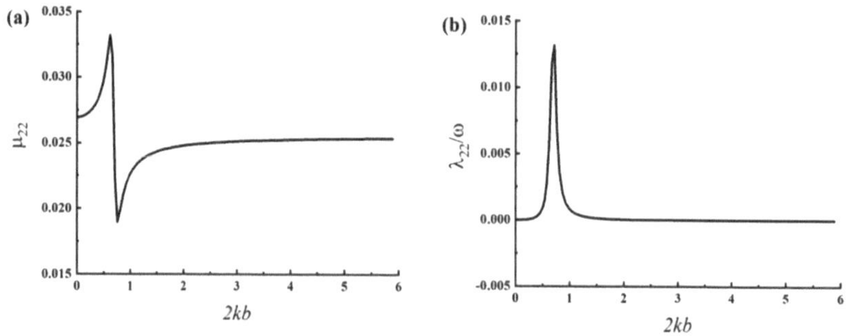

Figure 2. Analytical results of hydrodynamic coefficients of the moonpool under heave motion (by Zhou's program): (**a**) added mass; (**b**) damping coefficient.

It can be observed from Figure 3a that, under the sway motion, the first maximum of the added mass appears in the range of frequency $2kb = 0 \sim 1$. The positive and negative extrema of the added mass appear near the second resonant frequency of $2kb = 3.1$. It can be observed from Figure 3b that, under the sway motion, the peak of the damping coefficient also appears in the range of frequency $2kb = 0 \sim 1$, whereas the damping coefficient only fluctuates weakly near the second resonant frequency of $2kb = 3.1$, where the extremum of the added mass appears.

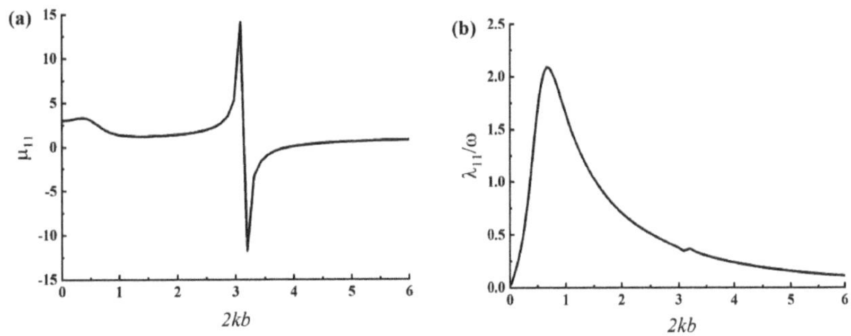

Figure 3. Analytical results of hydrodynamic coefficients of the moonpool under sway motion (by Zhou's program): (**a**) added mass; (**b**) damping coefficient.

In the code provided by Dr. Zhou, ω is given first, $\Delta\omega$ is taken as 0.1, and the wave number k is then obtained through iteration of the dispersion equation. The dispersion equation is expressed as follows:

$$\omega^2 = gk\tanh(kh) \tag{5}$$

where g is the gravitational acceleration. Furthermore, according to Equations (20)–(24) by Zhou et al. (2013) [6], the eigenvalues, α_n, β_n, and γ_n correspond to three different water depths, h, h_1, and h_2, respectively, and the subsequent solving is then carried out. Table 1 shows the calculated frequency, wave number, and increments of the wave number according to the method of Zhou. It can be observed from Table 1 that, with an increase in ω, the step size Δk gradually increases. Additionally, from a numerical point of view, Δk was greater than $\Delta\omega$. Although the given ω has an equal difference increment, k obtained by iteration does not change uniformly in an equal difference, and Δk is greater than $\Delta\omega$. This may lead to an inaccurate capture of the resonant frequency and thus, the variation in the damping coefficient near the resonant frequency is not captured.

Table 1. k and Δk under different ω.

ω	k	Δk
...	...	...
5.2	2.75637	0.10449
5.3	2.86340	0.10703
5.4	2.97248	0.10908
5.5	3.08359	0.11111
5.6	3.19674	0.11315
5.7	3.31193	0.11519
5.8	3.42915	0.11722
5.9	3.54842	0.11927
6.0	3.66972	0.12130
6.1	3.79307	0.12335
6.2	3.91845	0.12538

3. Solution Improvement and Comparative Analysis

3.1. Improvement of Analytical Solution

We adjust part of the analytical solution code of the hydrodynamic coefficients under sway motion, which is related to the solution of the wave number. The relevant parameters are the same as those described in Section 2. The program is changed to specify k first, then obtain ω using Equation (5), and then α_n, β_n and γ_n. The step size Δk was set to 0.1, and the comparisons between the adjusted new analytical results and original analytical results are shown in Figure 4. It can be observed from Figure 4a that, after the adjustment, the value of the positive extremum of the added mass increases significantly near the second resonant frequency of $2kb = 3.1$ compared with the original one. As shown in Figure 4b, however, the damping coefficients are almost the same.

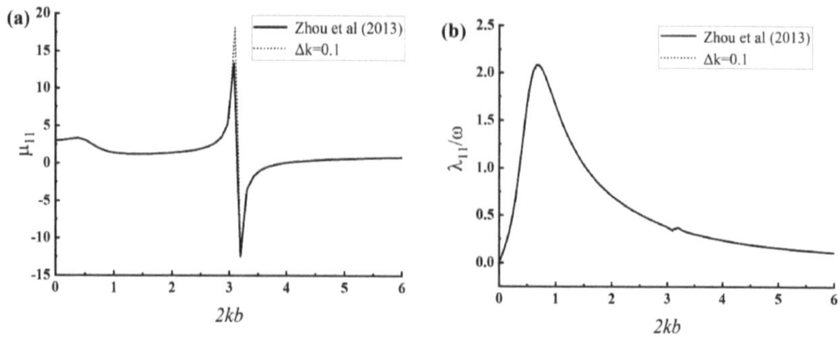

Figure 4. Comparison between the original analytical and improved results (Δk = 0.1) [6]: (**a**) added mass; (**b**) damping coefficient.

The Δk near the second resonant frequency ($2kb$ = 3.14~3.15) is more carefully adjusted, and Δk is set as 0.01, 0.001, and 0.0001, respectively. The hydrodynamic coefficients for the different Δk values are shown in Figure 5. As shown in Figure 5a, when Δk = 0.01, the values of the two extrema of the added mass, which are 380.040 and −89.876 at $2kb$ = 3.14 and 3.15, respectively, vary greatly compared with Δk=0.1. It can be observed from Figure 5b that, when Δk = 0.01, the damping coefficients, which are 0.00265 and 0.530 at $2kb$ = 3.14 and 3.15, respectively, no longer fluctuate weakly near the second resonant frequency; instead, it shows an obvious numerical change. As shown in Figure 5c, when Δk = 0.001, the two extrema of the added mass, especially the negative extreme values, tend to infinity, which are 787.393 and −10,580.496 at $2kb$ = 3.141 and 3.142, respectively. It can be observed from Figure 5d that, when Δk = 0.001, the minimum of the damping coefficient at $2kb$ = 3.140 is 0.00265, and the maximum tends to infinity and appears at $2kb$ = 3.142 with a value of 251.760. As shown in Figure 5e, when Δk = 0.0001, the two extrema of the added mass, which are 23,681.718 and −10,580.496 at $2kb$ = 3.1419 and 3.1420, respectively, tend to infinity. As shown in Figure 5f, when Δk = 0.0001, the damping coefficient has a minimum of 0.00014 at $2kb$ = 3.1402 and a maximum of 1129.184 at $2kb$ = 3.1419. Regardless of the value of Δk, the damping coefficient is always greater than 0, which is in line with the physical reality. As mentioned above, if Δk is more accurate, then the second resonant frequency is captured more accurately. The damping coefficient also appears as an extremum near the second resonant frequency, and the value approaches infinity.

To observe the variation of the hydrodynamic coefficients more carefully with an accuracy of Δk, the hydrodynamic coefficients for different values of Δk are amplified in local details. Figure 6 shows the comparisons among the details of the added mass and damping coefficient near the second resonant frequency for different sizes of Δk. As shown in Figure 6, when Δk = 0.0001, near the second resonant frequency, the curves of the added mass and damping coefficient change more smoothly, and the resonant frequency is captured more accurately.

Figure 7 shows the comparisons among the details of the added mass and damping coefficient in finite numerical intervals for different sizes of Δk. It can be observed from Figure 7 that, regardless of Δk, the results of the added mass and damping coefficient are almost identical in finite numerical intervals.

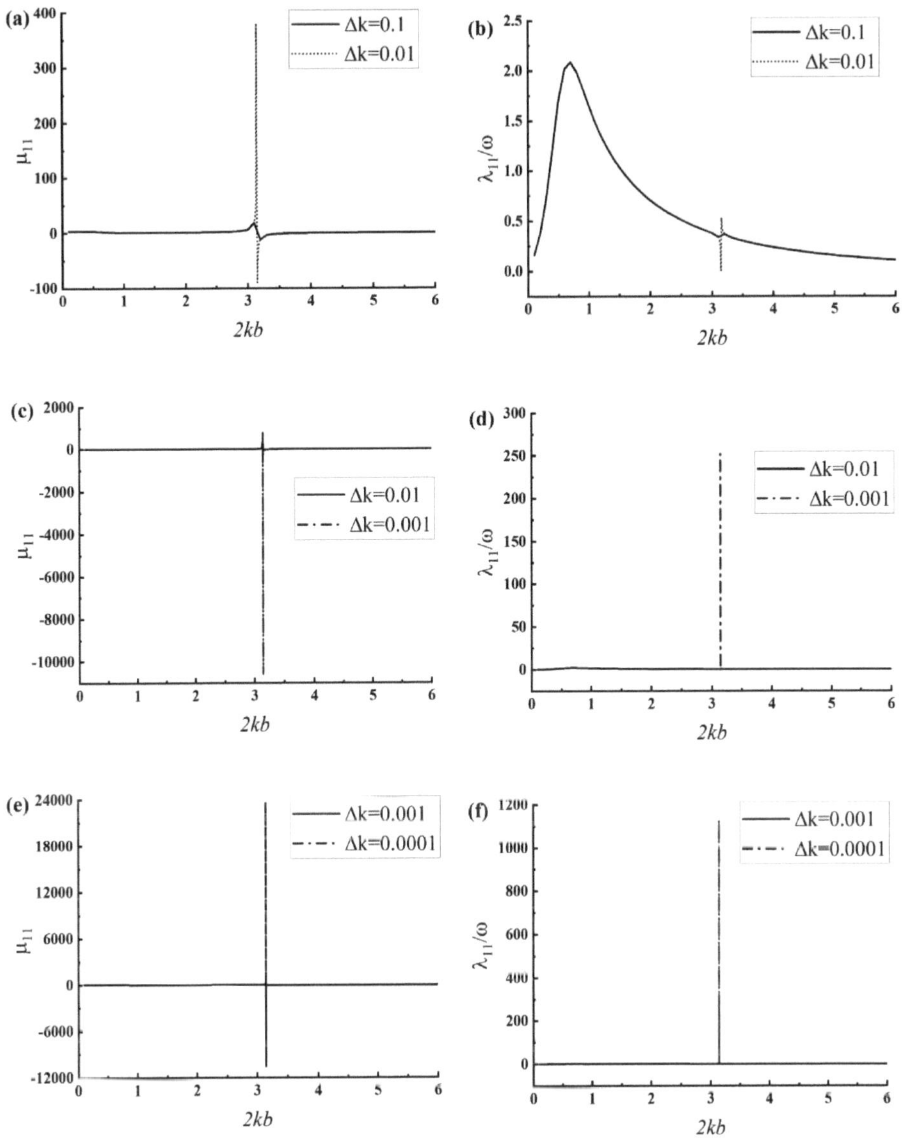

Figure 5. Comparisons of hydrodynamic coefficients with different increments of Δk: (**a**) added mass ($\Delta k = 0.1$, $\Delta k = 0.01$); (**b**) damping coefficient ($\Delta k = 0.1$, $\Delta k = 0.01$); (**c**) added mass ($\Delta k = 0.01$, $\Delta k = 0.001$); (**d**) damping coefficient ($\Delta k = 0.01$, $\Delta k = 0.001$); (**e**) added mass ($\Delta k = 0.001$, $\Delta k = 0.0001$); (**f**) damping coefficient ($\Delta k = 0.001$, $\Delta k = 0.0001$).

The analytical solutions discussed in this paper are based on the assumptions of an ideal fluid and a small motion amplitude. After the improvement of the solution process, from the above analyses and comparisons among the hydrodynamic coefficient results, it can be deduced that, although the values of the extrema of the two hydrodynamic coefficients at the second resonant frequency are different, their tendency to infinity is obvious.

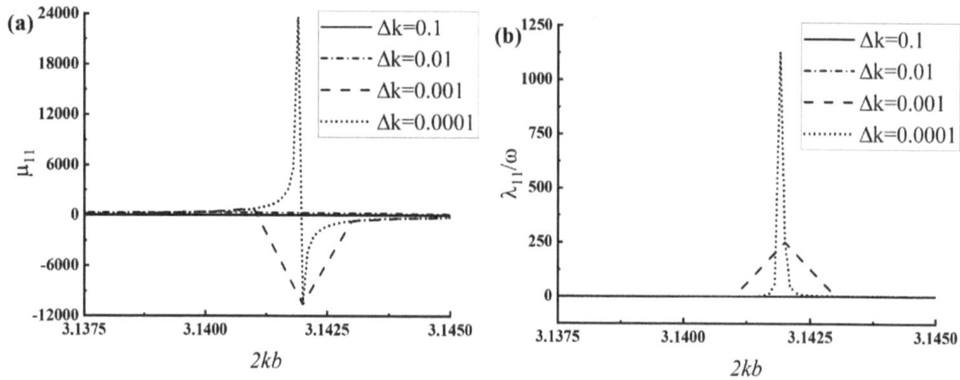

Figure 6. Comparisons of details of hydrodynamic coefficients near the second resonant frequency with different increments of Δk: (**a**) added mass; (**b**) damping coefficient.

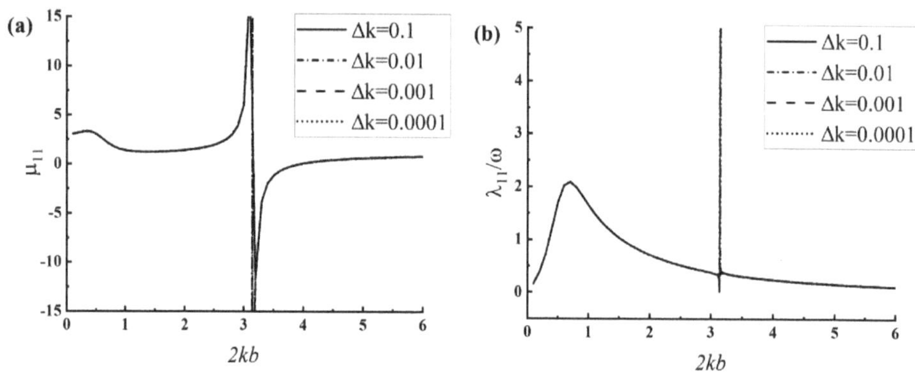

Figure 7. Comparisons of details of hydrodynamic coefficients in finite numerical intervals with different increments of Δk: (**a**) added mass; (**b**) damping coefficient.

The point of resonance in the analytical solution discussed in this study is a singularity. The point of resonance was profound; however, it is regarded as an infinitesimal "point" in geometry, it does not exist, and all physical laws fail at that point. Therefore, the specific values of the hydrodynamic coefficients at the resonance point are meaningless. Moreover, infinity does not correspond to the physical reality; thus, it is not necessary to examine the specific values of the hydrodynamic coefficients in depth studies. However, the sudden change in hydrodynamic coefficients near the second resonant frequency is very important and cannot be neglected.

3.2. Comparison to Far-Field Results

The analytical solution discussed above is based on the near-field method, that is, the fluid force is calculated by integrating the pressure on the moonpool along the wetted body surface. The damping coefficient can also be calculated based on the wave amplitudes at infinity, which is referred to as the far-field method. The damping coefficient of a two-dimensional rectangular moonpool with the far-field method derived by Zhou et al. (2013) [6] is expressed as

$$\frac{\lambda_{pq}}{\omega} = \frac{gC_g\left(A_q^- A_q^{-*} + A_q^+ A_q^{+*}\right)}{2h_1 b^{1+\delta_{3p}+\delta_{3q}} \omega^3} \tag{6}$$

where

$$C_g = \frac{g}{2\omega}\tanh kh\left(1+\frac{2kh}{\sinh 2kh}\right) \quad (7)$$

is the group velocity, A_q^+ and A_q^- are complex wave amplitudes at $x \to +\infty$ and $x \to -\infty$, respectively, and the superscript $*$ indicates a complex conjugate.

Figure 8 shows the comparisons between the damping coefficients of the moonpool under sway motion obtained using the two methods. This indicates that the results obtained with the two methods are in good agreement, which further proves that the improved analytical results are reasonable and correct.

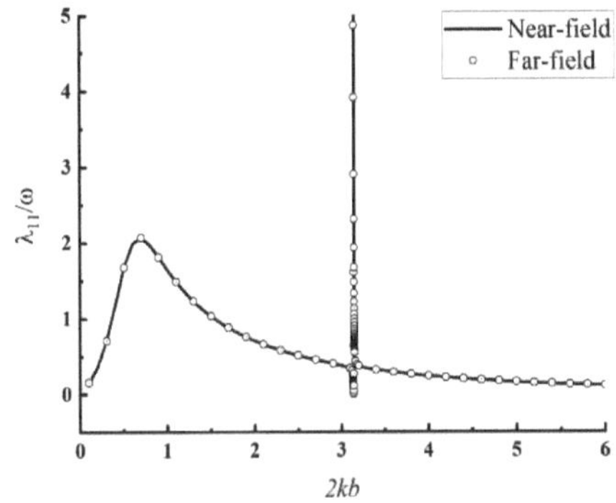

Figure 8. Comparison of damping coefficients with the near- and far-field methods.

3.3. *Comparisons of Results with Different Openings*

The hydrodynamic coefficients of the moonpool under sway motion with different a/b values are calculated to verify whether the improved analytical solution is applicable to the moonpool with different opening ratios. Assuming a/b as 0.2, 0.4, 0.6, and 0.8, respectively, the other parameters are the same as those in previous section. Figure 9 shows the comparisons among the hydrodynamic coefficients of the moonpool with different a/b values in finite numerical intervals. It can be observed that, under sway motion, the opening has little effect on the hydrodynamic coefficients, and the hydrodynamic coefficients have slight differences only at the first peak. For different a/b values, both the added mass and damping coefficients tend to be infinite near the second resonant frequency. Therefore, it is not accidental that the value of the damping coefficient tends to infinity at the resonant frequency; instead, it is universal.

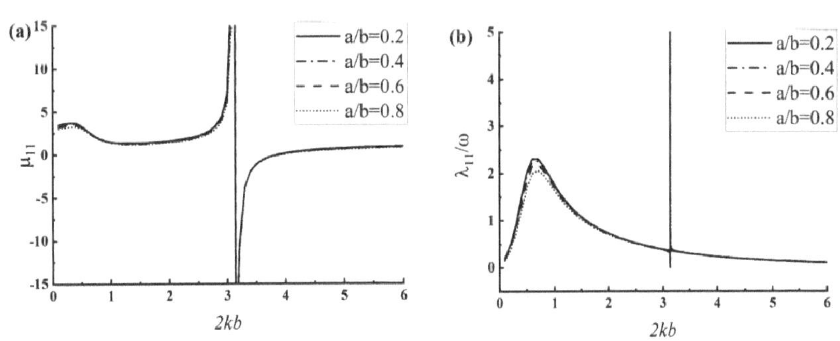

Figure 9. Comparison of hydrodynamic coefficients with different values of a/b: (**a**) added mass; (**b**) damping coefficient.

4. Conclusions

It is of great physical significance to discuss sudden changes in the hydrodynamic coefficients at the resonant frequency when a two-dimensional rectangular moonpool oscillates. In this study, the analytical solution of the hydrodynamic coefficients of a two-dimensional rectangular moonpool under sway motion based on the linear potential flow theory is improved. In the improved analytical solution, the phenomenon that the damping coefficient also appears as an extremum near the second resonant frequency was found. The variation characteristics of the hydrodynamic coefficients at the second resonant frequency of the moonpool, namely, the phenomenon in which the values of the added mass and damping coefficient tend to infinity at the second resonant frequency, is studied in detail. The improved analytical solution is verified by comparing the results of the near- and far-field methods and comparing the results under different opening ratios. In addition, we also simulated the forced oscillation motion of the moonpool using Fluent software, and the numerical results of the hydrodynamic coefficients are compared with the improved analytical solutions. Although there is no need to list the relevant calculation results, and the calculation method is also a commonly used method, the accuracy of the improved analytical solutions can be further illustrated from the perspective of numerical solutions. All the results show that the added mass and damping coefficient have extrema corresponding to each other at the resonant frequencies when the moonpool is under sway motion. The phenomenon in which the extremum appears cannot be neglected, although the detailed values of the extrema are insignificant for discussion.

Author Contributions: Conceptualization and methodology, Y.Z. and H.Z.; validation and formal analysis, Y.W., Z.Z. and H.Z.; writing—original draft preparation, Y.W., Z.Z. and H.Z.; writing—review and editing, Z.Z. and Yu Zhang; funding acquisition, Y.Z. and H.Z.. All authors have read and agreed to the published version of the manuscript.

Funding: This research was funded by National Natural Science Foundation of China (NSFC) (Grant Nos. 52201321 and 51679132). This work was also supported by Shanghai Frontiers Science Center of "Full Penetration" Far-Reaching Offshore Ocean Energy and Power.

Institutional Review Board Statement: Not applicable.

Informed Consent Statement: Not applicable.

Data Availability Statement: Not applicable.

Acknowledgments: This paper was completed after a thorough discussion with Wu G. X. His serious attitude toward scientific research was impressive. We express our sincere gratitude to Wu for taking time out of his busy schedule to answer the authors' questions.

Conflicts of Interest: The authors declare no conflict of interest.

References

1. Fukuda, K. Behavior of Water in Vertical Well with Bottom Opening of Ship, and its Effects on Ship-Motion. *J. Soc. Nav. Archit. Jpn.* **1977**, *1977*, 107–122. [CrossRef]
2. Molin, B. On the piston and sloshing modes in moonpools. *J. Fluid Mech.* **2001**, *430*, 27–50. [CrossRef]
3. Faltinsen, O.M. A Numerical Nonlinear Method of Sloshing in Tanks with Two-Dimensional Flow. *J. Ship Res.* **1978**, *22*, 193–202. [CrossRef]
4. Miao, G.P.; Ishida, H.; Saitoh, T. Influence of gaps between multiple floating bodies on wave forces. *China Ocean. Eng.* **2000**, *4*, 407–422.
5. Faltinsen, O.M.; Rognebakke, O.F.; Timokha, A.N. Two-dimensional resonant piston-like sloshing in a moonpool. *J. Fluid Mech.* **2007**, *575*, 359–397. [CrossRef]
6. Zhou, H.W.; Wu, G.X.; Zhang, H.S. Wave radiation and diffraction by a two-dimensional floating rectangular body with an opening in its bottom. *J. Eng. Math.* **2013**, *83*, 1–22. [CrossRef]
7. Zhang, H.-S.; Zhou, H.-W. Wave radiation and diffraction by a two-dimensional floating body with an opening near a side wall. *China Ocean. Eng.* **2013**, *27*, 437–450. [CrossRef]
8. Zhu, R.C.; Guo, H.Q.; Miao, G.P. A computational method for evaluation of added mass and damping of ship based on CFD theory. *J. Shanghai Jiaotong Univ. (Chin.)* **2009**, *43*, 198–203.

9. Heo, J.-K.; Park, J.-C.; Koo, W.-C.; Kim, M.-H. Influences of Vorticity to Vertical Motion of Two-Dimensional Moonpool under Forced Heave Motion. *Math. Probl. Eng.* **2014**, *2014*, 424927. [CrossRef]
10. Huang, H.Y.; Xu, X.; Zhang, X.S. Study on three-dimensional moonpool resonance of fixed and free-floating vessels. *Chin. J. Hydrodyn. (Chin.)* **2019**, *34*, 482–488.
11. Zhang, X.; Huang, H.; Song, X. On natural frequencies and modal shapes in two-dimensional asymmetric and symmetric moonpools in finite water depth. *Appl. Ocean. Res.* **2019**, *82*, 117–129. [CrossRef]

sustainability

MDPI

Article

The Effects of Meridian Surface Shape on the Pressure Pulsation of a Multi-Stage Electric Submersible Pump

Danyang Du [1], Yong Han [2,*], Yu Xiao [2], Lu Yang [1] and Xuanwei Shi [1]

[1] CNOOC EnerTech Drilling & Production Co. Ltd., Tianjin 300452, China
[2] National Research Center of Pumps, Jiangsu University, Zhenjiang 212013, China
* Correspondence: yonghan@stmail.ujs.edu.cn

Abstract: The influence mechanism of the internal pressure fluctuation propagation law of multi-stage submersible electric pump (ESP) is still unclear, which has been a major problem restricting the stable exploitation of deep-sea oil and gas. In order to investigate the effect of different meridian profiles on the pressure pulsation characteristics of three-stage submersible electric pumps, the unsteady Reynolds-averaged Navier–Stokes (URANS) method is used to numerically investigate it. The results show that the lower the pressure pulsation amplitude in the pump caused by the meridional shape that is more in line with the flow law, has a positive effect on the operation stability. The change of the shape of the meridian greatly affects the pressure pulsation law in the secondary and final pumps. The rotor–stator interaction causes the pressure pulsation amplitude of the monitoring point in the middle of the pump chamber to reach a peak value. By using continuous wavelet transform analysis, it is found that the regularity of 1–2 times frequency conversion is complicated due to multiple pulsation sources and low frequency propagation coupling between stages. At 3–6 times frequency, it is basically close to the pulsation rule of the blade frequency. The above research provides a basis for improving the operation stability of the ESP.

Keywords: electric submersible pump; pressure pulsation; meridian; numerical simulation

Citation: Du, D.; Han, Y.; Xiao, Y.; Yang, L.; Shi, X. The Effects of Meridian Surface Shape on the Pressure Pulsation of a Multi-Stage Electric Submersible Pump. *Sustainability* **2022**, *14*, 14950. https://doi.org/10.3390/su142214950

Academic Editor: Valeria Palomba

Received: 23 September 2022
Accepted: 5 November 2022
Published: 11 November 2022

1. Introduction

An electric submersible pump (ESP) is widely used in oil and gas production as an efficient artificial lift device [1]. However, the typical structure of cascaded supercharging also makes its running stability obviously insufficient. In the 1970s, the problem that the pressure pulsation in the pump affects the operation stability attracted the attention of scholars. At first, the research on the noise and cavitation of the pump began. At that time, the pressure pulsation was regarded as a random signal related to cavitation [2–4]. In the early 1980s, Japanese scholars pointed out that the disturbance between the centrifugal pump impeller and the volute separation tongue or the diffuser would lead to the generation of pressure pulsation [5–7]. Since then, in order to reduce vibration and noise and improve the stability of pump operation, scholars have conducted in-depth research on the pressure pulsation in the pump through theoretical analysis, numerical simulation and experimental measurement [8–10].

Dring et al. [11] pointed out that the rotor–stator interaction between the impeller and the diffuser can be divided into potential flow interference and wake interference. Potential flow interference is the potential flow effect of the inviscid fluid propagating upstream and downstream caused by the relative motion between the impeller and the diffuser and wake interference is a flow field fluctuation that only acts downstream due to the impeller blade wake impacting the inlet edge of the diffuser blades. Chalghoum et al. [12] found that the pressure near the volute tongue of a centrifugal pump is inversely proportional to the angle between the blade outlet and the volute tongue. Arndt et al. [13] found through the centrifugal pump pressure pulsation test that the increase in the number of diffuser blades would lead to a significant weakening of the impeller blade pressure pulsation. Brennen et al. [14] found that the maximum pressure pulsation intensity on the diffuser appeared near the suction leading edge of the blade, while the maximum pressure pulsation intensity on the impeller blade appeared at the trailing edge of the blade. Through experimental research, Khalifa et al. [15] found that reducing the flow rate of the double volute pump would significantly increase the pressure pulsation intensity in the pump. Solis et al. [16] carried out numerical calculation of pressure pulsation for centrifugal pumps with four different structures. Through comparative analysis, it was found that the amplitude of pressure pulsation at the passing frequency of the blade could be reduced by increasing the radial clearance.

Due to the obvious advantages of the CFD method in the study of ESPs, it has not only been used in the design process of ESPs but has also been widely used in the research on the stability of ESPs [17,18]. In the ESP, the hydraulically induced instability is the most complex component in the operation instability of the pump system [19,20]. Inside the ESP, due to the limited number of blades of the impeller, the internal flow field will be unevenly distributed, and then there will be strong dynamic and static coupling with the diffuser, which will induce the constant fluctuation of the pressure in the pump [21,22]. This phenomenon was first discovered by Dring and Joslyn et al. [11,23] in a pump visualization experiment, and was summarized as the phenomenon of rotor–stator interaction. Then, Iino et al. [24] successfully obtained the pressure variation law in an ordinary centrifugal pump through the instantaneous pressure measurement, and summarized the influence of the geometrical dimensions of the rotor components on it. In the follow-up research, Zhang et al. [25] analyzed the rotor–stator interaction and the pressure fluctuation phenomenon caused by numerical calculation method, and successfully established the relationship between the pressure fluctuation law and the number of dynamic and static rotor blades. On the basis of these studies, the researchers further analyzed the pressure pulsation in the pump and the instability effect caused by it through theoretical derivation, numerical calculation and experimental verification. Zhou et al. [26] found that the periodicity of pressure pulsation in the pump is related to the number of blades, but the intensity of pressure pulsation in each model is not related to the number of blades. Yang et al. [27] carried out multi-stage numerical calculation on the ESP based on the special structure of the ESP and found that there is a certain difference in the pressure pulsation in the ESP. The reasons were analyzed, and the inference that the operation instability in the ESP will have inter-stage influence is made. The structural design of the multi-stage ESP makes it easy to aggravate the unsteady state of the pump body when the liquid medium flows inside the lengthy flow channels, and the multi-stage series structure will inevitably superimpose the pressure fluctuation of the medium in the back-stage pump. Not only that, since the ESP is generally composed of impeller and diffuser, it is very easy to produce rotor–stator interaction between the two when the pump body is running, which induces adverse fluctuations in the liquid medium and causes vibration and noise in the pump.

In the past literature, only the static and dynamic interference mechanism in the pump is generally studied, and the influence of the superposition of inter-stage fluctuations caused by the special structure of the ESP is ignored. At the same time, the shape of the meridian surface of the impeller and diffuser is the key structure that affects the flow law in the pump. Based on ANSYS CFX 17.0 software, this paper conducts an unsteady study on the three-stage model of the multi-stage ESP. By setting the corresponding monitoring points at the same position in the calculation domain of the impeller, pump chamber and diffuser flow channel in the three-stage structure, the analysis summarizes the results. The pressure propagation law at the monitoring point, and the pressure pulsation coefficient is used to visualize the fluctuation difference analysis of the pressure pulsation at all stages. The reasons for the difference of the multi-level structure and the influence of different meridian plane shapes on the variation law of pressure pulsation were explored, in order to provide a reference for the stability optimization of the pump.

2. Physical Model

The research object is a medium specific speed ESP. The basic design parameters of the pump: speed n = 2850 r/min, flow rate Q = 80 m^3/h, single-stage head H = 20 m, specific speed n_s = 165.41, and the structural parameters: number of impeller blades Z = 6, the number of diffuser blades Z_2 = 7, the impeller inlet diameter D_1 = 80 mm, the impeller outlet diameter D_2 = 148 mm, the impeller outlet width b_2 = 22 mm, the impeller hub diameter d_h = 30 mm, the impeller blade wrap angle ϕ = 105°.

Based on Taguchi design method, the meridional surfaces of impeller and diffuser are parametrically decomposed, and the hydraulic performance optimization under multiple working conditions is carried out by effect analysis method and signal-to-noise ratio analysis method. The factor levels and the performance indexes of the corresponding schemes are selected based on the statistical results. Based on the analysis of variance, the linear equations of each factor and the performance indexes, such as head and efficiency, are fitted. Finally, according to the optimized design objectives, the optimization scheme with better performance is determined. This is described in detail in a previous investigation by Bai [28] and Yang [29]. The optimized meridional surface shape facilitates the reduction of flow losses. After the optimization of the meridian, on the premise that the head meets the requirements, the overall efficiency at the rated operating point is increased by 6.8%. The schematic diagram of the meridian of the optimized impeller and diffuser is shown in Figure 1. The impeller meridional surface avoids sudden contraction and expansion, which reduces the hydraulic losses. For the convenience of recording, the different meridional before and after optimization are named as scheme 1 (black) and scheme 2 (red).

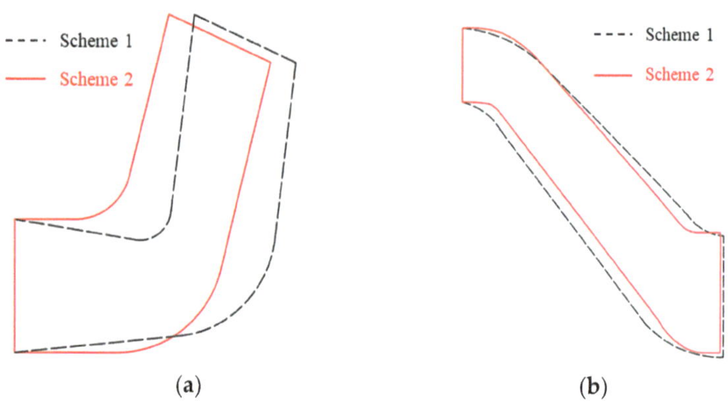

Figure 1. Comparison of different meridian profiles. (**a**) Impeller; (**b**) Diffuser.

3. Numerical Simulation

3.1. Computational Domain

Most of the application scenarios of ESPs are in deep liquid media, and they are often multi-stage superimposed mechanisms. Simulation cannot be limited to single-stage considerations. The multi-stage structure causes the liquid inflow at the impeller inlet of the pump body from the second stage to be on the basis of pre-swirl. In order to ensure that the simulation and experimental flow are as accurate as possible, the three-stage structure is generally simulated to replace the numerical simulation of the pump body with three or more stages. Except for the first and last two stages, the flow states of the remaining stages are mostly similar to the second stage, and the hydraulic performance can be regarded as the equivalent superposition of the lift. After assembling the three-dimensional model, the water body is divided, and the three-dimensional expansion diagram of the water body of the whole flow channel is obtained as shown in Figure 2. The structure of the shroud and hub in the water body of the pump chamber, which has little effect on the performance, will be simplified, and the pump chamber and the diffuser will be combined to save the calculation time.

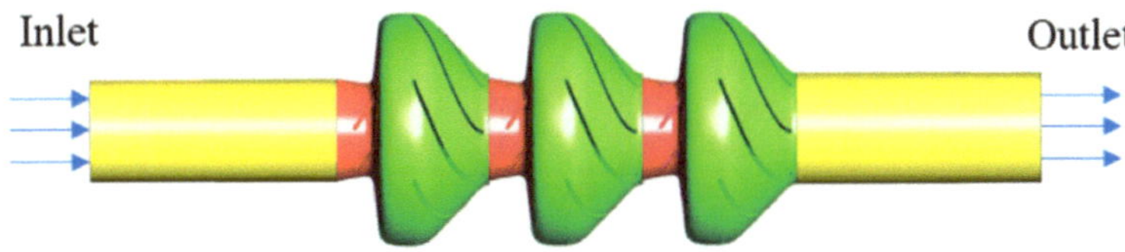

Figure 2. 3D water body assembly development.

3.2. Meshing and Independence Verification

The AMSYS ICEM software is used to perform full hexahedral structured mesh division for the inlet and outlet pipe sections in the computational domain to ensure that the quality of the inlet and outlet pipes is greater than 0.6 to meet the calculation requirements [30,31]. The impeller and diffuser of the ESP are meshed in TurboGird, a special mesh generation software for turbomachinery, and the aspect ratio and orthogonality of the mesh are strictly limited. For the negative volume distribution that occurs, the topological structure is used to generate a mesh structure with a mass greater than 0.3, and y+ is controlled to be less than 100. Figure 3 shows the mesh of the single-channel impeller and diffuser. The full-channel mesh can be obtained by rotating the single-channel grid by its period multiple in ANSYS ICEM [32,33]. Before numerical calculation, it is generally necessary to discretize the computational domain and divide it into multiple small grid units. However, due to the consideration of computer performance and calculation cycle, an appropriate and accurate grid number setting can maximize the use of existing computer performance [34]. As shown in Table 1, when the grid size is less than 2 mm, the prediction error is more obvious. As the number of grids continues to increase, the prediction performance gradually becomes stable, and the solution error fluctuation is less than 1%, which can be ignored. Considering the computer computing performance and time cost [35], the mesh size of the final selected model is 2 mm. The grid-independent analysis of the whole flow field is basically consistent with the trend of single flow channel, and the size is 2 mm, and the number of grids is the period multiple of the single flow channel of the corresponding component.

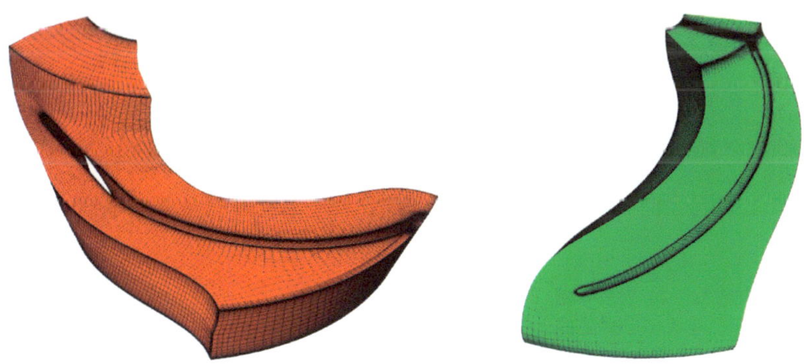

(a) (b)

Figure 3. Single runner grid. (**a**) Impeller; (**b**) Diffuser.

Table 1. Grid dependency analysis.

Size/mm	1.4	1.7	2.0	2.3	2.6
Impeller	573,678	454,745	349,804	279,843	184,696
Diffuser	590,164	468,813	354,508	286,473	189,041
Head/m	58.19	58.05	58.11	56.65	55.08
Efficiency/%	75.49	75.55	75.45	74.18	73.69

3.3. Boundary Conditions

After meshing, each calculation subdomain is imported into ANSYS CFX software in turn to simulate the multi-stage ESP. The inlet boundary condition is set to pressure inlet and the outlet is set to mass outflow. The impeller is set to rotating domain and adopts rotating coordinate system (MRF), and the speed is set to 2850 r/min. The inlet and outlet sections and the diffuser are set as static domains, the wall surface adopts no-slip boundary condition, the wall surface roughness is set to 20 μm, and the standard wall surface function is used for the near wall surface. The iterative convergence accuracy is set to 1×10^{-4}, and the discrete format adopts the first-order upwind style. During steady-state solution, the static interface adopts the default setting, and the dynamic interface is set to the frozen rotor type. In the unsteady solution, the steady calculation result is used as its initial value to improve the calculation efficiency, and the dynamic and static interface is changed to transient rotor stator. The transient numerical simulation impeller rotates 9 times, the total time is total time = 0.18947 s, the step size is 2° of impeller rotation as the unit time, $\Delta t = 1.16959 \times 10^{-4}$, and the initial time is set to 0 s. Solve a time step with a maximum number of inner loops of 10. A large number of literature studies have shown that the SST k-ω model has the characteristics of high accuracy and good flow separation and capture effect in the simulation of rotating machinery, so it is selected to simulate the internal flow characteristics of the ESP [36,37].

3.4. Monitoring Point

In order to summarize the variation law of pressure pulsation in the pump through computational simulation analysis, it is necessary to set up corresponding monitoring points in each computational domain. As shown in Figure 4a, it is the monitoring point in the impeller. In this chapter, five monitoring points Y1, Y2, Y3, Y4 and Y5 are set in the first-stage impeller, all of which are located on the middle section of the impeller flow channel. Among them, Y1 is located at the inlet of the impeller blade, Y5 is located at the outlet of the impeller blade, and the rest are equally distributed. As shown in Figure 4b, there are five monitoring points D1, D2, D3, D4, and D5 in the diffuser, of which D1 and D5 are located at the inlet and outlet respectively, and their distribution is consistent with the impeller. As shown in Figure 4c, the distribution of monitoring points in the pump chamber is also consistent. The position of the monitoring point in the secondary and final stage is the same as that of the first stage, and the labels increase in turn. The two-dimensional plane distribution map is shown in Figure 4d.

The pressure of each monitoring point in the multi-stage ESP increases with the number of stages, and its pressure magnitude also increases step by step. Therefore, in order to exclude the influence of the large magnitude change on the accuracy of the analysis, the monitored pressure is processed in a dimensionless way, and the pressure pulsation coefficient C_p is used to consider the fluctuation degree of the pressure pulsation in the pump [38], which is defined as follows:

$$C_p = \frac{p - \overline{p}}{\frac{1}{2}\rho u_2^2} \tag{1}$$

where p is the pressure value at the monitoring point in Pa; $\overline{p}$ is the mean value of the monitoring point pressure during one rotation in Pa; u_2 is the peripheral speed at the impeller outlet in m/s; ρ is the density of the liquid medium in kg/m^3.

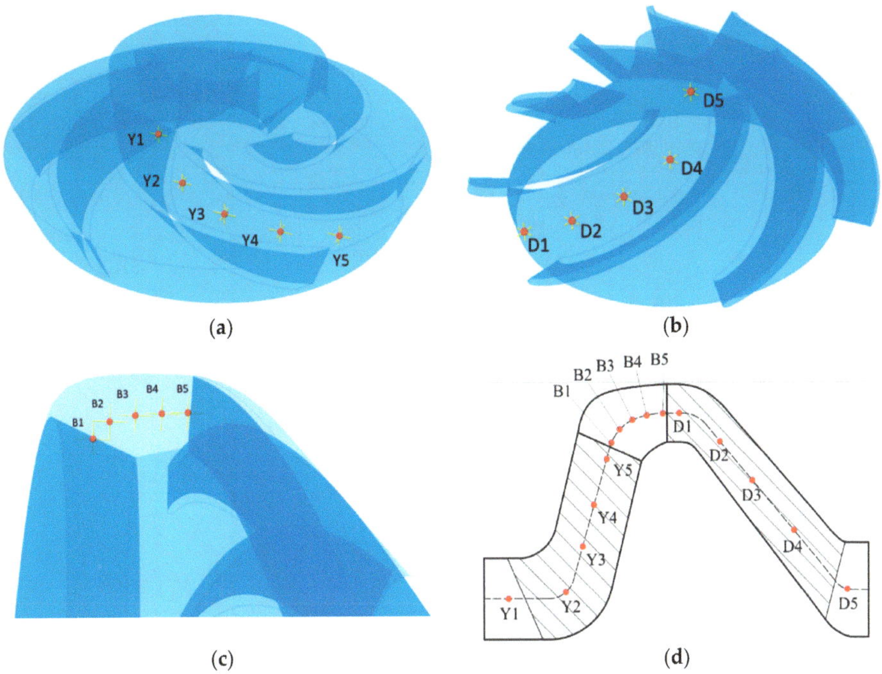

Figure 4. Distribution map of pressure pulsation monitoring points. (**a**) Impeller; (**b**) Diffuser; (**c**) Pump chamber; (**d**) Plane distribution map of monitoring points.

The time domain is to study the variation law of the pulsating signal at different times, and the frequency domain is to analyze the variation of the intensity distribution of the pulsating signal at different frequencies. The frequency domain analysis is the pulsation information obtained by transforming the time domain signal by the Fast Fourier Transform (FFT) method. Usually, the frequency domain analysis is based on the multiple change of the rotation frequency. The meaning of the rotation frequency f_{n} is as follows:

$$f_n = \frac{n}{60} \tag{2}$$

where n represents the rotational speed of the impeller in r/min; f_{n} represents the rotational frequency of the impeller in Hz. The speed of this paper is 2850 r/min, so the frequency of this paper is 47.5 Hz.

$$f_{\mathrm{D}} = f_{\mathrm{n}} \times Z_{\mathrm{D}} \tag{3}$$

where f_{D} represents the blade passing frequency of the diffuser in Hz; Z_{D} is the number of diffuser blades 7. Therefore, f_{D} is 332.5 Hz.

$$f_{\mathrm{Y}} = f_{\mathrm{n}} \times Z_{\mathrm{Y}} \tag{4}$$

where f_{Y} represents the blade passing frequency of the impeller in Hz; Z_{Y} is the number of impeller blades 6. Therefore, f_{Y} is 285 Hz.

3.5. Simulation Accuracy Verification

Yang et al. [27,39] compared the numerical simulation results of the external characteristics and interstage pressure pulsation characteristics of the three-stage ESP with the experimental results. The experimental loop and all instrumentation details are detailed in references. After analysis, because the simulation does not consider the friction loss of the front and rear cover plates of the impeller, the external characteristic results are slightly higher than the experimental data, as shown in Figure 5, but the error meets the requirements. The predicted value of pressure pulsation at the monitoring point in the middle of the pump chamber are consistent with the experimental trend, and both the main frequency signal and the fluctuation period can verify the accuracy of the numerical simulation, as shown in Figure 6.

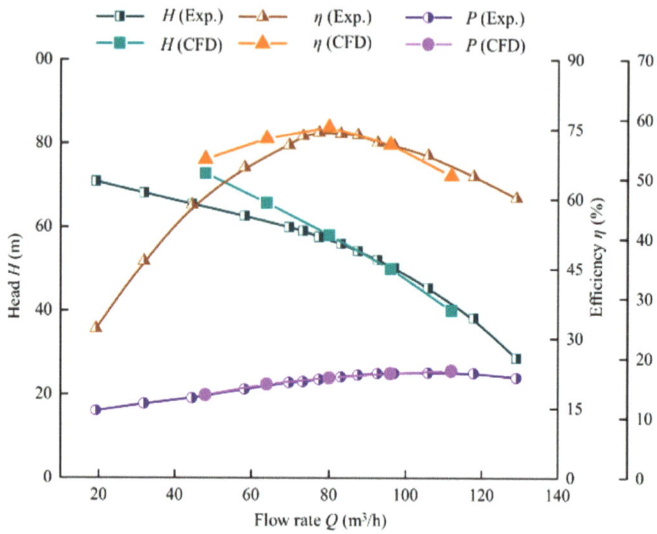

Figure 5. Comparison between simulation results and experimental results of external characteristics.

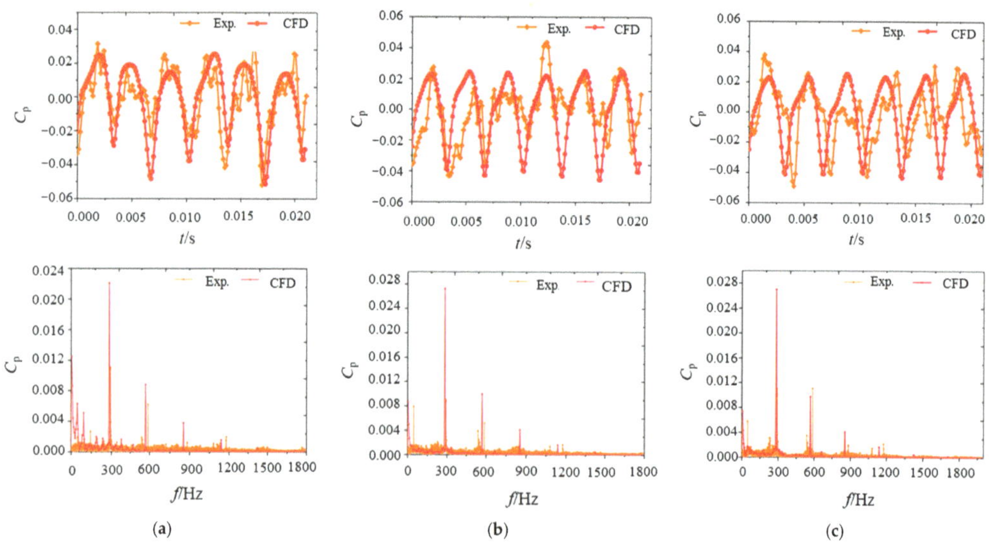

Figure 6. Comparison of pressure fluctuation experiment and simulation results under rated flow (up: time domain, down: frequency domain). **(a)** 1st stage; **(b)** 2nd stage; **(c)** 3rd stage.

4. Results and Discussion

4.1. Pressure Pulsation Inside the Impeller

Figure 7 is the distribution diagram of the internal pressure pulsation characteristics of the impeller under the rated flow condition. The flow of the liquid medium in the second stage of the multi-stage pump is relatively stable, so the analysis is mainly based on the pulsation characteristics in the secondary pump. Figure 7a depicts the comparison of the time domain characteristics of the secondary impeller in scheme 1 and scheme 2. From the fluctuation amplitude, it can be clearly found that the magnitude of the fluctuation of scheme 2 is greatly reduced, and the degree of pressure pulsation is significantly weakened. In terms of cycles, the curves of the two schemes have a 7-cycle law, which is consistent with the number of diffuser blades, because the liquid at the inlet of the secondary impeller flows out from the end of the first-stage diffuser. In scheme 1, the pulsation at the front of the impeller is weak, and the pulsation gradually increases with the flow direction to the peak in the middle of the impeller. In scheme 2, the pulsation is weaker and the peak occurs in the middle and tail of the impeller. The law of severe pulsation of the tail impeller should be mainly caused by the dynamic and static interference generated by the pulsation source in the pump chamber. Figure 7b is a comparison diagram of frequency domain characteristics. It can be seen that the frequency signal is larger at the low-frequency pulsation and 7 times the frequency, but the signal strength of scheme 2 is obviously weakened. Consistent with the above analysis, the main frequency of the impeller under the influence of dynamic and static interference occurs at the passing frequency of the diffuser, due to its close proximity to the upper diffuser. Second, the signal strength at multiple harmonics also decreases with increasing multiplier. The main reason for the large low-frequency pulsation intensity is the inter-stage coupling phenomenon caused by the superposition of inter-stage pulsation, which leads to the transmission of the first-stage low-frequency pulsation signal to the secondary. As shown in Figure 7c–e, since the inlet is a steady flow, the first-stage pulsation signal at the monitoring point Y1 at the impeller inlet is smaller than the other stages. In terms of the flow direction, scheme 1 shows the law that the pulsation intensity is large in the middle, and the pulsation is small at the beginning and the end. In the case of scheme 2, the intermediate monitoring point Y3 not only reduces the overall signal strength, but also makes its amplitude lower than that of the first-stage impeller in the unsteady state in the stable state of the second and last stages. From the perspective of inter-stage characteristics, the intensity of the pulsation in Scheme 1 increases sharply compared with the first stage, generally reaching its peak in the second stage, and slightly stable in the third stage. For scheme 2, the law at the inlet is similar to that of scheme 1, but the fluctuation range in the middle obviously has a flat trend, and the tail remains stable from the second stage impeller. In general, the stability between the two levels of the scheme is higher.

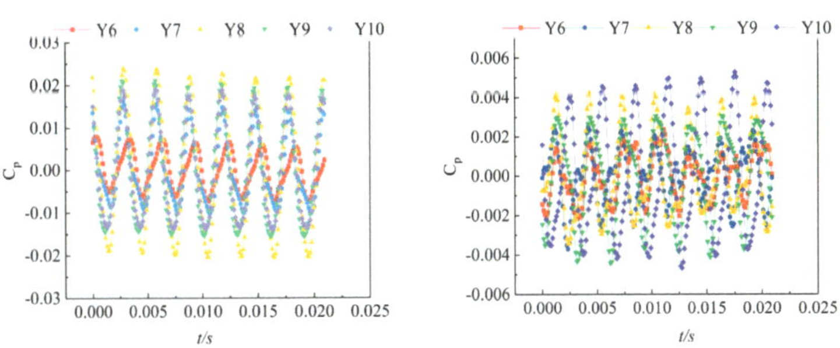

(a)

Figure 7. *Cont.*

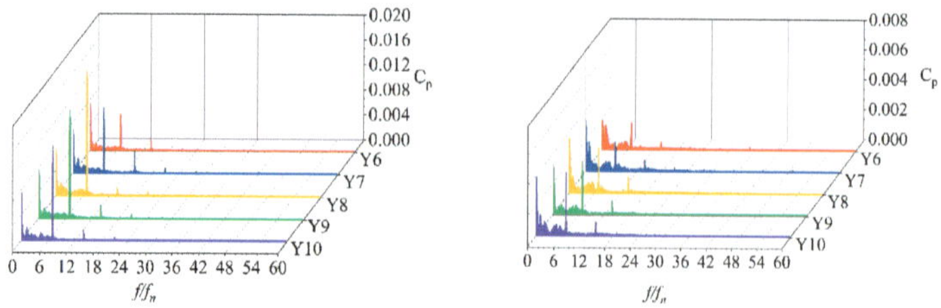

(**b**)

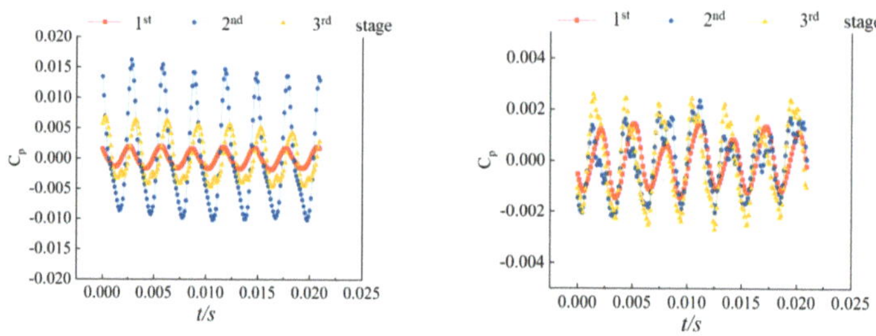

(**c**)

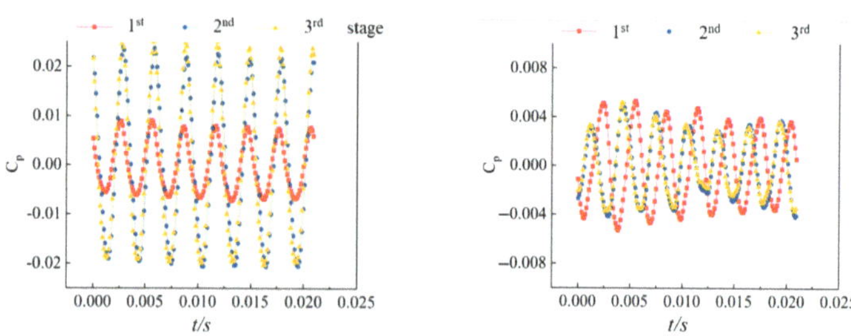

(**d**)

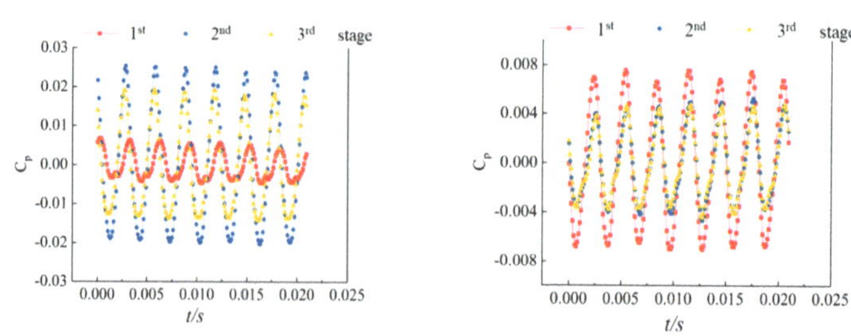

(**e**)

Figure 7. The influence of the shape of the meridian surface on the internal pressure pulsation characteristics of the impeller under rated operating conditions. (**a**) Time domain distribution of secondary impeller (left: scheme 1; right: scheme 2); (**b**) Frequency domain distribution of secondary impeller (left: scheme 1; right: scheme 2); (**c**) Inter-stage time domain distribution of impeller inlet monitoring points (left: scheme 1; right: scheme 2); (**d**) Inter-stage time domain distribution of monitoring points in the middle of the impeller (left: scheme 1; right: scheme 2); (**e**) Inter-stage time domain distribution of impeller outlet monitoring points (left: scheme 1; right: scheme 2).

4.2. Pressure Pulsation Inside the Chamber

Figure 8 is a distribution diagram of the pressure pulsation characteristics inside the pump chamber under the rated flow condition. The pump chamber is the pressure source of dynamic and static interference in the ESP, and the pulsation in each flow channel is deeply affected by it. Among them, the time domain diagram of the secondary pump chamber in Figure 8a, from the perspective of periodicity, is mostly six cycles, which are the same as the number of impeller blades. At the same time, it can be clearly found that the signal amplitude of the minimum value is abnormally large relative to the maximum value of the pulsating signal. This is because in the process of conveying the medium, the pressure drops caused by the jet is stronger than the pressure drops caused by the wake. The extremes of this pulsation all peak in the middle region of the pump chamber. It is worth mentioning that the pump chamber inlet of scheme 1 is greatly affected by the impeller pulsation law, thus showing a situation of seven cycles, but the monitoring point after the flow direction is greatly affected by the pulsation source. In terms of phase, due to the design of the structure, the pulsation characteristics before optimization are more complicated, and the regularity of the phase after optimization is more obvious. Figure 8b is the frequency domain distribution diagram. As far as the low-frequency pulsation is concerned, although there is an inter-stage coupling phenomenon, due to the rated operating point and the secondary pump chamber, the pulsation intensity is much lower than the intensity of the arterial-venous interference pulsation source itself. Scheme 2 optimizes the meridional surface structure, which greatly weakens the low-frequency pulsation region and further reduces the pulsation intensity of the interstage coupling. Additionally, because the monitoring points are relatively close to the pulsation source, the pulsation intensity is similar at the main frequency and its multiplier, and there is no obvious improvement. Figure 8c–e describes the distribution of the inter-stage pulsation characteristics of the pump chamber. It can be seen that due to the steady state, the pulsation characteristics are relatively obvious, and the difference between the stages is low, which is especially reflected in the optimized part of the pump chamber. In scheme 1, the pulsation of each monitoring point in the first stage of the pump chamber fluctuated slightly, and the other parts of the second stage and the last stage showed good regularity. The pulsation intensity of the inlet monitoring point B1 and the central monitoring point B3 is not obvious. Only at the final monitoring point B5, which is close to the diffuser, the signal intensity has some weakening trend.

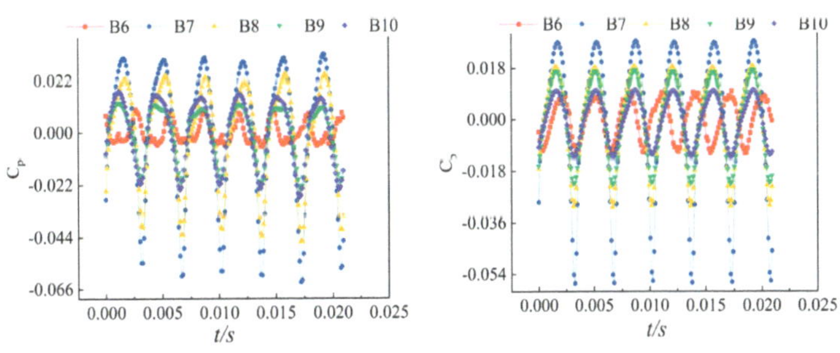

(**a**)

Figure 8. *Cont.*

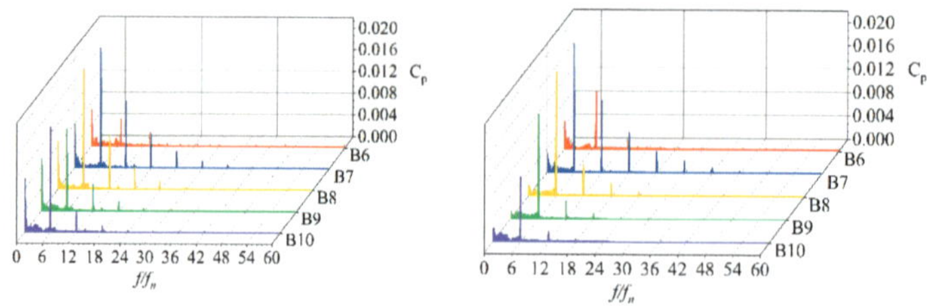

(b)

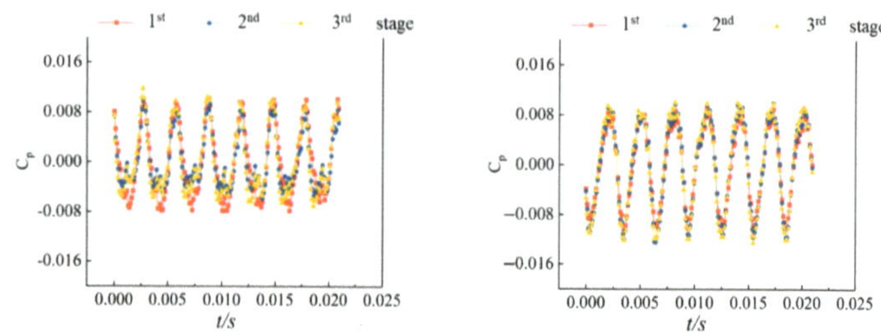

(c)

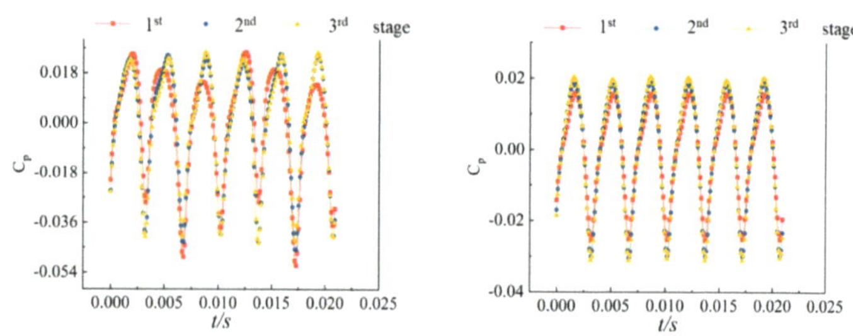

(d)

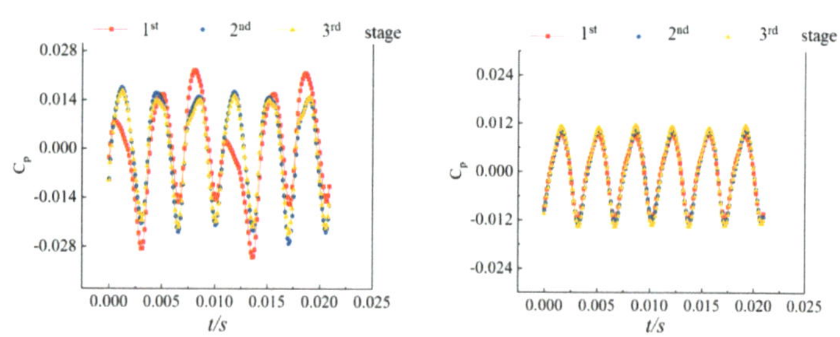

(e)

Figure 8. Influence of the shape of the meridian plane on the pressure pulsation characteristics inside the pump chamber under rated operating conditions. (**a**) Time domain distribution of secondary pump chamber (left: scheme 1; right: scheme 2); (**b**) Frequency domain distribution of secondary pump chamber (left: scheme 1; right: scheme 2); (**c**) Inter-stage time domain distribution of monitoring points at the inlet of the pump chamber (left: scheme 1; right: scheme 2); (**d**) Inter-stage time domain distribution of monitoring points in the middle of the pump chamber (left: scheme 1; right: scheme 2); (**e**) Inter-stage time domain distribution of monitoring points at pump chamber outlet (left: scheme 1; right: scheme 2).

4.3. Pressure Pulsation Inside the Diffuser

Figure 9 is the distribution diagram of the pressure pulsation characteristic inside the diffuser under the rated flow condition. Figure 9a is a time domain diagram of the secondary diffuser. First, the pulsation presents six cycles, which is equal to the number of impeller blades. Secondly, the pulsation intensity at the monitoring point gradually decreases with the flow direction, and reaches the lowest at the outlet of the diffuser. On the whole, the pressure pulsation intensity of scheme 2 tends to attenuate compared with scheme 1, and the phase change occurs at the monitoring point D10 at the outlet of the diffuser in scheme 2. The reason is that the monitoring point is the farthest from the pulsation source of the current stage, and is affected by the pulsation source of the two stages before and after at the same time, so there is a certain coupling, which eventually leads to the change of the time domain phase. Figure 9b is a comparison diagram in the frequency domain. Through the scale comparison, it can be found that the pulsation signal at the diffuser frequency of scheme 2 is relatively low, and all except the low frequency part at 1–2 times the rotation frequency is significantly reduced. Figure 9c–e are the comparison diagrams between pulsation stages. At the monitoring point D1 at the inlet of the diffuser, the extreme point of the first stage of scheme 1 fluctuates up and down, while scheme 2 not only reduces the overall amplitude, but also makes the tertiary pulsation develop regularly. The law of the monitoring point D3 in the middle of the diffuser is roughly close to that of D1. The second and last stages show better stability due to the pulsation under the rated flow condition. The second harmonic part of scheme 2 is significantly improved, and the periodicity is relatively complete. The magnitude of the overall signal intensity at the monitoring point D5 at the outlet of the diffuser becomes smaller, and the phase periodicity in the first and second stages of scheme 2 is enhanced to a certain extent, and the final stage may be affected by more sources of pulsation.

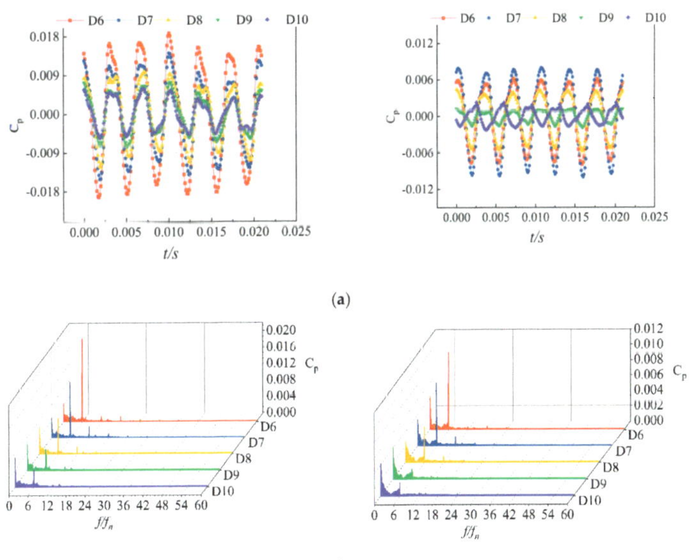

Figure 9. *Cont.*

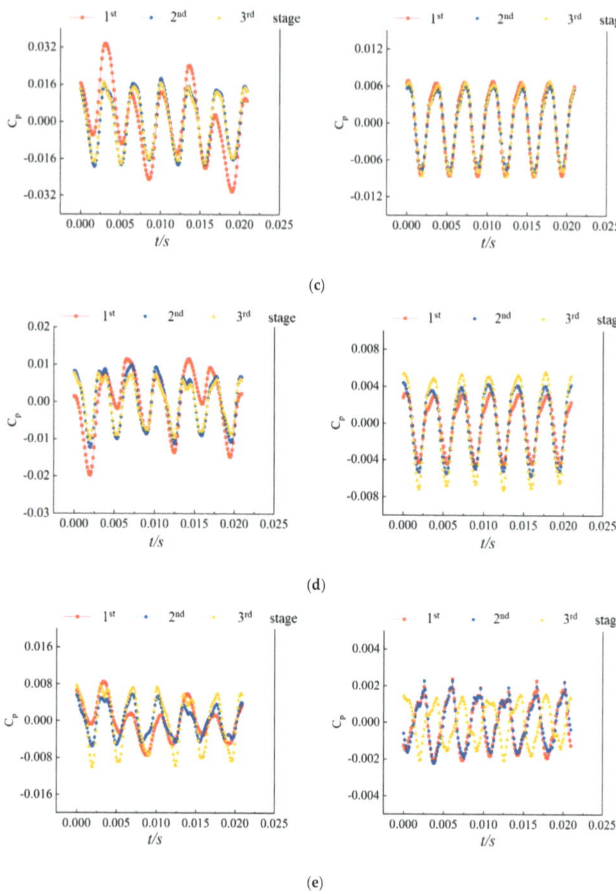

Figure 9. Influence of the shape of the meridian plane on the internal pressure pulsation characteristics of the diffuser under rated operating conditions. (**a**) Time domain distribution of secondary diffuser (left: scheme 1; right: scheme 2); (**b**) Secondary diffuser frequency domain distribution (left: scheme 1; right: scheme 2); (**c**) Inter-stage time domain distribution of diffuser inlet monitoring points (left: scheme 1; right: scheme 2); (**d**) Inter-stage time domain distribution of monitoring points in the middle of the diffuser (left: scheme 1; right: scheme 2); (**e**) Inter-stage time domain distribution of diffuser outlet monitoring points. (left: scheme 1; right: scheme 2).

4.4. Analysis of CWT

In the above, a qualitative analysis is carried out on the time-frequency characteristics in the ESP. By decomposing the equal-frequency pulsation signal of the blade frequency by FFT, the high-frequency variation law in the multi-stage ESP is analyzed. However, it cannot do anything about the low frequency signal part. Since the ESP is affected by many different pulsation sources, the changes of its coupled pulsation signal are extremely complex. In order to explore the variation law of low-frequency signal in the multi-stage ESP, CWT was used to analyze the middle point of the secondary impeller, diffuser and pump chamber of the multi-stage ESP. The essence of CWT is a filtering process of the original signal [40]. The detail signal at each scale is decomposed by energy through CWT, trying to discriminate and extract by the energy scale, and then analyze the low-frequency pulsation characteristics. The inner product of continuous wavelet transform is as follows:

$$WT_f(a,\tau) = \langle f(t), \psi_{a,x}(t) \rangle = \frac{1}{\sqrt{a}} \int_R f(t)\psi^* \left(\frac{t-\tau}{a} \right) dt \quad (5)$$

where the $f(t)$ transform is the CWT, the conjugate of the dilated and translated wavelet $\psi^*((t-\tau)/a)$ is employed and a and τ are the scale (or dilation) and translation parameters, respectively.

Figure 10 is a CWT pulsation distribution diagram of different meridian under rated flow conditions. From a longitudinal perspective, since the flow state of the ESP is relatively stable at the rated flow operating point, the low-frequency components from the impeller to the diffuser are very similar. However, only the pulsation components in the pump chamber are slightly blurred, which may be due to the middle point of the dynamic and static interference, and the clarity of the pulsation components at the monitoring point in the middle of the diffuser is enhanced. In terms of the overall horizontal comparison, the overall magnitude of the scale is one order of magnitude lower than before optimization, which is reflected in both high and low frequencies. The period and pulsation trend at 3–6 times the rotation frequency of the impeller is roughly similar, but the amplitude and phase are shifted. Due to the influence of multiple pulsation sources at the pump chamber, the pulsation oscillation is larger than that of the impeller diffuser. It is worth mentioning that the pulsation amplitude at 1 time the rotation frequency f_n of scheme 2 is almost 0, and the two meridian planes are affected. The difference in pressure pulsation is obvious. Moreover, the periodicity is more obvious at the 2 times the frequency of scheme 2, and the frequency of 3 times and above increases with the increase of the multiple, the lower the resolution, and the more obvious the phenomenon of pulsation superposition. The CWT pulsation in the diffuser has the same variation law as that of the pump chamber, especially at 1 times the rotational frequency, the magnitude is reduced by another level.

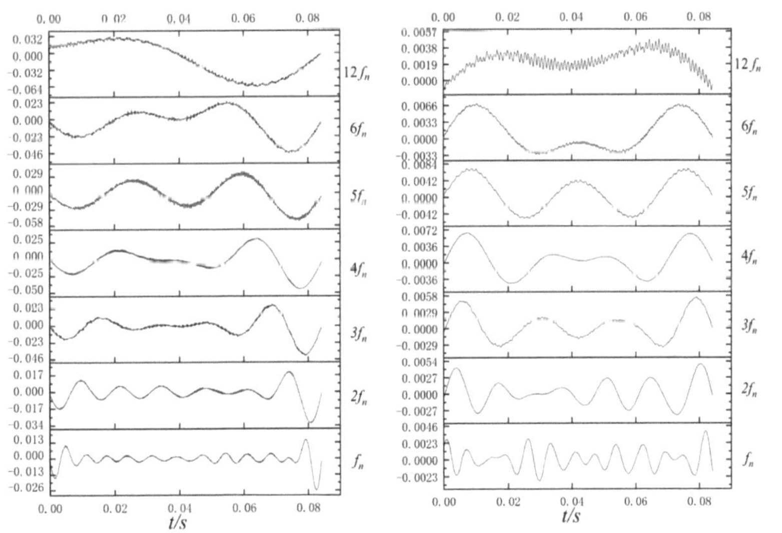

(**a**)

Figure 10. *Cont.*

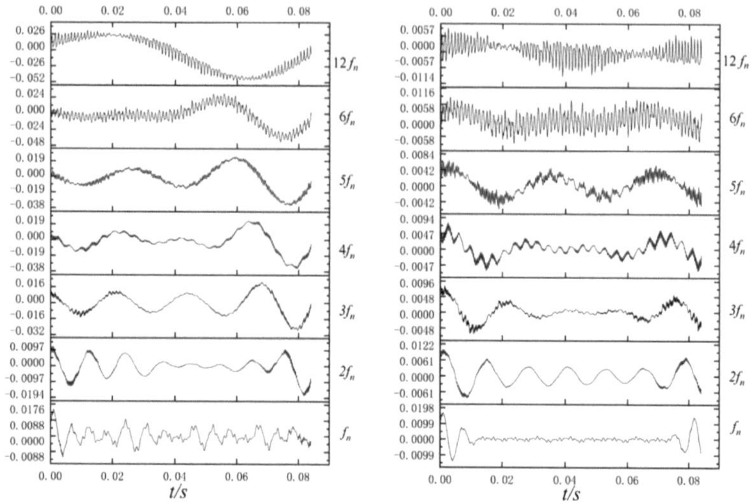

(b)

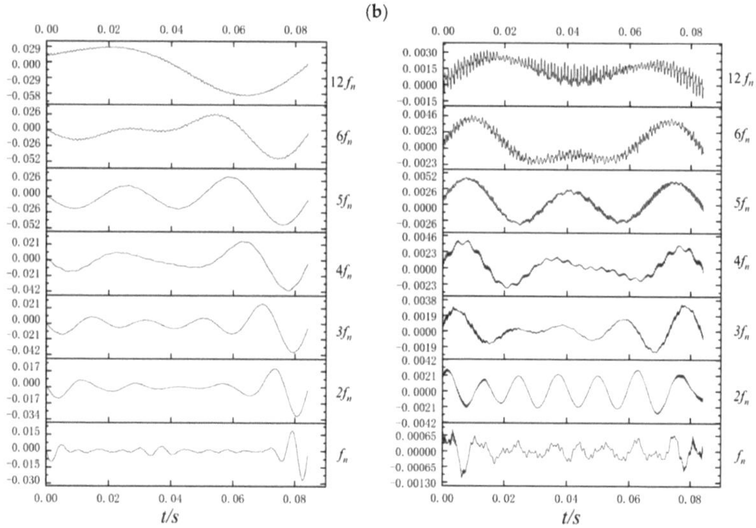

(c)

Figure 10. Wavelet pulsation characteristics of different meridian under rated flow conditions (left: scheme 1; right: scheme 2). (**a**) Impeller wavelet pulsation distribution; (**b**) Pump chamber wavelet pulsation distribution; (**c**) Diffuser wavelet pulsation distribution.

5. Conclusions

Transient numerical simulations of two multi-stage ESPs with different meridional shapes were carried out. At the same time, the instantaneous pressure changes at different positions in the impellers, pump chambers and diffuser at all stages were monitored. The main conclusions of this paper are as follows:

1. The shape of the meridian of scheme 2 is more in line with the flow law, so the pulsation intensity in each component in the pump is smaller than that of scheme 1. Optimizing the shape of the meridian has a contribution to reducing the pulsation strength and increasing the operational stability, which improves the hydraulic per-

formance. Transient pressure pulsation in the secondary and final stages of the pump is a key factor affecting the unsteady flow in the pump;

2. The pulsation amplitude peaks at the middle monitoring point in the pump chamber, which is caused by the rotor–stator interaction between the impeller and the diffuser. Because of the pulse source and pulse propagation coupling, the pressure pulsation period at the monitoring points in the impeller and pump cavity is the same as the number of diffuser blades, and the pressure pulsation period at the monitoring points in the diffuser is the same as the number of impeller blades;
3. The low-frequency pulsation signal after CWT is basically close to the pulsation law of the leaf frequency when the frequency is 3–6 times. At 1–2 times the rotation frequency, the regularity is relatively complicated due to various pulsation sources and low-frequency propagation coupling between stages.

Based on the above studies, we believe that by optimizing the shape of the impeller and diffuser meridian, it is possible to effectively reduce the pressure pulsation amplitude and improve the operational stability of multi-stage submersible electric pumps.

Author Contributions: Conceptualization, D.D. and Y.H.; methodology, Y.X.; software, Y.X.; validation, L.Y. and X.S.; investigation, Y.X.; resources, D.D. and L.Y.; writing—original draft preparation, Y.X.; writing—review and editing, Y.H.; visualization, D.D.; supervision, L.Y.; project administration, X.S.; funding acquisition, D.D. and X.S. All authors have read and agreed to the published version of the manuscript.

Funding: The work was sponsored by Scientific Research Projects of China National Offshore Oil Corporation (JTKY-SXSH-2020-HF-06).

Conflicts of Interest: The authors declare no conflict of interest.

References

1. Zhu, J.; Zhu, H.; Zhang, J.; Zhang, H.Q. A numerical study on flow patterns inside an electrical submersible pump (ESP) and comparison with visualization experiments. *J. Petrol. Sci. Eng.* **2019**, *173*, 339–350. [CrossRef]
2. Spence, R.; Amaral-Teixeira, J. A CFD parametric study of geometrical variations on the pressure pulsations and performance characteristics of a centrifugal pump. *Comput. Fluids* **2009**, *38*, 1243–1257. [CrossRef]
3. Guan, D.; Cong, X.; Li, J.; Niu, Z. Experimental test and theoretical modeling on the working characteristics of spherical water pump. *Flow Meas. Instrum.* **2022**, *85*, 102162. [CrossRef]
4. Zhou, L.; Hang, J.; Bai, L.; Krzemianowski, Z.; El-Emam, M.A.; Yasser, E.; Agarwal, R. Application of entropy production theory for energy losses and other investigation in pumps and turbines: A review. *Appl. Energy* **2022**, *318*, 119211. [CrossRef]
5. Guelich, J.; Bolleter, U. Pressure pulsations in centrifugal pumps. *ASME J. Vib. Acoust.* **1992**, *318*, 272–279. [CrossRef]
6. Li, D.; Qin, Y.; Wang, J.; Zhu, Y.; Wang, H.; Wei, X. Optimization of blade high-pressure edge to reduce pressure fluctuations in pump-turbine hump region. *Renew. Energy* **2022**, *181*, 24–38. [CrossRef]
7. Shim, H.S.; Kim, K.Y. Effects of the number of blades on impeller-volute interaction and flow instability of a centrifugal pump. *Proc Inst. Mech. Eng. A-J. Power Energy* **2022**, 09576509221094951. [CrossRef]
8. Zheng, Y.; Chen, Y.; Mao, X.; Wang, H.; Shi, W.; Kan, K.; Zhang, Y. Pressure pulsation characteristics and its impact on flow-induced noise in mixed-flow pump. *Trans. Chin. Soc. Agric. Eng.* **2015**, *31*, 67–73.
9. Shi, L.; Yuan, Y.; Jiao, H.; Tang, F.; Cheng, L.; Yang, F.; Zhu, J. Numerical investigation and experiment on pressure pulsation characteristics in a full tubular pump. *Renew. Energy* **2021**, *163*, 987–1000. [CrossRef]
10. Guan, D.; Wu, J.H.; Jing, L.; Hilton, H.H.; Lu, K. Kinematic modeling, analysis and test on a quiet spherical pump. *J. Sound Vib.* **2016**, *383*, 146–155. [CrossRef]
11. Dring, R.; Joslyn, H.; Hardin, L.; Wagner, J. Turbine rotor-stator interaction. *ASME J. Eng. Power* **1982**, *104*, 729–742. [CrossRef]
12. Chalghoum, I.; Kanfoudi, H.; Elaoud, S.; Akrout, M.; Zgolli, R. Numerical Modeling of the Flow Inside a Centrifugal Pump: Influence of Impeller–Volute Interaction on Velocity and Pressure Fields. *Arab. J. Sci. Eng.* **2016**, *41*, 1–14. [CrossRef]
13. Arndt, N.; Acosta, A.; Brennen, C.; Caughey, T. Experimental investigation of rotor-stator interaction in a centrifugal pump with several vaned diffusers. *ASME J. Turbomach.* **1990**, *112*, 98–108. [CrossRef]
14. Brennen, C.; Franz, R.; Arndt, N. Rotor/Stator Unsteady Pressure Interaction. 1988. Available online: https://www.researchgate.net/publication/30759128_RotorStator_Unsteady_Pressure_Interaction (accessed on 22 September 2022).

15. Khalifa, A.; Al-Qutub, A.; Ben-Mansour, R. Experiments on pressure fluctuations in a high pressure-double volute centrifugal pump under part load conditions. In Proceedings of the 4th International Conference on Thermal Energy: Theory and Applications, Abu-Dhabi, United Arab Emirates; 2009. Available online: https://www.researchgate.net/publication/281902977_Experiments_on_Pressure_Fluctuations_in_a_High_Pressure-Double_Volute_Centrifugal_Pump_under_Part_Load_Conditions (accessed on 22 September 2022).
16. Solis, M.; Bakir, F.; Khelladi, S.; Noguera, R. Numerical study on pressure fluctuations reduction in centrifugal pumps: Influence of radial gap and splitter blades. *Int. Sch. Res. Not.* 2011, 2011, pp. 1–14. Available online: https://downloads.hindawi.com/archive/2011/479594.pdf (accessed on 22 September 2022).
17. Rutter, R.; Sheth, K.; O'Bryan, R. Numerical flow simulation and validation of an electrical submersible pump//Fluids Engineering Division Summer Meeting. *Am. Soc. Mech. Eng. Digit. Collect.* **2013**, *55546*, V01AT03A005.
18. Bai, L.; Zhou, L.; Han, C.; Zhu, Y.; Shi, W. Numerical study of pressure fluctuation and unsteady flow in a centrifugal pump. *Processes* **2019**, *7*, 354. [CrossRef]
19. Muggli, F.A.; Holbein, P.; Dupont, P. CFD calculation of a mixed flow pump characteristic from shutoff to maximum flow. *J. Fluids Eng.* **2002**, *12*, 798–802. [CrossRef]
20. Wang, H.; Hu, Q.; Yang, Y.; Wang, C. Performance Differences of Electrical Submersible Pump under Variable Speed Schemes. *Int. J. Simul. Model.* **2021**, *20*, 76–86. [CrossRef]
21. Maitelli, C.W.S.P.; Bezerra, F.V.M.; Mata, W. Simulation of flow in a centrifugal pump of esp systems using computational fluid dynamics. *Braz. J. Pet. Gas* **2010**, *4*, 001–009.
22. Hammer, S.; Phan, D.; Peter, J.; Werder, T.; Meyer, R.; Liebich, R.; Thamsen, P. Active Flow Control by Adaptive Blade Systems in Periodic Unsteady Flow Conditions. 2014. Available online: http://www.depositonce.tu-berlin.de/handle/11303/197 (accessed on 22 September 2022).
23. Joslyn, D.; Dring, R. Three-Dimensional Flow in an Axial Turbine: Part 1—Aerodynamic Mechanisms. *J. Turbomach.* **1992**, *114*, 61–70. [CrossRef]
24. Iino, T.; Kasai, K. An analysis of unsteady flow induced by interaction between a centrifugal impeller and a vaned diffuser (1st report, Measurement of pressure fluctuations in pump impellers). *Trans. Jpn. Soc. Mech. Eng. Part B* **1985**, *51*, 3748–3753. [CrossRef]
25. Zhang, M.; Wang, H.; Tsukamoto, H. Unsteady Hydrodynamic Forces due to Rotor-Stator Interaction on a Diffuser Pump with Identical Number of Vanes on the Impeller and Diffuser. *Trans. Jpn. Soc. Mech. Eng. Part B* **2005**, *127*, 2020–2027. [CrossRef]
26. Zhou, L.; Shi, W.; Lu, W. Performance analysis on deep-well centrifugal pump guide vanes based on numerical simulation. *Trans. Chin. Soc. Agric. Eng.* **2011**, *27*, 38–42.
27. Yang, Y.; Zhou, L.; Shi, W.; He, Z.; Han, Y.; Xiao, Y. Interstage difference of pressure pulsation in a three-stage electrical submersible pump. *J. Petrol. Sci. Eng.* **2021**, *196*, 107653. [CrossRef]
28. Bai, L.; Yang, Y.; Zhou, L.; Li, Y.; Xiao, Y.; Shi, W. Optimal design and performance improvement of an electric submersible pump impeller based on Taguchi approach. *Energy* **2022**, *252*, 124032. [CrossRef]
29. Yang, Y.; Zhou, L.; Hang, J.; Du, D.; Shi, W.; He, Z. Energy characteristics and optimal design of diffuser meridian in an electrical submersible pump. *Renew. Energy* **2021**, *167*, 718–727. [CrossRef]
30. Kan, K.; Zhang, Q.; Zheng, Y.; Xu, H.; Xu, Z.; Zhai, J.; Muhirwa, A. Investigation into Influence of Wall Roughness on the Hydraulic Characteristics of an Axial Flow Pump as Turbine. *Sustainability* **2022**, *14*, 8459. [CrossRef]
31. Zuo, Z.; Tan, L.; Shi, W.; Chen, C.; Ye, J.; Francis, E.M. Transient Characteristic Analysis of Variable Frequency Speed Regulation of Axial Flow Pump. *Sustainability* **2022**, *14*, 11143. [CrossRef]
32. Zhang, Y.; Zhang, Z.; Zheng, J.; Zhang, J.; Zheng, Y.; Zang, W.; Lin, X.; Fernandez-Rodriguez, E. Experimental investigation into effects of boundary proximity and blockage on horizontal-axis tidal turbine wake. *Ocean. Eng.* **2021**, *225*, 108829. [CrossRef]
33. Zhang, Y.; Zang, W.; Zheng, J.; Cappietti, L.; Zhang, J.; Zheng, Y.; Fernandez-Rodriguez, E. The influence of waves propagating with the current on the wake of a tidal stream turbine. *Appl. Energy* **2021**, *290*, 116729. [CrossRef]
34. Han, Y.; Zhou, L.; Bai, L.; Shi, W.; Agarwal, R. Comparison and validation of various turbulence models for U-bend flow with a magnetic resonance velocimetry experiment. *Phys. Fluids* **2021**, *33*, 125117. [CrossRef]
35. El-Emam, M.A.; Zhou, L.; Yasser, E.; Bai, L.; Shi, W. Computational methods of erosion wear in centrifugal pump: A state-of-the-art review. *Arch. Comput. Method. E.* **2022**, *29*, 1–26. [CrossRef]
36. Deyou, L.; Hongjie, W.; Gaoming, X.; Ruzhi, G.; Xianzhu, W.; Zhansheng, L. Unsteady simulation and analysis for hump characteristics of a pump turbine model. *Renew. Energy* **2015**, *77*, 32–42. [CrossRef]
37. Gao, B.; Zhang, N.; Li, Z.; Ni, D.; Yang, M. Influence of the blade trailing edge profile on the performance and unsteady pressure pulsations in a low specific speed centrifugal pump. *J. Fluids Eng.* **2016**, *138*, 051106. [CrossRef]
38. Zhao, X.; Xiao, Y.; Wang, Z.; Luo, Y.; Cao, L. Unsteady flow and pressure pulsation characteristics analysis of rotating stall in centrifugal pumps under off-design conditions. *J. Fluids Eng.* **2018**, *140*, 021105. [CrossRef]
39. Yang, Y.; Zhou, L.; Han, Y.; Hang, J.; Lv, W.; Shi, W.; He, Z.; Pan, B. Pressure pulsation investigation in an electrical submersible pump based on Morlet continuous wavelet transform. *Proc. Inst. Mech. Eng. Part C: J. Mech. Eng. Sci.* **2021**, *235*, 6069–6079. [CrossRef]
40. Zhu, J.; Zhu, H.; Zhang, J.; Zhang, H.Q. Intelligent fault diagnosis of hydraulic piston pump based on wavelet analysis and improved alexnet. *Sensors* **2021**, *21*, 549. [CrossRef] [PubMed]

sustainability

MDPI

Article

Effect of Mooring Parameters on Dynamic Responses of a Semi-Submersible Floating Offshore Wind Turbine

Baolong Liu [1,2] and Jianxing Yu [1,2,*]

1 State Key Laboratory of Hydraulic Engineering Simulation and Safety, School of Civil Engineering, Tianjin University, Tianjin 300354, China
2 Tianjin Key Laboratory of Port and Ocean Engineering, School of Civil Engineering, Tianjin University, Tianjin 300354, China
* Correspondence: yjx2000@tju.edu.cn

Abstract: Based on a new semi-submersible floating offshore wind turbine (FOWT), a coupling aero-hydro-flexible model was established to study its dynamic behaviors, as well as the corresponding mooring system, under complicated sea scenarios. The aerodynamic load, the wave load, the current load, and the mooring load were taken into consideration. To further investigate the influence of the mooring parameters on the floating system, the diameter and the total length of mooring lines, which are the most critical parameters in mooring line design, were chosen to be analyzed. Particularly, five diameters and seven lengths were adopted to establish the FOWT mooring system, and a time-domain simulation was carried out for each cases. Based on the numerical simulations, their influences on the mooring system stiffness and the dynamic responses of FOWT were studied. The results show that the diameter has little influence on the static shape of the mooring line. The mooring system stiffness can be effectively increased by reducing the length and increasing the diameter of mooring lines. Moreover, the surge motion of floating foundation can be effectively controlled by increasing the mooring line diameter and decreasing mooring line length under the rated sea scenario. From this aspect, the dynamic response features of the FOWTs could be improved.

Keywords: mooring system; dynamic response; floating offshore wind turbine; semi-submersible platform

Citation: Liu, B.; Yu, J. Effect of Mooring Parameters on Dynamic Responses of a Semi-Submersible Floating Offshore Wind Turbine. *Sustainability* 2022, *14*, 14012. https://doi.org/10.3390/su142114012

Academic Editors: Yuquan Zhang, Jianhua Zhang, Ling Zhou, Bin Huang and Firoz Alam

Received: 16 August 2022
Accepted: 26 October 2022
Published: 27 October 2022

1. Introduction

With the traditional fossil energy crisis and environmental pollution becoming serious, renewable energy development has become one of the key approaches to overcome this predicament. Wind energy, as an easily accessible and huge-reserve renewable energy, has achieved rapid development in the past decades. Compared with onshore wind, the offshore wind speed is higher and its turbulence intensity is lower. In addition, offshore wind turbines occupy little farming resources, which releases the negative impact on the environment. Therefore, it is suitable for large-scale development and utilization. At present, with the exploitation of the offshore resources in in the China coastal shallow-water areas becoming saturated, the development of offshore wind resources has gradually moved to the far-reaching sea area, which water depths exceed 50 m.

A complete offshore wind turbine system usually consists of a turbine, a tower, a platform, and a mooring system. As the structure for energy conversion, the turbine extracts energy from the surrounding flow, which are divided into water turbines and wind turbines according to the different fluids. Based on the hydrodynamic and structural dynamics analysis, there are many research studies on water turbines and wind turbines. Zhang et al. [1,2] studied the influence of the support structure and bypass flow on the turbine wake through model experiments, expanding the knowledge of the turbine wake under waves and currents [3]. In offshore wind farms, multiple wind turbines are usually installed in one area in order to obtain power more effectively, which leads to a wake interaction. Rezvane et al. [4] proposed an innovative method to study the inflow velocity of wind turbines after considering the multiple wake effects. Giovanni et al. [5] studied the influence of turbulence on the wake dynamic mode of NREL-5MW turbine and the wind field layout is optimized according to the analysis wake model.

The present floating foundations of the FOWT are derived from the experience of offshore oil and gas platforms. The common floating platform types include semi-submersible, Spar, and TLP [6]. At present, scholars worldwide have carried out relevant research on the dynamic responses of floating wind energy facilities. For example, based on the computational fluid dynamics (CFD) method, Li et al. [7] calculated the nonlinear hydrodynamic parameters including the second-order difference-frequency wave load, the radiation wave load of the semi-submersible wind turbine platform. They found that the dynamic performance of a floating platform predicted by the modified hydrodynamic coefficient was in better agreement with the experimental results. Jiang et al. [8] proposed a stepped Spar type FOWT platform, and verified its applicability in both 50 m and 100 m water-depth conditions, according to a model test and numerical analysis. A conceptual design of reduced-draft FOWT was proposed by Zhang et al. [9], to operate in a 70 m water-depth area, and its dynamic behaviors were predicted by in their in-house numerical simulation tool. It was found that increasing the diameter of the anti-motion structure in the bottom can effectively reduce the FOWT pitch motion, as well as the mooring tension, thus improving the safety and robustness of the FOWT system. Beshbichi et al. [10] studied the dynamic response of a two-rotor wind turbine mounted on a spar floating platform by comparing it with the baseline OC3 single-rotor design. According to their results, the two-rotor design has an advantage of mass saving, but a disadvantage in motion, that is the low-frequency yaw response is increased at the rated operating wind speed. Ren et al. [11] calculated the hydrodynamic parameters and motion responses of a TLP FOWT in the case of tendon failure. The numerical prediction is based on the software WAMIT and ANSYS/AQWA. It was found that the heave, pitch, and roll motions of the FOWT were all affected with tendon failure scenarios. In addition, the natural frequency of the platform and the tension of the unbroken tendon were also affected.

As a permanent mooring structure, the FOWT maintains the energy generation in the predetermined operating area under the restriction of the mooring system to prevent the drift motion under the action of wind and waves. To some extent, the generation efficiency of the FOWT depends on the performance of the mooring system. Therefore, the mooring system of FOWT has been studied extensively. Hsu et al. [12] used commercial software to analyze the dynamic mooring load of a semi-submersible FOWT under the operating and extreme conditions. They found that the impact load might be observed in the mooring line under extreme conditions. The aerodynamic–hydrodynamic–mooring coupling dynamic analysis models were successively established by Bae et al. [13] and Li et al. [14], respectively. The dynamic response of the DeepCwind semi-submersible FOWT and the Hywind Spar FOWT were analyzed with a single mooring line broken. It was found that the mooring line failure would cause a large drift motion of the floating platform, while the mooring system is not redundancy. Yang et al. [15] demonstrated the transient dynamic response of a new multi column FOWT in the case of tendon fracture, and found that the tendon fracture connecting between the buoyancy tank and the ballast tank would have a significant influence on the pitching motion. In order to reduce the dynamic response of the semisubmersible FOWT, Giulio Ferri et al. [16], proposed a configuration optimization method of the platform and mooring system by establishing an effective frequency domain simulation model. The results show that the optimized structure has better performance and smaller platform size.

Moreover, the parametric studies have been conducted to investigate the influence of the catenary mooring line parameter on the dynamic performance of the floaters. For example, Ghafari and Dardel [17] established the numerical model of a moored semi-submersible platform model and the effects of mooring line stretch and buoy size are studied in both time and frequency domain. They found that the wider stretch and smaller buoy size could reduce the platform motions, but enlarge the tensions in the mooring lines. Wu et al. [18] numerically investigated the effect of three different mooring line types and two auxiliary mooring system on the floating current turbine which was adopted to capture the hydrokinetic energy from Kuroshio current. Their results showed that the system robustness could be achieved when the mooring mass density could exceed the corresponding thresholds. Ali et al. [19] performed sensitivity analysis on the mooring system of a deep-water turret-moored FPSO. The influences of water depth, diameter, and the hydrodynamic coefficients of the mooring line were studied. It could be observed from their results that the water depth and mooring line diameter had significant effects on surge motion reduction, but the hydrodynamic coefficients had little impact on the FPSO motion. Huang and Yang [20] optimized the mooring system of OC4 semi-submersible FOWT depending on the shallow water area with five different water depths ranging from 50 m to 100 m, based on the design rules or guidelines issued by Det Norske Veritas and American Petroleum Institute. Both the ultimate strength and fatigue damage of the catenary mooring lines were discussed via different met-ocean data. Particularly, the cost of each mooring system was carefully examined and compared. Ja'e et al. [21] further conclude the design method and the optimization techniques of the mooring system. They found that some evolutionary algorithms showed great potential and had been used for the mooring line parameter optimization works.

In the present study, a new-type semi-submersible floating platform with pontoon and three columns was adopted. For safety reasons, three groups of three mooring lines were designed, unlike the Hywind Scotland and Windfloat Atlantic. The prototype FOWT was established in the South China Sea. Moreover, previous research on the mooring system of FOWTs usually focuses on the performance of the particular mooring system under different sea conditions. In this work, more attentions are paid to the mooring line parameters' influence on the dynamic behaviors of FOWT. We hope to provide reference for the design of other floating platforms and their corresponding mooring systems. Therefore, considering different mooring line diameters and lengths, which are the most critical parameters of the mooring line design, the effect of the mooring system on the stiffness and the dynamic response of the floating wind turbine system was analyzed based on the numerical simulation method. A coupled model with wind turbine, floating platform, and mooring system was conducted to perform the dynamic response simulations of a new semi-submersible FOWT via different mooring lines.

In the following section, the numerical algorithms are briefly described. The FOWT system is clarified in Section 3. The simulation results of the FOWT with different mooring system are carefully discussed in Section 4. Finally, the conclusions are drawn in Section 5.

2. Methodology

2.1. Aerodynamic Loads

The blade element momentum theory is used to calculate the aerodynamic loads on the wind turbine. It is assumed that the blade is discretized into a plurality of independent blade elements along the length direction of the blade, and the airfoil remains unchanged along the same element. When the pressure difference is generated after the wind flow passes through the rotor, the axial and tangential induced velocities at each blade element can be calculated through the momentum loss of the wind flow, and then the thrust dT and torque dQ on the blade element can be determined by the following formula,

$$\begin{aligned} dT &= \tfrac{1}{2}\rho W^2 BC(C_l\cos\phi + C_d\sin\phi)dr \\ dQ &= \tfrac{1}{2}\rho W^2 BC(C_l\sin\phi - C_d\cos\phi)rdr \end{aligned} \quad (1)$$

where ρ is the air density, W is the wind speed, B is the number of the blades, C is the chord length of the blade element profile, C_l and C_d represent the lift and drag coefficient of the blade element at the corresponding angle of attack, ϕ is the angle of incidence, and r is the distance between the blade element and the hub center. The aerodynamic load on a single blade can be obtained by integrating the distributed load along the blade span direction.

2.2. Wave Loads

In this study, the three-dimensional potential flow theory was used to calculate the hydrodynamic loads on the floating platform. This theory was developed from the linear potential flow theory, and is also one of the common tools to solve the hydrodynamic coefficient of the floating body without forward speed.

According to the linear potential flow theory, the velocity potential Φ of the floating body in the flow field is composed of three parts: the undisturbed incident velocity potential Φ_w, the diffraction velocity potential Φ_d under the assumption that the floating body is stationary, and the radiation velocity potential Φ_j caused by the six degree of freedom motion of the floating body. These three velocity potential terms all satisfy Laplace equation. By combining with the boundary conditions of the flow field, the velocity of the floating body can be obtained, and then the wave force on the wet surface of the floating body can be obtained by numerical integration.

At the same time, because the second-order wave load easily generates the second-order drift force of the low-frequency resonance of the mooring system, the Newman approximation method was used to deal with the second-order differential frequency wave load on the floating platform. Hereby, the wave loads could be achieved based on the following equations,

$$F_{wave\_1}(t) = \mathrm{Re}\left[\sum_{i=1}^{M} \eta_i F_1(\omega_i)\right] \tag{2}$$

$$F_{wave\_2d}(t) = \mathrm{Re}\left[\sum_{i=1}^{M}\sum_{j=1}^{M} \eta_i \eta_j^* F_{2d}(\omega_i, \omega_j)\right] \tag{3}$$

where η_i and η_j^* donate the elevation and its conjugation of i-th wave component in complex domain, $F_1(\omega)$ is the linear wave load transfer function, and the $F_{2d}(\omega_i,\omega_j)$ is difference-frequency quadric transfer function calculated by Newman approximation method.

2.3. Current Loads

The main types of ocean current include shear flow and uniform flow. The sea condition in this study is uniform flow. In uniform flow, the flow velocity at different points on the same streamline is equal, the flow velocity distribution on each cross-section is the same, and the average cross-section flow velocity is constant. The flow load acting on the surface of the floating body is estimated by the drag force formula as follows,

$$F_{cur} = \frac{1}{2} C_D \rho_W A V_{cur}^2 \tag{4}$$

where, ρ_W is the density of sea water, C_D is the drag force coefficient related to the shape, A is the force area of the floating body in the direction of positive projection of velocity, and V_{cur} is the relative velocity of the body.

2.4. Mooring Loads

The internal forces of mooring lines mainly affected by axial tension can be calculated by the finite element method as shown in Figure 1. The bar elements are expressed by the overall Lagrange formula. The expression is based on the Bergan method and is modified according to the comprehensive cross-section force and small strain theory. Based on the small strain theory, it can be deduced that the axial force N of the unit is:

$$N = \frac{L - L_0}{L_0}(EA) \tag{5}$$

where L_0 is the initial stress-free unit length and EA is the axial stiffness. The tangential stiffness relation of the unit is derived by incremental form of virtual work principle:

$$\Delta S_{int} = (k_G + k_M)\Delta v \tag{6}$$

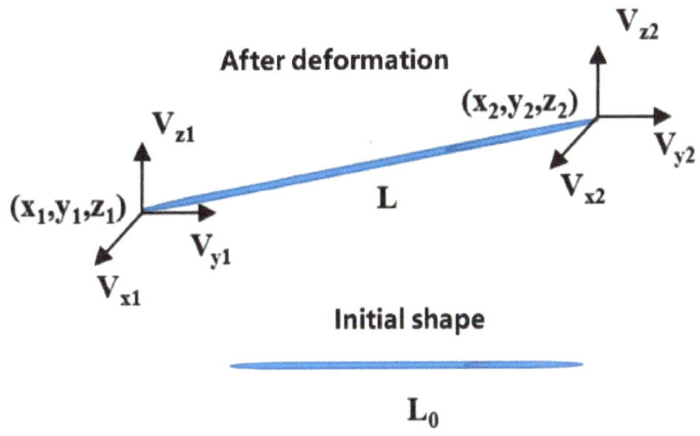

Figure 1. Model of bar element.

In this equation, ΔS_{int} is the increment of internal force vector, and k_G and k_M are the geometric stiffness matrix and the material stiffness matrix, respectively. Δv is the increment of displacement vector.

2.5. Equation of Motion in the Time Domain

For a FOWT, the platform motion governing equation in the time domain can be written as following,

$$(\mathbf{M}+\mathbf{A}_\infty)\ddot{x}(t)+\int_0^t h(t-\tau)\dot{x}(\tau)d\tau+\mathbf{D}f(\dot{x})+\mathbf{K}(x)x=q(t,x,\dot{x}) \quad (7)$$

where **M** is the body mass matrix, **A** is the frequency-dependent attached water quality matrix, *h(t)* is the retardation function, **D** is the nonlinear damping matrix and **K** is the restoring stiffness matrix of the floating body. $x, \dot{x}, \ddot{x}$ represent the displacement, velocity, and acceleration vector of the floating body, respectively. q is the external excitation load, including the first and second order wave loads, nonlinear restoring forces provided by the mooring lines, and aerodynamic loads on the wind turbines.

The added mass and potential flow damping calculated by the three-dimensional potential flow theory are frequency-dependent, while the wave frequency components are complex under irregular wave conditions, so the added mass and potential flow damping corresponding to a certain frequency cannot be directly determined and used in the time domain equation. On this basis, the added mass and potential flow damping relating to the frequency are transformed into convolution function by convolution integration. Fourth-order Runge–Kutta numerical method was used to solve the motion response of floating platform.

3. Description of the New-Type Semi-Submersible FOWT

3.1. The Semi-Submersible Floating Platform

In previous studies on semi-submersible FOWT, the OC4 DeepCWind FOWT are usually adopted. However, in this study, we adopted a novel semi-submersible floating foundation which was composed of columns, heave plates, lower floating bodies, and square cross braces. The whole structure is arranged in an equilateral triangle for better stability. The basic structure type of the floating platform is shown in Figure 2, and the main parameters are shown in Table 1.

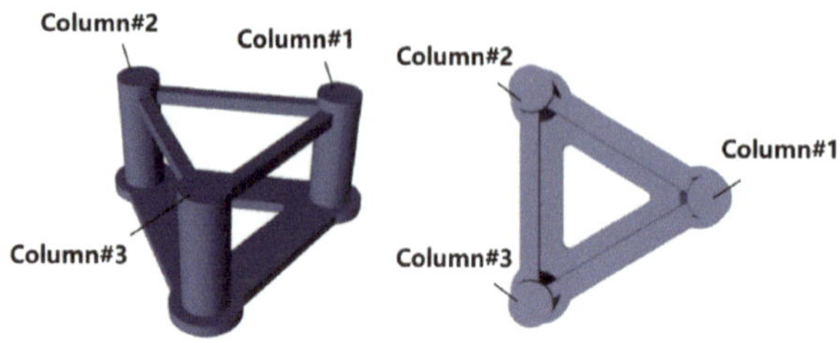

Figure 2. Floating platform.

Table 1. Main parameters of the floating platform.

Parameter	Value
Total length	71.96 m
Total width	80.0 m
Molded depth	33.0 m
Column diameter/spacing	13.0 m/60.0 m
Pontoon width/height	14.0 m/3.5 m
Heave plate diameter/height	20.0 m/3.5 m
Square cross brace length/width/height	47 m/3.5 m/3.5 m
Draft	18 m
Weight	5621.066 t

3.2. The Mooring System

Nine mooring lines were used in the present study, which were divided into 3 groups. Each group, which consisted of three mooring lines, was connected with the column and the lines were arranged in parallel. The configuration of mooring system is shown in Figure 3.

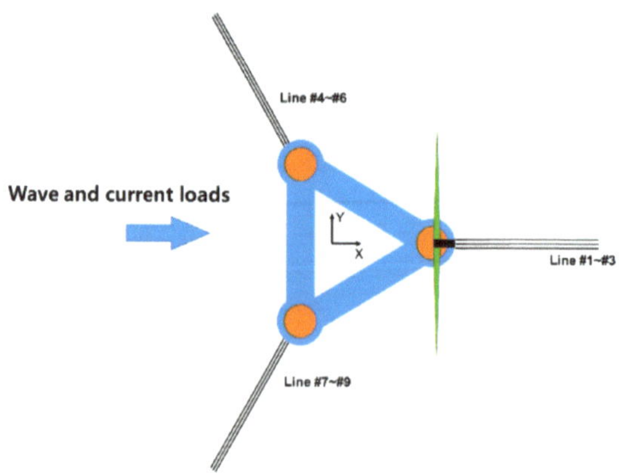

Figure 3. Configuration of mooring system.

A single mooring line was composed of three sections. The profile is displayed in Figure 4. The upper section of the splash zone was a R3S anchor chain with a length of 20 m. The middle suspension section was R3 anchor chain with a length of 131 m; the lowest lying section was M2 anchor chain with a length of 280 m. The main parameters of each section are shown in Table 2. It should be noted that the diameter means the nominal diameter of the chains, so the thickness of the chain is same as the diameter, which is required by the international standard ISO-1074. In fact, each component and its corresponding length are designed based on the guideline approved by the China Classification Society [22]. This design meets the requirements of both buoy motion and extreme tension in the lines, like the design work based on the API or DNV guidelines in another citation [20]. These results could be observed in other publications [23,24]. It should be noted that in those works, the mooring lines are simplified as two-part cable. The Upper part S3 whose minimum broken tension is larger is merged into S2. Specifically, when analyzing the effect of mooring line diameter, since a single mooring line is composed of three sections of anchor chains of different materials, the diameter of the three sections will increase/decrease the corresponding change value on the original design parameters. When analyzing the effect of mooring line length, since the tension of the upper mooring line is greater, the length of the S3 section is defined to change, and the position of the fairlead and anchor point remains unchanged.

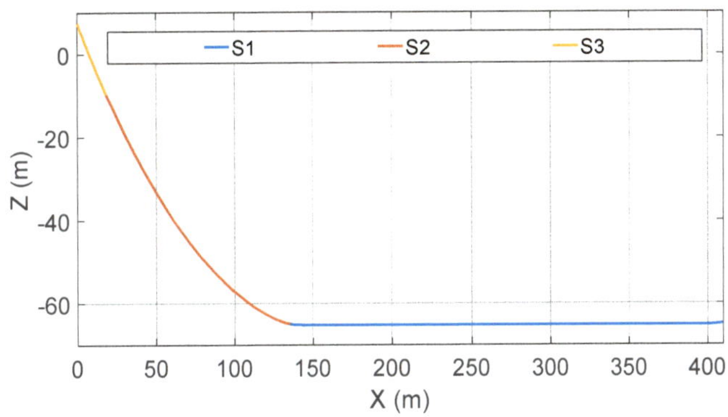

Figure 4. Profile of the mooring line.

Table 2. Main parameters of the mooring system.

Component	S1 (Lowest)	S2 (Middle)	S3 (Upper)
Level	M2	R3	R3S
Nominal Diameter (mm)	208	122	122
Length (m)	280	131	20
Dry weight (kg/m)	861	296.2	296.2
Minimum broken tension (kN)	24,140	11,365	12,690
Axial stiffness (N)	3.69×10^9	1.27×10^9	1.27×10^9

3.3. The Wind Turbine

NREL 5MW wind turbine [25] was adopted as the wind turbine in this study, but the tower was modified for the corresponding wind farm. To be specific, the tower column with height of 78 m, top diameter of 4.77 m, and bottom diameter of 6.96 m was redesigned according to the actual environmental conditions of offshore wind farms in China. In the numerical model, the tower column was divided into 30 sections which the diameter changes linearly, and the beam element was adopted to model the blades and tower. In other words, both bending and torsional moments on these slender structures were taken into consideration. Their dynamic behaviors were simulated by the Riflex module in the SIMA package. However, the nacelle and the rotor hub were modelled as the rigid bodies which include the six-DOF motion responses, respectively. The corresponding oscillations were predicted by the SIMO package. The models of wind turbine and tower column can be observed in Figure 5. Moreover, the semi-submersible foundation was modelled as the rigid body while the bar elements were adopted to establish the numerical model of all nine mooring lines. The geometry modelling and the hydrodynamic coefficient prediction of the floater were finished in SESAM, and the coupled model was established in SIMA.

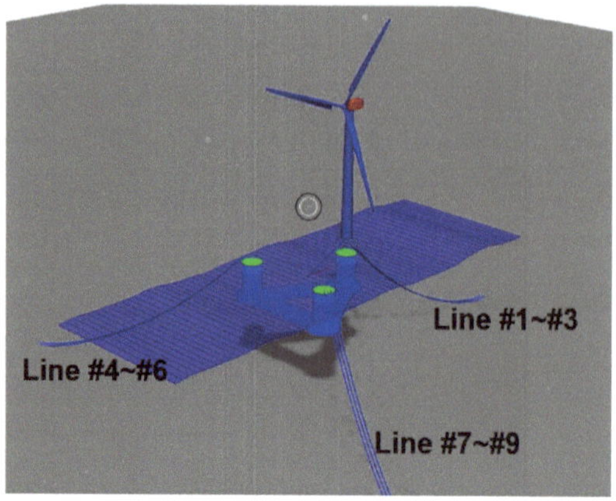

Figure 5. Dynamic simulation model in SIMA.

4. Results and Discussion

In this section, two major mooring parameters are chosen to be analyzed. One is the mooring line diameter which could significantly affect the construction cost and minimum breaking loads. The other is the mooring line length. This parameter usually affects the pretension as well as the body motion. Firstly, the static restoring stiffness of the mooring system is calculated based on the finite element algorithm with different body motions. This is followed by the dynamic response simulations under the rated operation sea scenario. To perform these simulations, SIMA, one of the SESAM numerical packages is adopted. The established numerical model is shown in Figure 5. Due to the limitation of the manuscript, we hope to brief and quick display the most significant results of the FOWT, the buoy motion and the mooring tension (like the original manuscript shows), which are directly and mainly determined by the mooring system, other than spreading out all the results to make the manuscript like an engineering report. Therefore, we select two major mooring line parameters: the diameter and the length. To be specific, five sensitivity analysis cases of the mooring line diameter with a 10 mm step and seven sensitivity analysis cases of the mooring line length with a 1 m step are adopted to perform the 3 h CPU-time simulations. To avoid the stochastic influence caused by the irregular wave and the unsteady wind, we choose 10 random seeds in every case, and the statistic results displayed in the manuscript has been chosen from the same random seed.

4.1. Static Stiffness Characteristics of the Mooring System

The stiffness is one of the important design parameters of the mooring line for the large drifts in the horizontal plane of the FOWT are mainly limited by the mooring system. Among many parameters, the diameter and length of mooring line are the most important parameters to be determined in mooring design. Therefore, the finite element method described in Section 2.4 is used to analyze the mooring line mooring type and stiffness of mooring system in this section.

4.1.1. Effect of Mooring Line Diameter on Mooring Stiffness

The change of mooring line diameter will not only change the weight of mooring line, but also change the tensile stiffness of mooring line. Based on the basic principles of material mechanics and the empirical formulas obtained from classification tests, it is assumed that the weight and tensile stiffness of the mooring line are proportional to the square of the diameter. On this basis, five different diameters of mooring lines are selected for stiffness characteristics analysis. The static configuration of a single mooring line, surge sway and yaw stiffness curves of the mooring system are shown in Figure 6a–d, respectively.

It can be found from the calculation results in Figure 6a that the static shape of a single mooring line does not change when the mooring line diameter changes. However, from the stiffness curves in Figure 6b,c, it can be seen that both surge and sway stiffness of the mooring system increases with the increasing diameter, and significant nonlinearity was found when the displacement is more than 5 m which ensuring the motion of the semi-submersible FOWT in a limited range. Moreover, according to the yaw restoring moment curves, we could observe the linearity in the stiffness. Particularly, it could be seen that we the restoring stiffness of the horizontal motions is positive relative to the mooring line diameter. In order words, the larger-diameter lines could provide larger restoring force/moment for the FOWT because of the heavier weight. However, the cost of the thicker lines is undoubtedly more expensive.

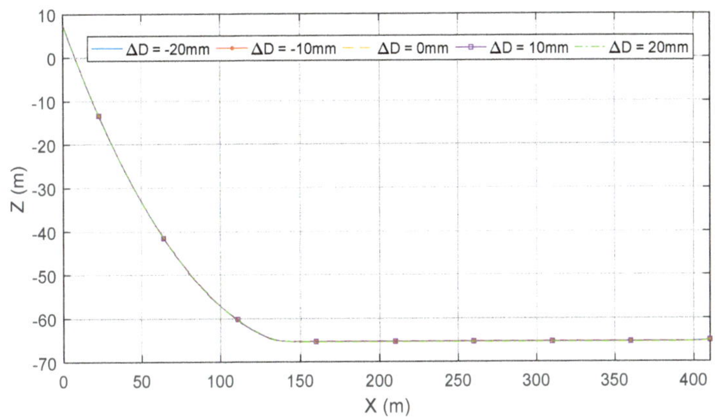

(**a**)

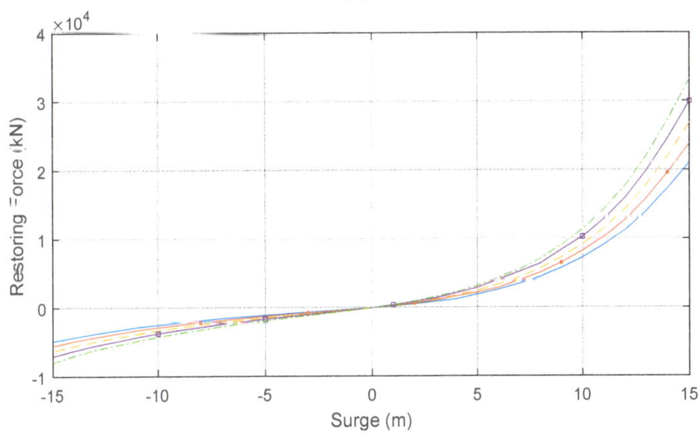

(**b**)

Figure 6. *Cont.*

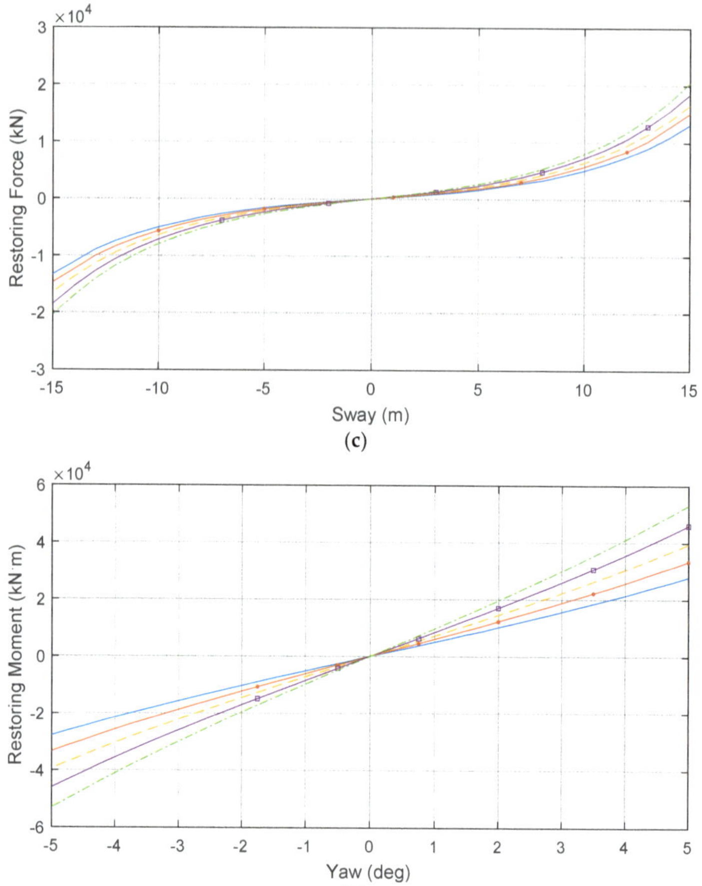

(**c**)

(**d**)

Figure 6. Effect of diameter on the static mooring stiffness. (**a**) Static mooring line configuration; (**b**) Surge Stiffness; (**c**) Sway Stiffness; (**d**) Yaw Stiffness.

4.1.2. Effect of Mooring Line Length on Mooring Stiffness

Choosing the appropriate length of mooring line is not only beneficial to guarantee positioning performance of mooring system but also to avoid excess project investment with using the too long mooring line. Therefore, the stiffness analysis of the different length is carried out with the position of the mooring pipe and anchor unchanged. On this basis, six more mooring lines in different length are selected for the stiffness analysis. The static type of the single mooring line, surge sway and yaw stiffness curves of the mooring system are shown in Figure 7a–d, respectively.

The calculation results show that the static type of the single mooring line in water changes significantly with the length changes. On the one hand, the position touching the ground of the mooring line is closer to the fairlead with the increase in length, which leads to the increase in the angle between lifting section and horizontal plane. On the other hand, the curvature of the lifting section increases with the increase in length, both of which leads to the decrease in the horizontal component of the mooring line tension. This is also confirmed by the stiffness curves of the mooring system shown in Figure 7b–d, i.e., the stiffness of the mooring system in the DOFs of surge, sway, and yaw, increases significantly as the length of the mooring line decreases. Moreover, when the mooring length is too short, the horizontal restoring stiffness is extremely large. Although this is beneficial to the positioning effect, it may cause other issues such as the large tension which will be discussed later based on the dynamic response predictions.

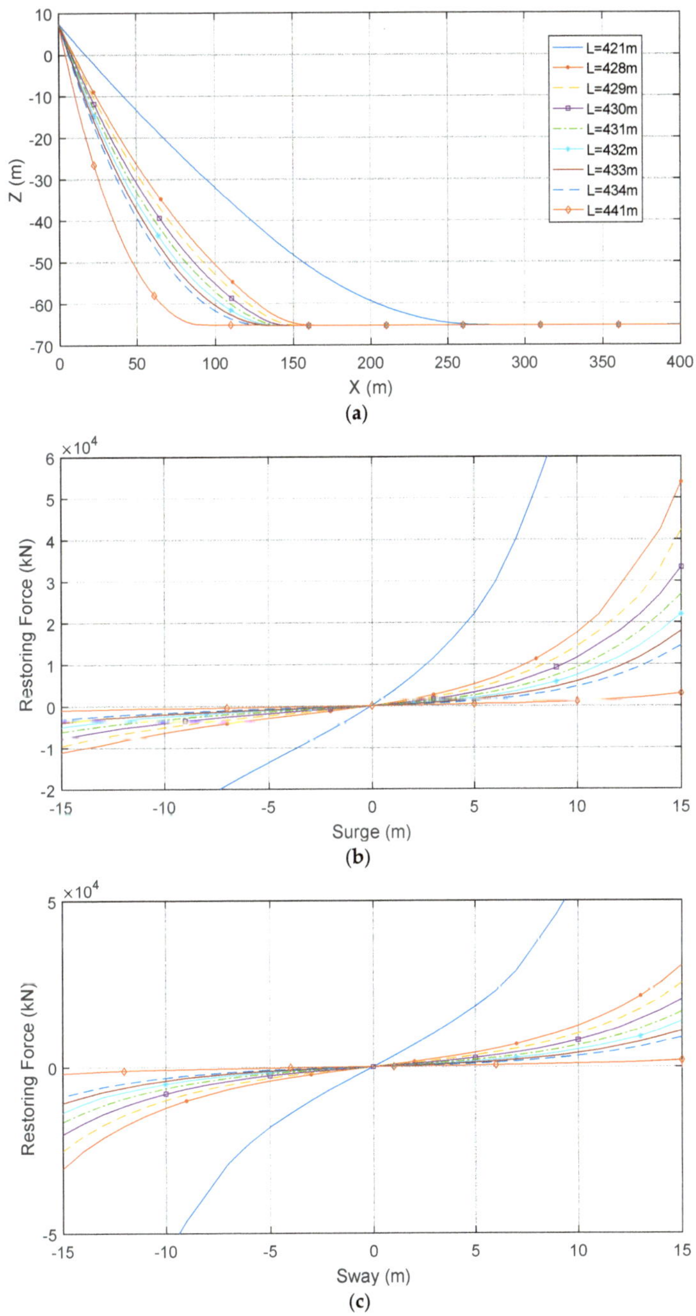

Figure 7. *Cont.*

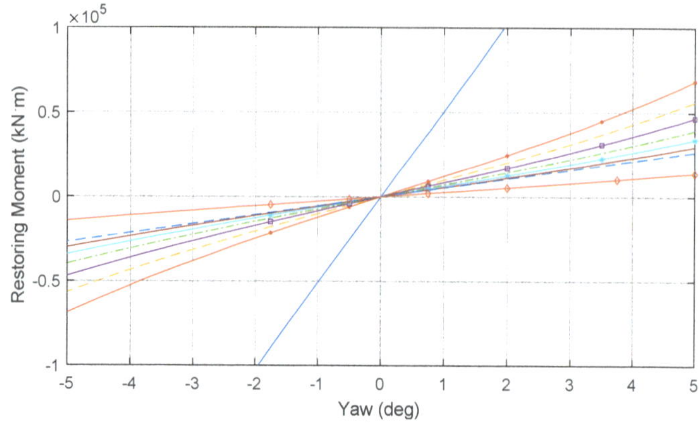

(**d**)

Figure 7. Effect of length on the static mooring stiffness; (**a**) Static mooring line configuration; (**b**) Surge Stiffness; (**c**) Sway Stiffness; (**d**) Yaw Stiffness.

4.2. Dynamic Response Analysis of the Wind Turbine System

4.2.1. Free Decay Tests

Based on the three-dimensional potential flow theory described in Section 2.2, the hydrodynamic parameters of the floating platform are calculated in the frequency domain. Then we established the wind turbine-floating buoy-mooring line coupled system in the time domain, and firstly performed the free decay tests to calculate the natural frequencies of the buoy. The relaxation curves are displayed in Figure 8 and the corresponding natural periods of the platform motion are listed in Table 3. It can be observed from the results that the natural period of the floating platform in this study is large, which far away from the peaked wave period of the operational sea condition (10.0 s), so it could be derived that the buoy oscillations will be seldom affected by the resonance caused by the high-frequency wave load.

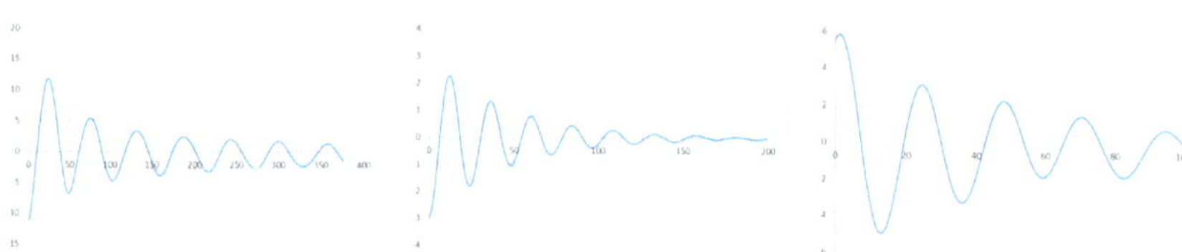

(**a**) (**b**) (**c**)

Figure 8. Time histories of the free decay tests; (**a**) Surge; (**b**) Heave; (**c**) Pitch.

Table 3. Natural periods of the FOWT.

DOF	Natural Period (s)
Surge	53.33
Heave	24.25
Pitch	26.43

4.2.2. Environmental Conditions and Simulation Results

Based on the static stiffness analysis of the mooring system, the time domain calculation of the dynamic response characteristics of the FOWT is carried out under different mooring parameters using the integrated analysis model. The rated environmental conditions are selected considering the combined action of wind, wave, and current load in which the JONSWAP spectrum is used for wave propagation along the X-axis in the positive direction, with a significant wave height of 3.0 m and a peak period of 10.0 s. The API wind spectrum is used to simulate the turbulent wind with an average wind speed of 11.4 m/s and a constant surface current of 1.0 m/s. Each environmental load is incident along the direction of 0 degrees. Therefore, three degrees of freedom motions in the longitudinal plane are mainly analyzed, i.e., surge, pitch, and heave motions. The response of semi-submersible platform motion and mooring line tension under different conditions are obtained by 3 h simulation in the time domain. Specifically, the results of the tension are chosen from the upper chain connected with the fairlead in Line#2, because of the largest tension in the whole mooring system. An example of the time histories under the original mooring parameters is shown in Figure 9. The unstable range in the first 500 s of each time history curve is removed and the response characteristics of the motion and mooring line tension in the stable range are analyzed.

4.2.3. Effect of Mooring line Diameter on Dynamic Response

Statistical results of the platform motion under different mooring line diameters are shown in Figure 10. In the bar chart, the results of each bar are expressed as the minimum, mean, maximum, and standard deviation of the corresponding statistical variable under the relevant condition.

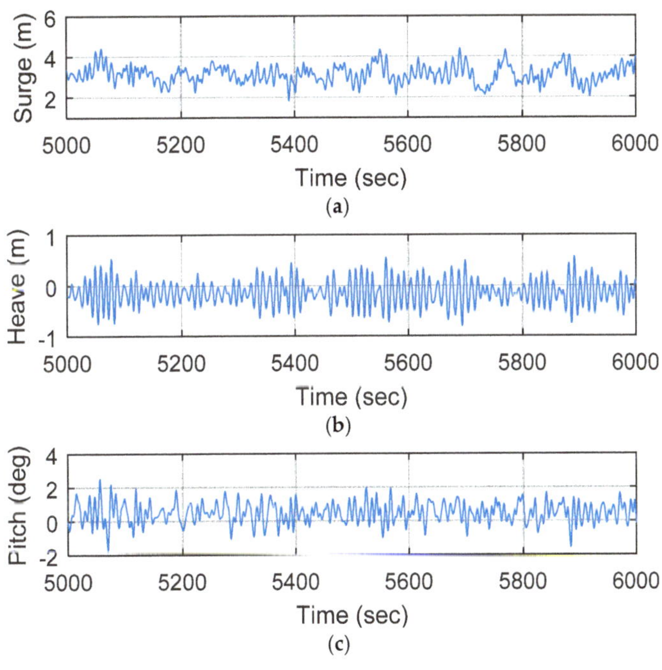

Figure 9. *Cont.*

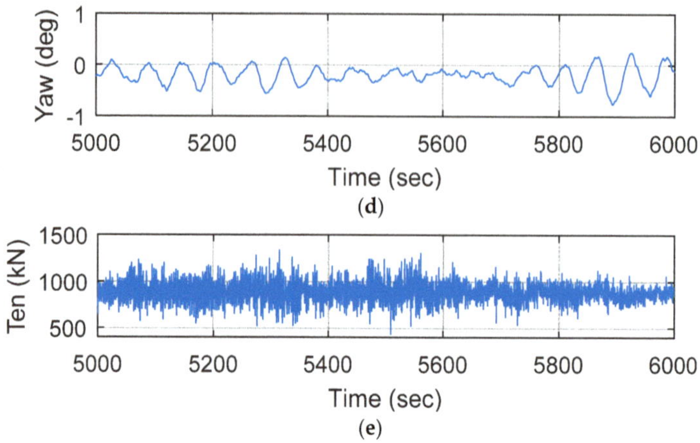

Figure 9. Time histories of dynamic responses of the semi-submersible FOWT; (**a**) Surge; (**b**) Heave; (**c**) Pitch; (**d**) Yaw; (**e**) Mooring line tension.

From the statistical results of platform motion in Figure 10a–c, it can be found that the mooring line diameter exerts a great influence on the surge and heave motion of the platform, which the surge motion decreases obviously with the increase in mooring line diameter. It can be acknowledged from the static stiffness characteristics of the mooring system that the stiffness of surge increases significantly with the increase in the mooring line diameter. Therefore, the surge motion decrease with the increase in the mooring system stiffness under the same conditions, which results in this phenomenon. At the same time, due to the increase in mooring line diameter, the mooring line mass increases, and the heave motion of the platform decreases accordingly. On the other hand, according to the statistic results of yaw motions, we could observe that the fluctuations of the yaw motion are generally decreasing with the increasing mooring line diameter. It also illustrates that the thicker and heavier mooring line could provide larger restoring stiffness which could be derived from the static results in Figure 6. The effect of mooring system on pitch motion is not significant, which shows a slightly decreasing trend with the increase in diameter. This is because the contribution of mooring system to the pitch stiffness is much smaller than that of the floating platform itself, while the coupling effect between the surge and pitch motion attributes to the slightly decreasing trend of the pitch motion. In addition, the statistical results of the tension of the mooring line shown in Figure 10e show that the tension of each mooring line increases with the increase in the diameter which due to the corresponding increase in the weight and the pre-tension of mooring line with the increase in the diameter.

4.2.4. Effect of Mooring Line Length on Dynamic Response

The statistic results of platform motion under different mooring line lengths are shown in Figure 11.

Comparing the results of Figures 10a and 11a, it can be found that the effect of the length of mooring line on the surge motion is more significant than that of the diameter. With the increase in the length of mooring line, the mean and amplitude of the surge of the floating platform are enlarged, which is due to the decrease in the stiffness of mooring system caused by the increase in length. It can be found from Figure 11b that the pitch motion of the platform is less affected by the length of the mooring line. As can be seen from the statistical results of the pitch motion shown in Figure 11c, the change in the pitch amplitude is relatively small, but it shows an enlarged trend with the growth of mooring line length. This is also observed in the statistical yaw fluctuations in Figure 11d. In addition, the comparative results of Figures 10e and 11e shows that the tension of each mooring line decreases as the length of the mooring line increases, which is different from the effect of changing the diameter of the mooring line. The reason is the mooring line becomes more relaxed with the increase in mooring line length, so the pre-tension as well as the dynamic tension of the mooring line tends to decrease under the environmental loads such as wind and wave. However, it should be noted that when the mooring line is too long, it will become too loose to moor the floating buoy, and the large drift may occur as the 441 m case shows. On the other hand, the shorter mooring line could significantly release the horizontal motions, but the line will become too tight. This will not only lead to the large mooring tension which may increase the fracture risk, but also will increase the set-down heave motion, as the results of the 421 m case displays. Therefore, it is necessary to choose a suitable mooring length during the design.

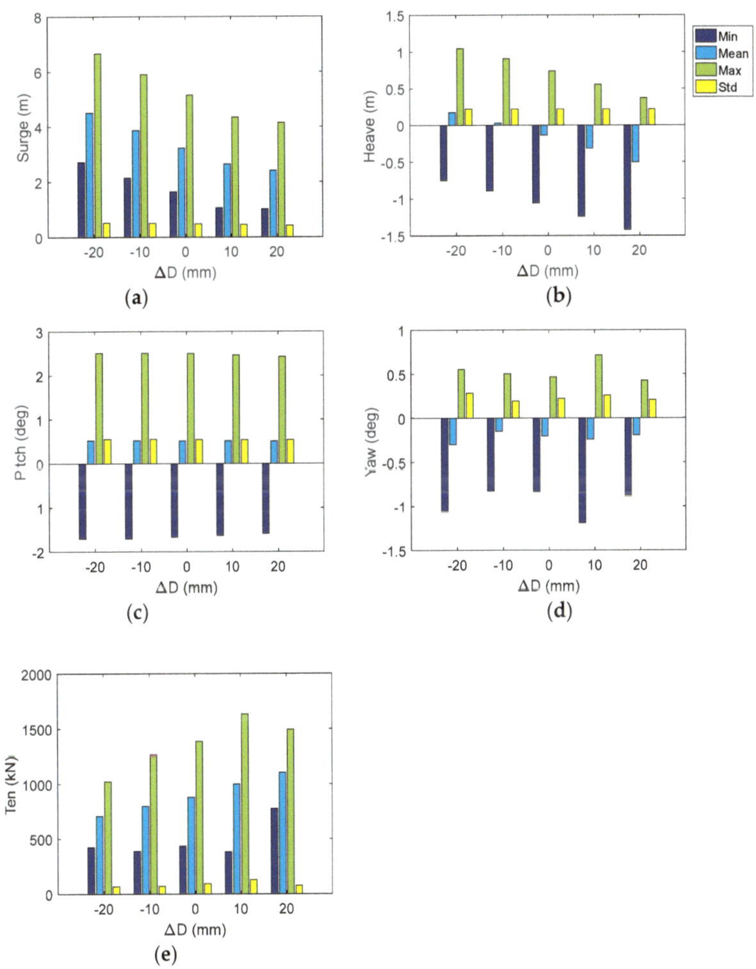

Figure 10. Statistic results dynamic response of the FOWT with different-diameter mooring lines; (**a**) Surge; (**b**) Heave; (**c**) Pitch; (**d**) Yaw; (**e**) Mooring line tension.

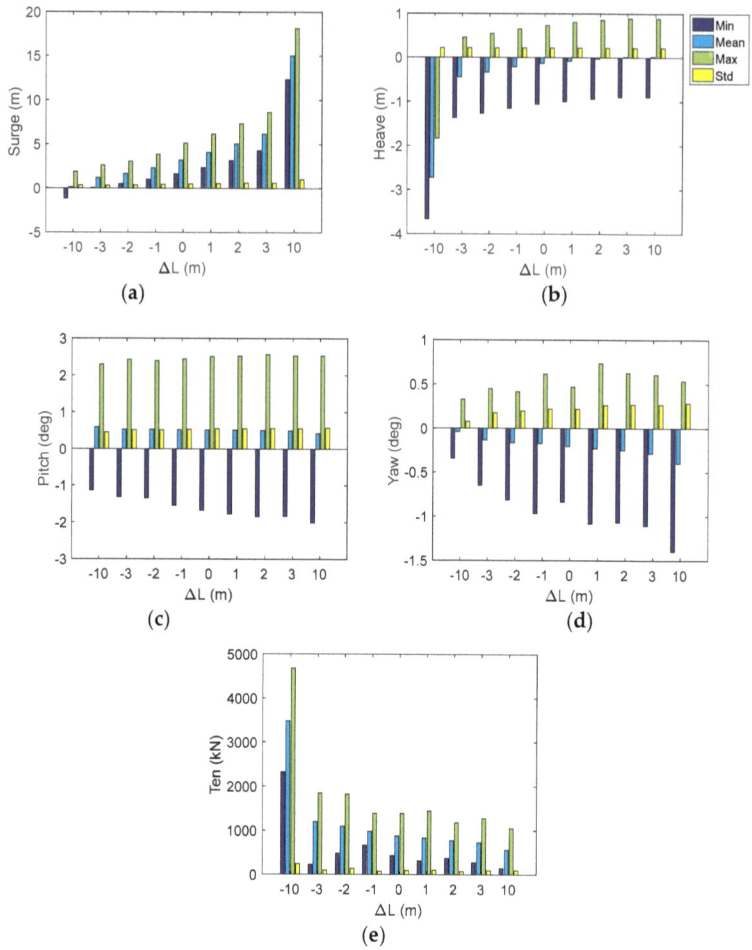

Figure 11. Statistic results dynamic response of the FOWT with different-length mooring lines; (**a**) Surge; (**b**) Heave; (**c**) Pitch; (**d**) Yaw; (**e**) Mooring line tension.

On the other hand, we observed an interesting phenomenon in the statistical results of mooring line loads in Figures 10e and 11e. The mean value of the mooring line tension varies linearly with the diameter and length, while the change in the maximum value is slightly different from the mean value. To be specific, in some particular cases, such as the changing diameter is changing length is −3 m, −1 m, 1 m, 2 m, and 3 m, we found the minimum tension is largest in the case of −1 m, while the maximum tension was observed in the case of −2 m. In fact, due to the water depth is relative shallow (65 m), heave and pitch will also influence the extreme mooring tension. According to the results in Figure 11b, it shows that the heave motion is raised by the longer mooring line, and it means the larger vertical distance from the fairlead to the anchor and leads to the larger tension in all mooring lines. However, due to the surge motion is reduced by the longer line, the tension in line #2 will be decreased because of the smaller horizontal distance between both ends. Due to these two inconsistent factors, the mooring tension shows this interesting phenomenon. Simultaneously, the extreme tension will also be slightly influenced by the stochastic extreme wave load.

5. Conclusions

Through the research on the critical mooring design parameters such as diameter and length, the static configuration of the separate mooring line and restoring stiffness of the whole mooring system are analyzed by the finite element method. The dynamic responses of the semi-submersible FOWT are calculated based on the numerical model of wind turbine-platform-mooring system, and the following conclusions are drawn. The diameter of mooring line has little effect on the static type, but the length will significantly influence the static shape and the restoring loads in the mooring line. This is caused by the different pretension among the mooring system with different length. In addition, the stiffness of mooring system can be effectively improved by reducing the length of mooring line and increasing the diameter of mooring line. On the other hand, the surge motion of FOWT can be effectively controlled by increasing the diameter of the mooring line and decreasing the length of the mooring line under the rated operational sea scenario of the semi-submersible FOWT, while the extreme tensions in the lines will both influenced by all DOF motions of the buoy, due to the relative shallow water depth.

In present simulations, the mooring parameters are chosen based on the prototype engineering design and the changes are chosen in a small range, to show the influence and its tendency of the dynamic response of FOWT when the mooring line length and diameter change. However, it should be admitted that this is an initial step to analyze the influence in the specified range. In the following works, we will adopt the optimization algorithms (such as NAGA-III algorithm [26] and evolutionary algorithm [21]) to further investigate the global optimized mooring parameters. On the other hand, in the present work, two representative parameters, length and diameter, are adopted to performed the numerical simulations. However, there are many other parameters such as pretension, line numbers, and materials [27] In the following study, we will make additional investigations on other parameters, in order to fully grasp the key mooring parameter and its influence, so that we could provide reference for the design of other floating platforms and their corresponding mooring systems.

On the other hand, although the simulations are performed under a specific sea scenario, the results may have broader suggestion on the engineering practical. However, it should be admitted that there are many other DLCs suggested by the IEC61300 are not included in present work. We will further investigate the accidental and gust DLCs in our following investigations, so that we could fully capture the whole dynamic feature of this floating renewable energy system.

Author Contributions: Conceptualization, B.L. and J.Y.; methodology, B.L.; software, B.L.; validation, B.L. and J.Y.; formal analysis, B.L.; investigation, B.L.; resources, B.L.; data curation, B.L.; writing—original draft preparation, B.L.; writing—review and editing, B.L.; visualization, B.L.; supervision, B.L.; project administration, B.L.; funding acquisition, J.Y. All authors have read and agreed to the published version of the manuscript.

Funding: This research was funded Fund of the State Key Laboratory of Ocean Engineering, Shanghai Jiao Tong University (No. GKZD010081).

Institutional Review Board Statement: Not applicable.

Informed Consent Statement: Not applicable.

Data Availability Statement: Not applicable.

Conflicts of Interest: The authors declare no conflict of interest.

References

1. Zhang, Y.; Zhang, Z.; Zheng, J.; Zhang, J.; Zheng, Y.; Zang, W.; Lin, X.; Fernandez-Rodriguez, E. Experimental investigation into effects of boundary proximity and blockage on horizontal-axis tidal turbine wake. *Ocean Eng.* **2021**, *225*, 108829. [CrossRef]
2. Zhang, Y.; Zhang, J.; Lin, X.; Wang, R.; Zhang, C.; Zhao, J. Experimental investigation into downstream field of a horizontal axis tidal stream turbine supported by a mono pile. *Appl. Ocean. Res.* **2020**, *101*, 102257. [CrossRef]

3. Zhang, Y.; Zang, W.; Zheng, J.; Cappietti, L.; Zhang, J.; Zheng, Y.; Fernandez-Rodriguez, E. The influence of waves propagating with the current on the wake of a tidal stream turbine. *Appl. Energy* **2021**, *290*, 116729. [CrossRef]
4. Mirsane, R.S.; Farschad, T. An innovative method of investigating the wind turbine's inflow speed in a wind farm due to the multiple wake effect issue. *Energy Convers. Manag.* **2022**, *269*, 116077. [CrossRef]
5. De Cillis, G.; Cherubini, S.; Semeraro, O.; Leonardi, S.; De Palma, P. The influence of incoming turbulence on the dynamic modes of an NREL-5MW wind turbine wake. *Renew. Energy* **2022**, *183*, 601–616. [CrossRef]
6. Qu, X.; Li, Y.; Tang, Y.; Hu, Z.; Zhang, P.; Yin, T. Dynamic response of spar-type floating offshore wind turbine in freak wave considering the wave-current interaction effect. *Appl. Ocean Res.* **2020**, *100*, 102178. [CrossRef]
7. Li, H.; Bachynski-Polić, E.E. Validation and application of nonlinear hydrodynamics from CFD in an engineering model of a semi-submersible floating wind turbine. *Mar. Struct.* **2021**, *79*, 103054. [CrossRef]
8. Jiang, Z.; Wen, B.; Chen, G.; Xiao, L.; Li, J.; Peng, Z.; Tian, X. Feasibility studies of a novel spar-type floating wind turbine for moderate water depths: Hydrodynamic perspective with model test. *Ocean Eng.* **2021**, *233*, 109070. [CrossRef]
9. Zhang, J.; Li, Y.; Tang, Y.; Qu, X.; Huang, J. Analysis on dynamic response of new type reduced draft floating foundation wind turbine. *Acta Energ. Sol. Sin.* **2021**, *42*, 378–383. (In Chinese)
10. El Beshbichi, O.; Xing, Y.; Ong, M.C. Dynamic analysis of two-rotor wind turbine on spar-type floating platform. *Ocean Eng.* **2021**, *236*, 109441. [CrossRef]
11. Ren, Y.; Venugopal, V.; Shi, W. Dynamic analysis of a multi-column TLP floating offshore wind turbine with tendon failure scenarios. *Ocean Eng.* **2022**, *245*, 110472. [CrossRef]
12. Hsu, W.; Thiagarajan, K.P.; Manuel, L. Extreme mooring tensions due to snap loads on a floating offshore wind turbine system. *Mar. Struct.* **2017**, *55*, 182–199. [CrossRef]
13. Bae, Y.H.; Kim, M.H.; Kim, H.C. Performance changes of a floating offshore windturbine with broken mooring line. *Renew. Energy* **2017**, *101*, 364–375. [CrossRef]
14. Li, Y.; Zhu, Q.; Liu, L. Transient response of a SPAR-type floating offshore wind turbine with fractured mooring lines. *Renew. Energy* **2018**, *122*, 576–588. [CrossRef]
15. Yang, Y.; Bashir, M.; Michailides, C.; Mei, X.; Wang, J.; Li, C. Coupled analysis of a 10 MW multi-body floating offshore wind turbine subjected to tendon failures. *Renew. Energy* **2021**, *176*, 89–105. [CrossRef]
16. Ferri, G.; Marino, E.; Bruschi, N.; Borri, C. Platform and mooring system optimization of a 10 MW semisubmersible offshore wind turbine. *Renew. Energy* **2022**, *182*, 1152–1170. [CrossRef]
17. Ghafari, H.; Dardel, M. Parametric study of catenary mooring system on the dynamic response of the semi-submersible platform. *Ocean Eng.* **2018**, *153*, 319–332. [CrossRef]
18. Wu, J.T.; Chen, J.H.; Hsin, C.Y.; Chiu, F.C. Dynamics of the FKT system with different mooring lines. *Pol. Marit. Res.* **2019**, *26*, 20–29. [CrossRef]
19. Ali, M.O.A.; Ja'e, I.A.; Hwa, M.G.Z. Effects of water depth, mooring line diameter and hydrodynamic coefficients on the behaviour of deepwater FPSOs. *Ain Shams Eng. J.* **2020**, *11*, 727–739.
20. Huang, W.H.; Yang, R.Y. Water depth variation influence on the mooring line design for FOWT within shallow water region. *J. Mar. Sci. Eng.* **2021**, *9*, 409. [CrossRef]
21. Ja'e, I.A.; Ali, M.O.A.; Yenduri, A.; Nizamani, Z.; Nakayama, A. Optimisation of mooring line parameters for offshore floating structures: A review paper. *Ocean Eng.* **2022**, *247*, 110644.
22. China Classification Society. *Guidance Notes of the Floating Offshore Wind Turbine Platform*; GD29-2021; China Classification Society: Beijing, China, 2021. (In Chinese)
23. Li, H.; Zhang, K.; Dong, Y. Sensitivity analysis of dynamic response parameters of offshore floating wind turbine. *Hydropower New Energy* **2022**, *36*, 23–29. (In Chinese)
24. Zhao, Z.; Fan, L.; Kuang, X.; Zhou, S.; Cheng, H. Experimental research on coupling dynamic response of semi-submersible floating offshore wind turbine. *Shipbuild. China* **2022**, *63*, 198–206. (In Chinese)
25. Jonkman, J.; Butterfield, S.; Musial, W.; Scott, G. *Definition of a 5-MW Reference Wind Turbine for Offshore System Development*; No. NREL/TP-500-38060; National Renewable Energy Lab. (NREL): Golden, CO, USA, 2009.
26. Zhang, P.; Li, Y.; Tang, Y.; Zhang, R.; Qu, X. Study on the Dynamic Behaviours of an Articulated Offshore Wind Turbine under the Severe Sea State. In Proceedings of the 9th Conference on Computational Methods in Marine Engineering (Marine 2021), Online, 2–4 June 2021.
27. Xu, S. Mooring Design and Analysis for Offshore Platforms and Wave Energy Converters. Ph.D. Thesis, University of Lisbon, Lisbon, Portugal, 2021.

sustainability

MDPI

Article

Turbulence Characteristics in the Mixing Layer of a Submerged Cavitating Jet at High Reynolds Numbers

Yongfei Yang [1,*], Gaowei Wang [1], Weidong Shi [1,*], Wei Li [2], Leilei Ji [2] and Hongliang Wang [3]

1 School of Mechanical Engineering, Nantong University, Nantong 226019, China
2 Research Center of Fluid Machinery Engineering and Technology, Jiangsu University, Zhenjiang 212013, China
3 School of Aeronautical and Mechanical Engineering, Changzhou Institute of Technology, Changzhou 213032, China
* Correspondence: yyf2020@ntu.edu.cn (Y.Y.); wdshi@ujs.edu.cn (W.S.)

Abstract: In this paper, unsteady and time-averaged turbulence characteristics in a submerged cavitating jet with a high Reynolds number are studied using large eddy simulation. The simulation is validated by comparing the vapor distribution using CFD and a high-speed photography experiment. The results indicate that the currently used numerical method can predict the evolution of the cavitation cloud in the jet accurately. The instantaneous and time-averaged flow fields of the submerged jet with three different cavitation numbers are studied. Comparing the frequency spectral of jets with different cavitation numbers, it is found that, for a fixed location, the frequency increases with the decrease in the cavitation number. Comparing the vorticity distribution at different streamwise locations, the instability process of the ring-shapes vortexes is revealed. Comparing the shape of the cavitation cloud and the vortexes in the jet finds that their spatial distribution and the temporal evolution are similar, indicating that the dynamic characteristics of the vortex and the cavitation affect each other. For the currently investigated cavitating jets, the Reynolds number increases with the decrease in the cavitation number. However, the spreading rate is lower for the jet with higher Reynolds numbers here. This is means that the momentum exchange between the jet and submerging water is reduced by the cavitation phenomenon.

Keywords: cavitating jet; LES; mixing layer; vortex

Citation: Yang, Y.; Wang, G.; Shi, W.; Li, W.; Ji, L.; Wang, H. Turbulence Characteristics in the Mixing Layer of a Submerged Cavitating Jet at High Reynolds Numbers. *Sustainability* 2022, 14, 11963. https://doi.org/10.3390/su141911963

Academic Editors: Yuquan Zhang, Jianhua Zhang, Ling Zhou and Bin Huang

Received: 16 August 2022
Accepted: 19 September 2022
Published: 22 September 2022

1. Introduction

In recent years, water jet technology has been widely used in various industrial fields, and water jets injected into pure water environments are called submerged jets. Cavitation is one of the main features of submerged jets, which is a phase change phenomenon that reduces the liquid pressure to the local saturated vapor pressure and transforms the liquid phase into a gas phase [1–3]. The high temperature and pressure generated at the moment of cavitation bubble collapse can cause serious damage to metallic components in hydraulic machinery, but the cavitation bubbles generated by high-pressure water jets can be used for surface strengthening of metallic materials under appropriate conditions [4,5]. The cavitation behavior in submerged jets involves complex and variable physical phenomena, such as turbulence, high temperature and pressure, and phase changes. Therefore, a comprehensive understanding of the turbulence characteristics and cavitation properties of submerged jets is necessary to provide a theoretical basis for the in-depth application of water jet technology.

The current research on submerged jets is mainly based on both experimental studies and numerical simulations. In recent years, many scholars have carried out related works, mainly using High Speed Photography (HSP) to observe the cavitation phenomenon in the jet and Particle Image Velocimetry (PIV) to measure the jet flow field. Zhang Y et al. filled the gap in the literature by experimentally visualizing the internal effects of waves and turbulence in the wake of a tidal turbine. In order to achieve this, it is crucial to quantify the average wake and turbulence properties, based on the turbine power and wave properties such as the wave energy and its height. Contour plots of velocity deficit, turbulence intensity and wake anisotropy are given for different flow positions [6–8]. Keiichi Sato et al. conducted a high-speed photographic study of a cavitation jet from a contraction–expansion nozzle and obtained the low-frequency and high-frequency signals of vacuolar cluster shedding by the statistics of the cavity length with time, and also captured the diffusion collapse of the vacuolar cloud as it impinges on the material surface using a high-speed photographic camera from the vertical direction [9,10]. Yang Y et al. captured the non-constant flow characteristics of high-pressure cavitation jets at three different angles with HSP experiments, and extracted information on their growth, shedding and collapse from the high-speed images. Finally, the optimal distance and impact performance of cavitation jet reinforcement were investigated [11,12]. Nakano et al. observed the evolution of vacuoles near the submerged jet nozzle using HSP and found that, in the early stage, most of the vacuoles were within the starting vortex and connected to each other in a vortex ring. The vacuoles within the vacuole ring grew rapidly near the nozzle and then gradually decreased in size. In the later stages of jet development, individual vacuoles formed, either connected together or randomly distributed in the jet shear layer [13]. PIV technology is widely used for flow field measurements because of its non-contact measurement properties and its ability to capture transient flow field information with high accuracy. Toshihiro Sawamura et al. performed velocity measurements of the cavitation of the submerged jet using PIV to obtain the average velocity distribution in the region where cavitation exists and found that, in the average velocity distribution, the center of the jet is higher than the edge and the higher region occurs in a part of the edge [14]. Shridhar et al. tested the cavitation vacuoles within the near-flow field of the jet using PIV; they found that the cavitation appeared first in the vortex ring and predicted the probability of cavitation vacuole formation by the distribution pattern, intensity and strain of the cavitation gas nuclei, and the predicted values were in good agreement with the experimental results [15]. Gopalan et al. studied the cavitation phenomenon in submerged water jets using PIV, and the PIV observations showed that the cavitation bubbles were mainly generated at the center of the vortex in turbulent flows [16]. Mao N et al. used Time-resolved Particle Image Velocimetry (TR-PIV) to measure the submerged jet flow field and orthogonal decomposition (POD) to extract the large-scale energy-containing vortex structure, and the results show that, as the pump pressure increases, shearing between the jet edge and ambient water increases, entrainment of the surrounding fluid is enhanced, and the cavitation cloud profile fluctuates dramatically under high-pressure conditions [17].

Since high-pressure submerged water jets are usually accompanied by cavitation, and the concentration of vacuoles increases when the cavitation number of the jet is low, it is difficult to accurately measure the internal flow field by means of visual measurements such as PIV; therefore, it is important to develop numerical simulation techniques for high-pressure submerged water jets. Martin M et al. conducted numerical simulations and experimental studies for submerged rotating jets, and various RANS models were used to calculate the flow field to obtain the velocity distribution at different locations of the jet [18]. Yang Y used three different turbulence models to represent the RANS and RANS-LES hybrid models and compared the probabilities of these models to simulate microscale vortex structures. By analyzing the relationship between small-scale vortices and cavitation formation, the probability of these models to simulate strong shear stress high-pressure cavitation jets is discussed [19,20]. Thus, LES is a feasible method to simulate the jet flow field. Yong Wang et al. used large eddy simulation to simulate the heat transfer process of the impinging jet and obtained the accurate heat transfer process of the jet cooling device by numerical calculation; they proposed a numerical simulation-based design method for the jet cooling device [21]. Y. C. Wang et al. used large eddy simulation combined with the Schnerr–Sauer cavitation model to study the high-Reynolds-number three-dimensional flow structure of a submerged jet in an axisymmetric cavity. The computational results capture the transient vortex structure and the three-dimensional flow structure in the shear layer. The growth of the jet's shear layer is not linearly diffused by the impactor [22,23].

In this paper, unsteady and time-averaged turbulence characteristics in a submerged cavitating jet with high Reynolds numbers are studied using large eddy simulation. The simulation is validated by comparing the vapor distribution using CFD and a high-speed photography experiment. Both instantaneous and time-averaged flow fields of the submerged jet with three different cavitation numbers are studied.

2. Experiment and CFD Method

2.1. Experiment Apparatus and Method

The unsteady characteristics and velocity field of the jet are tested using high-speed photography and particle image velocimetry (PIV). The experimental apparatus is shown in Figure 1. The pressure of the system is supplied by a plunger pump. The rated pressure is 50 MPa, and the rated flow rate is 15 L/min. The main part of the test bench is composed of upper and lower water tanks, and the upper tank is made of transparent material polymethyl methacrylate. During the experiment, the upstream pressure of the nozzle can be changed by controlling the rotating speed of the pump. For the high-speed photography, an Olympus Ispeed Camera is used to capture the image of the cavitation cloud, with a frequency of 20,000 fps. Under this frame rate, images with a pixel size of 352 × 272 can be obtained, which are finally analyzed by image-processing algorithms in MATLAB.

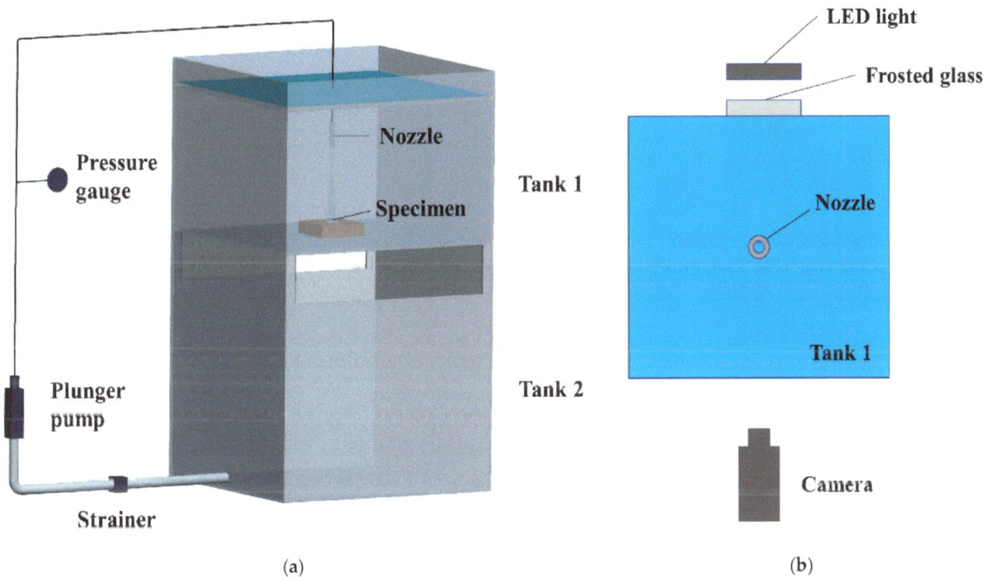

Figure 1. Experimental apparatus for cavitation jet. (**a**) High-pressure jet system; (**b**) High-speed photography.

2.2. Governing Equations and LES Approach

To reveal the structure of the vortex and the velocity fluctuations in the shear layer, the LES approach is used for the simulation of the submerged cavitation jet under high Reynolds numbers. Continuity and momentum equations are used as the governing equations, and the mixture model is used for the mixed-phase flow field calculation, and its control equations are as follows [24]:

$$\frac{\partial \rho}{\partial t} + \frac{\partial (\rho \overline{u}_j)}{\partial x_j} = 0 \quad (1)$$

$$\frac{\partial(\rho \overline{u}_j)}{\partial t} + \frac{\partial(\rho \overline{u}_i \overline{u}_j)}{\partial x_j} = -\frac{\partial \overline{p}}{\partial x_i} + \frac{\partial}{\partial x_i}\left(\mu \frac{\partial \overline{u}_i}{\partial x_i}\right) - \frac{\partial \tau_{ij}}{\partial x_j} \quad (2)$$

where u_i is the velocity component in direction i, p is the pressure for both vapor and liquid phases, and ρ and μ denote the mixture density and viscosity, respectively, which are defined as follows:

$$\rho = \sum_{k=1}^{n} \alpha_k \rho_k \quad (3)$$

$$\mu = \sum_{k=1}^{n} \alpha_k \mu_k \quad (4)$$

In Equation (2), τ_{ij} represents the subgrid-scale stresses (SGS), which are defined as follows:

$$\tau_{ij} = \rho(\overline{u_i u_j} - \overline{u}_i \overline{u}_j) \quad (5)$$

After the filtering operation, the SGS are unknown and need modeling. The current simulation is solved using Fluent and the subgrid-scale turbulence models employ the Boussinesq hypothesis. The subgrid-scale turbulent stresses are calculated as follows [25]:

$$\tau_{ij} - \frac{1}{3}\tau_{kk}\delta_{ij} = -2\mu_t \overline{S}_{ij} \quad (6)$$

where $\overline{S}_{ij}$ is the tensor of the rate of strain for the resolved scale, which is defined as:

$$\overline{S}_{ij} = \frac{1}{2}\left(\frac{\partial \overline{u}_i}{\partial x_j} + \frac{\partial \overline{u}_j}{\partial x_i}\right) \quad (7)$$

In Equation (7), μ_t is the subgrid-scale turbulence viscosity and it is closed by the wall-adapting local eddy viscosity (WALE) model in this simulation. In the WALE model, the subgrid-scale turbulence viscosity is modeled as follows:

$$\mu_t = \rho L_s^2 \frac{\left(S_{ij}^d S_{ij}^d\right)^{3/2}}{\left(\overline{S}_{ij}\overline{S}_{ij}\right)^{5/2} + \left(S_{ij}^d S_{ij}^d\right)^{5/4}} \quad (8)$$

where L_s and S_{ij}^d in the WALE model are defined as:

$$L_s = min\left(kd, C_w V^{1/3}\right) \quad (9)$$

$$S_{ij}^d = \frac{1}{2}\left(\overline{g}_{ij}^2 + \overline{g}_{ji}^2\right) - \frac{1}{3}\delta_{ij}\overline{g}_{kk}^2,\ \overline{g}_{ij} = \frac{\partial \overline{u}_i}{\partial x_j} \quad (10)$$

where k is the von Kármán constant and C_w is the default WALE constant set as the default value 0.325.

2.3. Physical Cavitation Model

When the cavitation phenomenon is considered in the simulation, the transport function of the vapor volume fraction is used [26]:

$$\frac{\partial}{\partial t}(\alpha \rho_v) + \nabla \cdot \left(\alpha \rho_v \vec{v}_v\right) = R_e - R_c \quad (11)$$

where α is the volume fraction of the vapor and R_e and R_c represent the vaporization and condensation rate, respectively, which can be obtained from the cavitation model.

Currently, different cavitation models are deduced by researchers and most of them are based on the Rayleigh–Plesset function. In this paper, the Zwart–Gerber–Belamri model is used to calculate the mass transfer between the vapor and the liquid phases. The mass transfer equation of the cavitation model is shown as follows:

$$R_e = F_{\text{vap}} \frac{3\alpha_{\text{nuc}}(1-\alpha_v)\rho_v}{R_B} \sqrt{\frac{2}{3} \frac{P_v - P}{\rho_l}} \text{ when } P \leq P_v \tag{12}$$

$$R_c = F_{\text{cond}} \frac{3\alpha_v \rho_v}{R_B} \sqrt{\frac{2}{3} \frac{P - P_v}{\rho_l}} \text{ when } P > P_v \tag{13}$$

where R_B is the radius of the homogenous bubbles, α_{nuc} is the volume fraction of non-condensable gas, F_{vap} is the vaporization rate coefficient, and F_{cond} is the condensation rate coefficient.

2.4. Simulation Setup

The simulation domain and conditions are set according to the experiment. A nozzle with a throat diameter of 2 mm is simulated, and the fluid domain with enough size is created to avoid the influence of the boundaries on the jet flow. As shown in Figure 2, the calculation domain has a diameter of 100 *d* and a length of 250 *d*. The present analysis is focused on the turbulence characteristics of the shear layer, which is located in the region of the outlet of the nozzle. The fluid domain is discretized into hexahedral grids using ANSYS ICEM, as shown in Figure 2. Since the focus is on a deep understanding of the flow in the mixing layer, the mesh in the shear layer closed to the nozzle outlet is refined. The total grid cell number for the whole domain is 39 million.

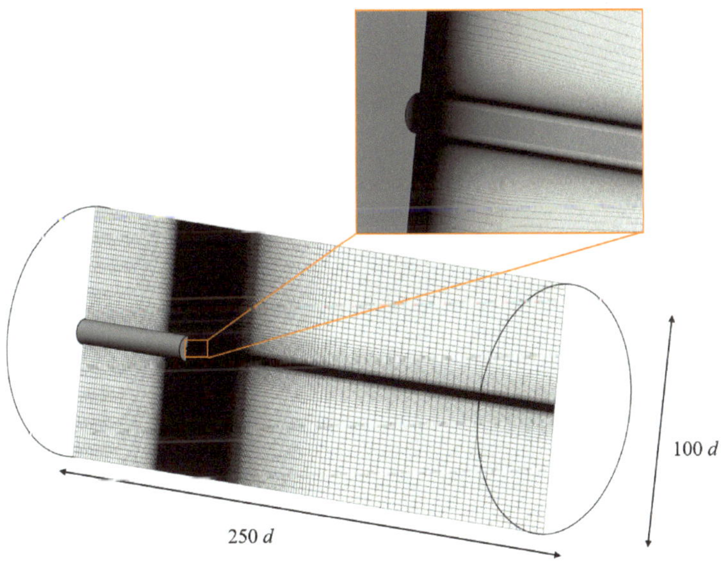

Figure 2. Mesh and sizes of the computational domain.

The numerical simulation contains three main types of boundary conditions. At the upstream inlet of the nozzle, the boundary is set as the pressure inlet, and the total pressure is set as 1–3 MPa, regarding the working conditions of the jet. The outlet of the domain is treated as the pressure outlet, and the gauge pressure is set as 0 Pa. The other boundaries are set as the wall. The time step is 100 μs and the residuals for the continuity equation, and the momentum and volume fraction transport equations, are 10^{-3}. The maximum iteration number for each time step is set as 30. The transient simulation was calculated for 1000 time steps initially, and then the velocity and pressure at the monitored point were stored for

1000 more time steps. Physical parameters in the whole domain were averaged for the last 1000 time steps to obtain the time-averaged value for the velocity and pressure fields.

3. Results and Discussion

3.1. Validation of CFD Results

Simulation of the cavitating jet is carried out according to the experiment, and the three working conditions of the jet are investigated both numerically and experimentally. Figure 3 shows the comparison between the high-speed image and the simulation results. Cavitation performance of a submerged jet is mainly affected by the nozzle geometry and the working conditions of the jet. The commonly used parameter to describe the working conditions is cavitation number, which is defined as follows:

$$\sigma = \frac{p_c - P_v}{\frac{1}{2}\rho v^2} \tag{14}$$

where p_c is the pressure at the downstream of the nozzle, P_v is the vapor pressure of the fluid, ρ is the density and v is the jet velocity. For the jet with a high value of p_c, the cavitation number can be defined approximately as:

$$\sigma = \frac{p_c}{\varphi \Delta p} \approx \frac{p_c}{\Delta p} \tag{15}$$

where φ is the flow coefficient, which is usually in the range of 0.97 to 0.98, and Δp is the pressure difference between the upstream and downstream sides of the nozzle throat.

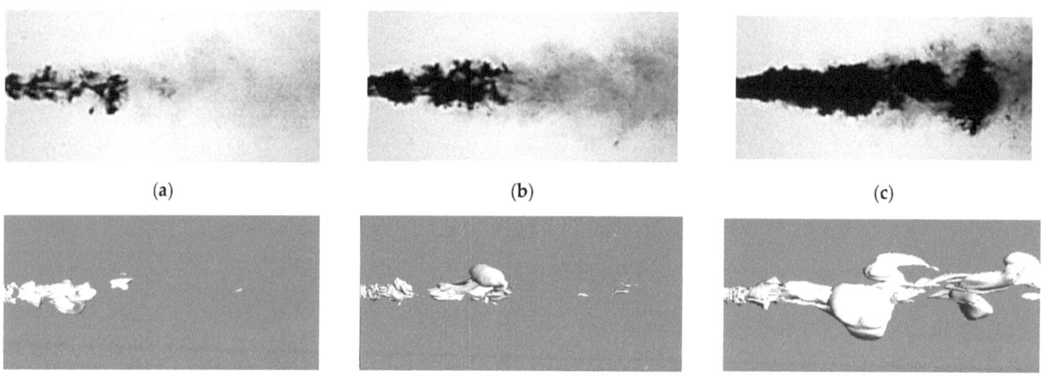

Figure 3. Cavitation cloud distribution by experiment and numerical prediction. (**a**) Exp. (σ = 0.1); (**b**) Exp. (σ = 0.05); (**c**) Exp. (σ = 0.033); (**d**) CFD (σ = 0.1); (**e**) CFD (σ = 0.05); (**f**) CFD (σ = 0.033).

Figure 3 shows the experiment results of the high-speed photography and the LES simulation of the jet for three different cavitation numbers, namely, σ = 0.1, 0.05 and 0.033. The images show that the length of the cavitation clouds increases obviously with the decrease in the cavitation number. When the cavitation number is 0.1, the cavitation bubbles only appear around the nozzle exit, while the bubble cloud length is almost doubled for the case when the cavitation number is 0.033. Figure 3d–f is the iso-surface of the vapor phase in the simulation result. By comparing the experimental and simulated results, it is found that the tendency and the scale of the cavitation cloud for CFD and high-speed photography are almost the same.

The time-averaged images of the high-speed images and the simulation results are compared in Figure 4. As shown in the images, gray-scale values of 5000 high-speed images are averaged, which represent the volume fraction of the vapor phase. For the simulation results, volume fraction values of more than 1000 time steps are calculated statistically, which is shown in Figure 4d–f. By comparing the time-averaged value of the cavitation cloud from experimental images and numerical simulation, it is found that the length of the jet is basically the same. The scale of the jet for both radial and axial directions are similar for experimental and numerical results. The results indicate that the currently used numerical method can predict the evolution of the cavitation cloud in the jet accurately.

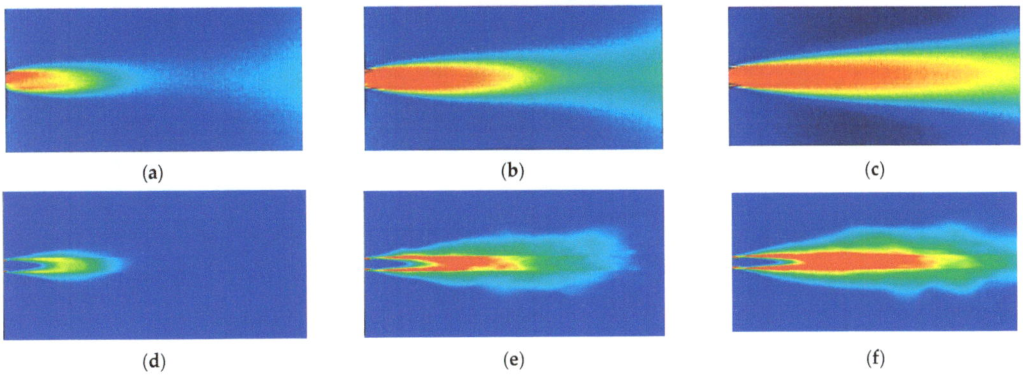

Figure 4. Time-averaged results of the cavitation clouds. (**a**) Exp. (σ = 0.1); (**b**) Exp. (σ = 0.05); (**c**) Exp. (σ = 0.033); (**d**) CFD (σ = 0.1); (**e**) CFD (σ = 0.05); (**f**) CFD (σ = 0.033).

3.2. Shear Layer Instability

When the high-speed jet is submerged in the relatively stationary fluid, velocity changes from hundreds of meters per second to zero in a small range of distance spatially in the radial direction. Under such conditions, the high-momentum fluid is affected by the surrounding fluid with low momentum by viscose forces. Figure 5 shows the transient velocity field of the submerged high-pressure jet in the region close to the nozzle exit, and the related cavitation number is σ = 0.05. It can be seen that the velocity decreases fast along the streamwise direction, and the velocity in the mixing layer fluctuates dramatically. At the same time, vortical structures can be found in the mixing layer, and the spatial scale of the vortexes increases gradually along the axial direction. This tendency is in accordance with the characteristic of shear layer Kelvin–Helmholtz instability [27]. The Kelvin–Helmholtz instability is a flow instability that arises due to shear stress along the interface between two different fluids and results in a concentration of vorticity along a velocity discontinuity in the tangent line.

To show the vortex structures in the mixing layer more clearly, the plane streamline in the mixing layer near the nozzle exit is plotted, which is colored by the velocity magnitude. Vortexes in the mixing layer can be seen clearly in Figure 6. Smally scaled vortexes started to appear at the boundary layer of the nozzle throat outlet, which is mainly caused by the counter flow between the streamwise main flow and the reversed flow from the region close to the end wall of the nozzle exit. The small-scale vortexes grow gradually, since the surrounding stationary fluid is entrained by the main flow. Meanwhile, the vortexes merge with each other in the early stages of the evolution and then break into small-scale vortexes under the effect of pressure and velocity fluctuations caused by the turbulence. Since the cavitation performance of the submerged high-pressure jet depends on the vortexes in the shear layer, it is of great importance to investigate the structure of the vortexes and the turbulence characteristics in the mixing layer. To observe the location of the vortexes and the spatial distribution of the velocity magnitude, it can be found that the velocity gradient in the mixing layer is not smooth, and each wave of the fluctuation corresponds to a large-scale vortex. In certain regions downstream of the nozzle exit, the wavelength of the velocity fluctuation increases, and the scale of the vortex increases in the same tendency.

This indicates that the velocity fluctuation can be caused by the interaction and evolution of the vortexes in the mixing layer under the high shear stress between the high-speed and stationary fluid.

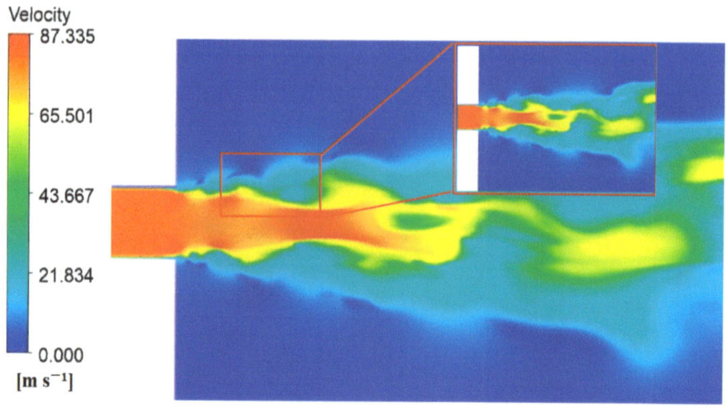

Figure 5. Velocity distribution of submerged cavitation jet (σ = 0.05).

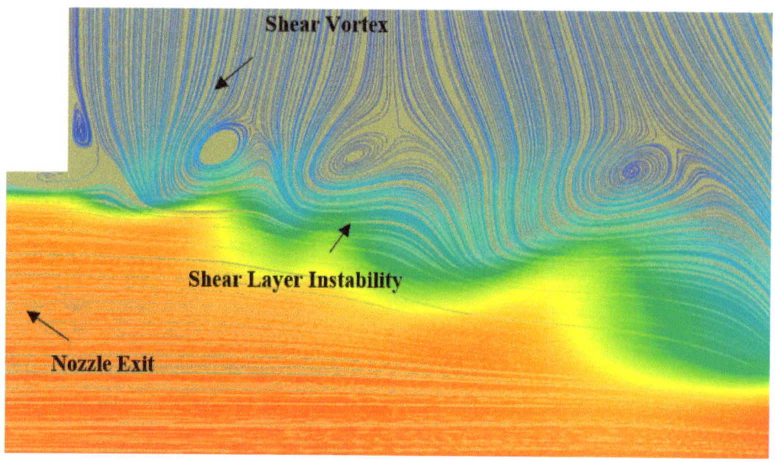

Figure 6. Shear layer instability in the submerged high-pressure jet (σ = 0.05).

According to the findings, the cavitating jet always sheds in a certain frequency, particularly in the case of self-excited oscillating cavitating jets. The exiting frequency and the theory play an important role in the design of the nozzle for the self-excited oscillating jet. To reveal the unsteady characteristic of the jet, streamwise velocity variations at different locations are monitored. The monitor points are located along the center line of the jet, from the nozzle exit to 10 d downstream. Figure 7 shows the time domain variation of the axial component of the velocity for three different cavitation numbers. It can be seen from the figures that the variation amplitude of the velocity increases obviously along the streamwise direction for each cavitation number. It indicates that the turbulence structures and the intensity are smaller in the region close to the nozzle, which increases gradually as the vortexes merge with each other and diffuse during motion. Comparing the monitored signals for the three different cavitation numbers, it can be seen that the velocity increases with the decrease in the cavitation number, since the pressure difference between the nozzle throat is increased. At the point x/d = 2, the variation of the velocity is relatively small for the three cavitation numbers, which means that the instability structures here are relatively small. Comparing the transient velocity at $x/d = 2.5$, it can be found that high-amplitude velocity fluctuations occur in the case with a high cavitation number, while the velocity variation still remains low in the case with lower cavitation numbers. When the cavitation number is $\sigma = 0.1$, at the point $x/d = 2.5$, the velocity varies from 20 m/s to 50 m/s, and the profile of the axial velocity in the time domain is similar to that at the further downstream point $x/d = 3$. This means that the turbulence is fully developed at the position $x/d = 2.5$, and the vortex structures become isotropic from this point. When the cavitation number is decreased to $\sigma = 0.05$, the transient axial velocity at $x/d = 2.5$ only shows several peaks, indicating large-scale and intensive instability starts to happen at this position. When $\sigma = 0.033$, the instability happens further downstream, where $x/d = 3$.

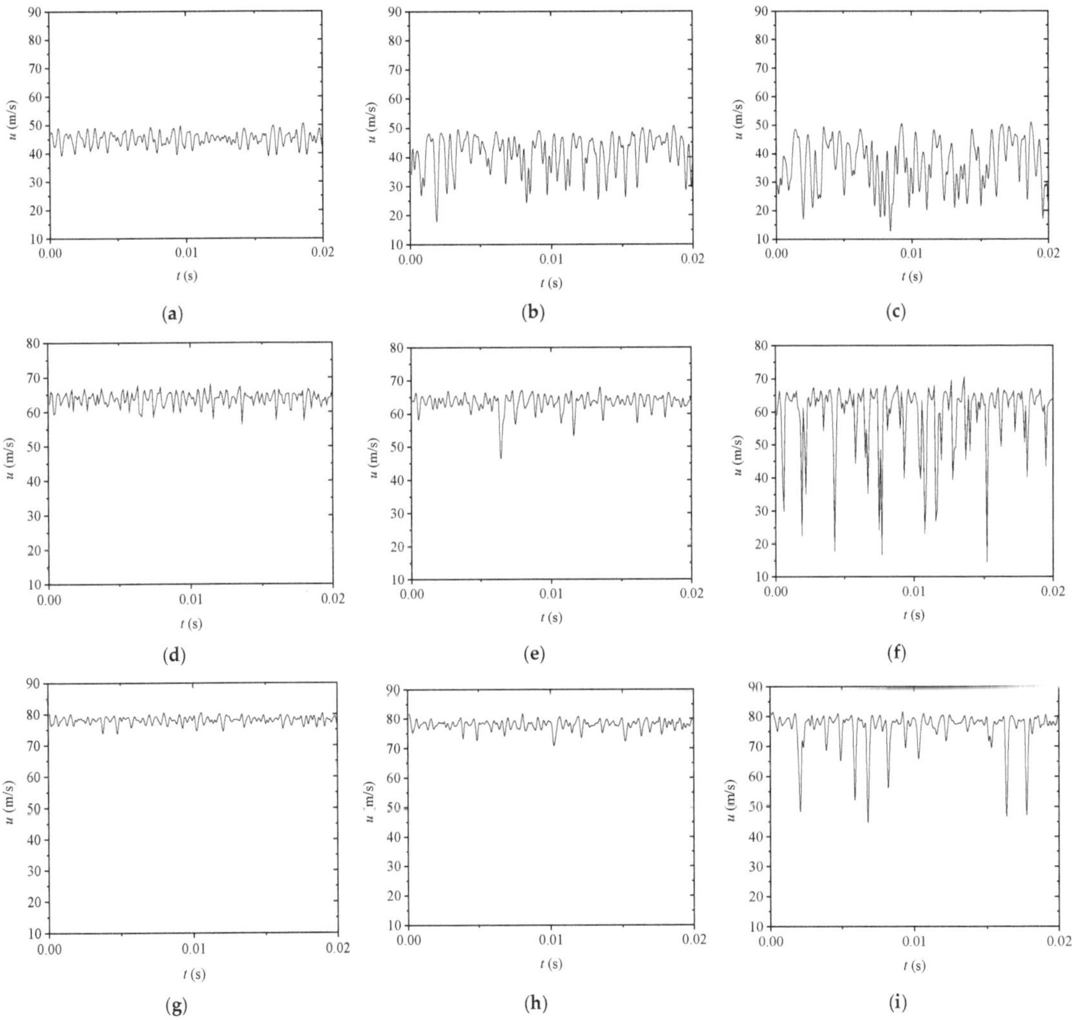

Figure 7. x-component velocity signals and the corresponding spectra inside the mixing layer. (**a**) $\sigma = 0.1$, $x/d = 2$; (**b**) $\sigma = 0.1$, $x/d = 2.5$; (**c**) $\sigma = 0.1$, $x/d = 3$; (**d**) $\sigma = 0.05$, $x/d = 2$; (**e**) $\sigma = 0.05$, $x/d = 2.5$; (**f**) $\sigma = 0.05$, $x/d = 3$; (**g**) $\sigma = 0.033$, $x/d = 2$; (**h**) $\sigma = 0.033$, $x/d = 2.5$; (**i**) $\sigma = 0.033$, $x/d = 3$.

To analyze the frequency characteristic of the turbulence flow in the cavitating jet, the monitored transient streamwise velocity was transformed to frequency field by FFT. Figure 8 shows the frequency of the temporal signal of the velocity component along the center line for different cavitation numbers. It can be seen from the figures that the main frequency of the velocity oscillation decreases as the monitor point moves downstream. The oscillation of the streamwise velocity reflects the vortex structures in the flow field. The scale of the vortexes increases and has a momentum exchange with the surrounding fluid in the process while moving downstream, which causes the decrease in the main frequency of the velocity magnitude. Comparing the frequency spectral of jets with different cavitation numbers, it can be found that, for a fixed location, the frequency increases with the decrease in the cavitation number. This is because, in the case of low cavitation numbers, the pressure differences between the inlet and outlet of the nozzle are larger, which produces jets with higher initial velocity. The jet with higher initial velocity contains more momentum, which decreases more slowly.

3.3. Vortexes and Cavitation in the Mixing Layer

As discussed above, the velocity in the jet varies in the complicated mode with the effect of the turbulence. It contains both low- and high-frequency signals, which may be caused by different types of vortexes with different size. To reveal the temporal evolution of the vortex structure in the mixing layer, vortexes near the nozzle outlet around one period are plotted. Figure 9 shows the Q-criterion iso-surface for the jet with a cavitation number of $\sigma = 0.05$, which is colored by the velocity magnitude. At the time when $t = 0$ s, the vortex in the fluid close to the nozzle is relatively weak and almost no vortex structure can be found in the nozzle. At $t = 60$ μs, a vortex ring is formed at the nozzle outlet. Spanwise vortexes dominate in the region $x/d = 0$ to 1.5, while a streamwise vortex shows up at around $x/d = 2$. The vortex dissipates gradually further downstream under the effect of the viscose force. At $t = 120$ μs, the newly formed vortex ring is broken into small-scale vortexes under the strong shear stress, and the streamwise vortex mostly disappears. After that, the region with strong vortexes is reduced, and the next period starts at around $t = 300$ μs. From the frequency distribution of the temporal spectral of the velocity, a wide band signal can be found. This can be related to the formation, distortion and breakage of the vortex ring structures.

To discuss the formation of the instability of the vortex ring, the vorticity magnitude on the plane perpendicular to the axis at different positions is displayed in Figure 10. It can be seen from the figures that, at the position just behind the nozzle outlet, the vortex is distributed in a circular region. The shape of the region with a high vorticity value is more regular in the case with a lower cavitation number, indicating that higher jet velocity makes the ring vortexes more stable. The same tendency can be observed on the plane at $x/d = 1.5$ and 3 for different cavitation numbers. Comparing the vorticity distribution at different streamwise locations, the instability process of the ring-shaped vortexes is revealed. Under the effect of the viscose force of the fluid, the second instability happens on the ring-shaped vortexes. The vortex ring breaks into several parts while moving downstream, due to the entrainment of the low-momentum fluid. The streamwise vortex structures show up accompanying the second instability of the vortex ring. In addition, it can be seen from the figure that the jet is affected by the shear vortex at the nozzle outlet to form a small vortex band, and the vortex gradually grows with the diffusion of the jet and is accompanied by dissipation.

Cavitation always happens in a high-pressure submerged jet when the jet velocity is high enough. The theory of the shear-induced Rankin vortex model can be used to explain the appearance of cavitation in the jet, and the vortex–bubble interaction is found to have an effect on the fluid and bubble dynamics to some degree. Figure 11 shows the iso-surface of the vapor volume fraction, from which the dynamic behavior of the cavitation cloud is reflected. From the figures, it is found that the vapor is mainly formed at the nozzle outlet. This is in accordance with the Rankin vortex theory. When the high-speed jet flows out from the nozzle, the high shear stress is formed in the mixing layer and a series of small-scale vortexes develops due to the K–H instability. The pressure in the Rankin vortex center decreases fast and reaches the vapor pressure of the fluid under the ambient temperature, and the cavitation occurs at the same time. A ring-shaped cavitation cloud is found at the nozzle outlet at the time t = 60 μs, and a vortex ring is created at the same location at the same time, according to Figure 11b. Then, the ring-shaped cavitation breaks into small parts under the effect of the vortex instability. Comparing the shape of the cavitation cloud and the cortexes in the jet, it is found that their spatial distribution and the temporal evolution are similar, indicating that the dynamic characteristics of the vortex and the cavitation affect each other.

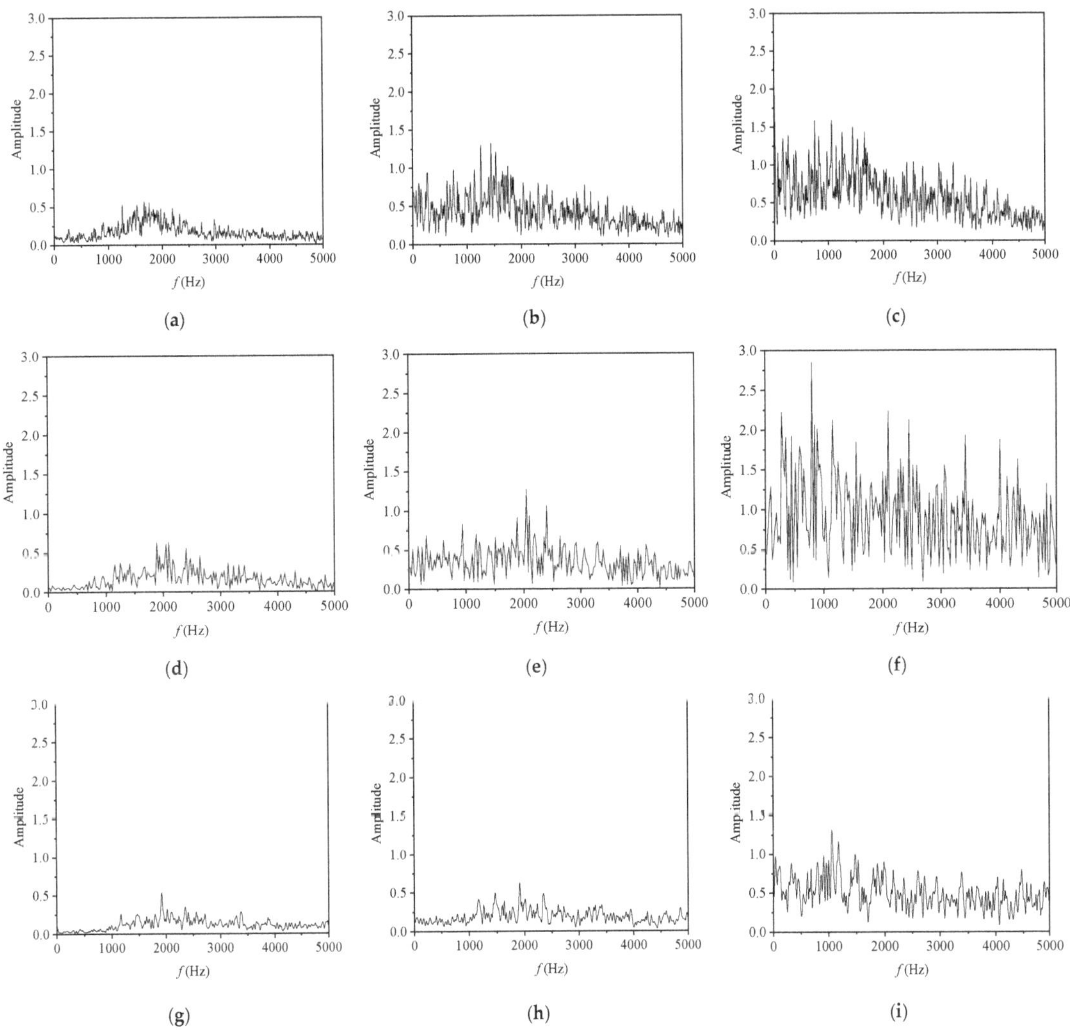

Figure 8. The frequency of the temporal signal of the velocity component along the center line. (**a**) σ = 0.1, x/d = 2; (**b**) σ = 0.1, x/d = 2.5; (**c**) σ = 0.1, x/d = 3; (**d**) σ = 0.05, x/d = 2; (**e**) σ = 0.05, x/d = 2.5; (**f**) σ = 0.05, x/d = 3; (**g**) σ = 0.033, x/d = 2; (**h**) σ = 0.033, x/d = 2.5; (**i**) σ = 0.033, x/d = 3.

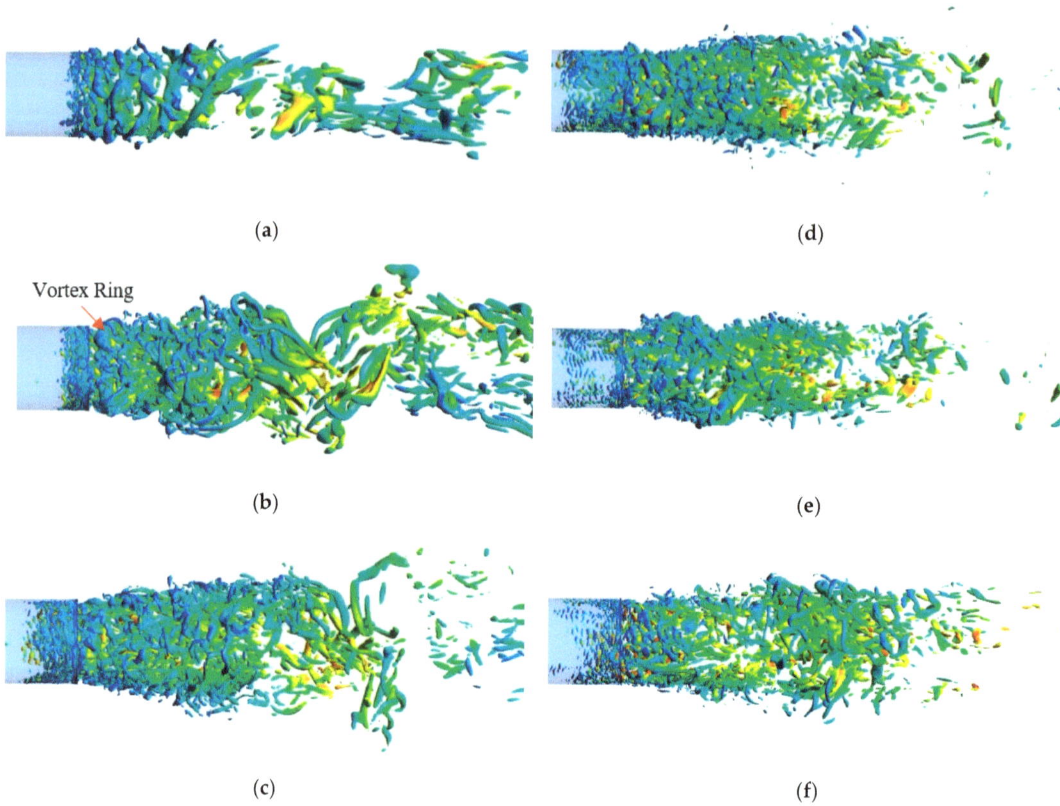

Figure 9. Vortex structure and evolution in the mixing layer (σ = 0.05). (**a**) t = 0 s; (**b**) t = 60 μs; (**c**) t = 120 μs; (**d**) t = 180 μs; (**e**) t = 240 μs; (**f**) t = 300 μs.

3.4. Time-Averaged Characteristics

To analyze the time-averaged characteristics of the jet, 1000 time steps of the transient simulation were averaged and statistically processed. Figure 12 shows the axial velocity along the center line and the radial direction at different x/D locations. As shown in Figure 12a, the velocity remains almost unchanged for a distance in the region close to the nozzle outlet, which is called the core region of the jet. It is found that the core region of the jet is extended as the cavitation number is decreased, since the momentum and the initial velocity are higher for the condition with a lower cavitation number. According to the research literature about submerged jets without cavitation, the velocity profile of the jet along the jet axis shows a similarity law, and the profile changes little when the Reynolds number is changed [28]. In the current research, the submerged jets are cavitating jets, and the variation of the non-dimensioned velocity profile changes with the variation of the cavitation number. One reason for this change is the difference of the vapor volume fraction. It is well known that the density and the viscosity of the vapor phase is obviously lower than that of the liquid phase, and the appearance of the cavitation may reduce the moment exchange between the high-speed jet and the surrounding fluids. Figure 12b presents the velocity distribution along the radial direction at different x/D locations, from which the jet's spreading rate can be compared. Trushar B. Gohil et al. [29] concluded that the radial variation shows a top-hat profile at the location close to the nozzle exit, and this profile is changed gradually at the downstream locations; the spreading rate is higher for the case with a high Reynolds number. In the current cavitating jets, a top-hat profile is also found. For the currently investigated cavitating jets, the Reynolds number increases with the decrease in the cavitation number. However, the spreading rate is lower for the jet with a higher Reynolds number here. This means that the momentum exchange between the jet and submerging water is reduced by the cavitation phenomenon.

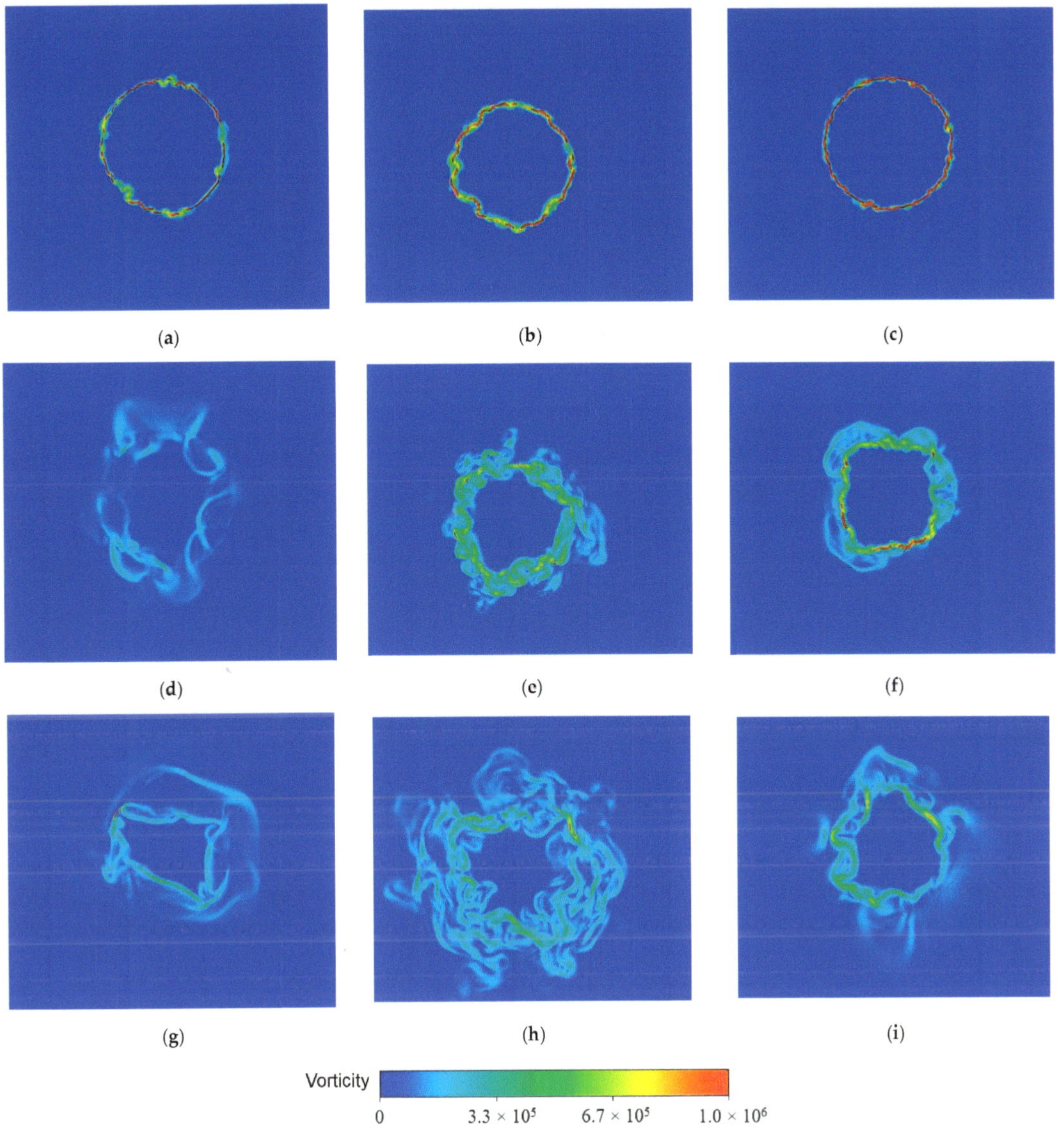

Figure 10. Contour shape of the vorticity on the cross-section at different streamwise locations. (**a**) $\sigma = 0.1$, $x/d = 0.05$; (**b**) $\sigma = 0.05$, $x/d = 0.05$; (**c**) $\sigma = 0.033$, $x/d = 0.05$; (**d**) $\sigma = 0.1$, $x/d = 1.5$; (**e**) $\sigma = 0.05$, $x/d = 1.5$; (**f**) $\sigma = 0.033$, $x/d = 1.5$; (**g**) $\sigma = 0.1$, $x/d = 3$; (**h**) $\sigma = 0.05$, $x/d = 3$; (**i**) $\sigma = 0.033$, $x/d = 3$.

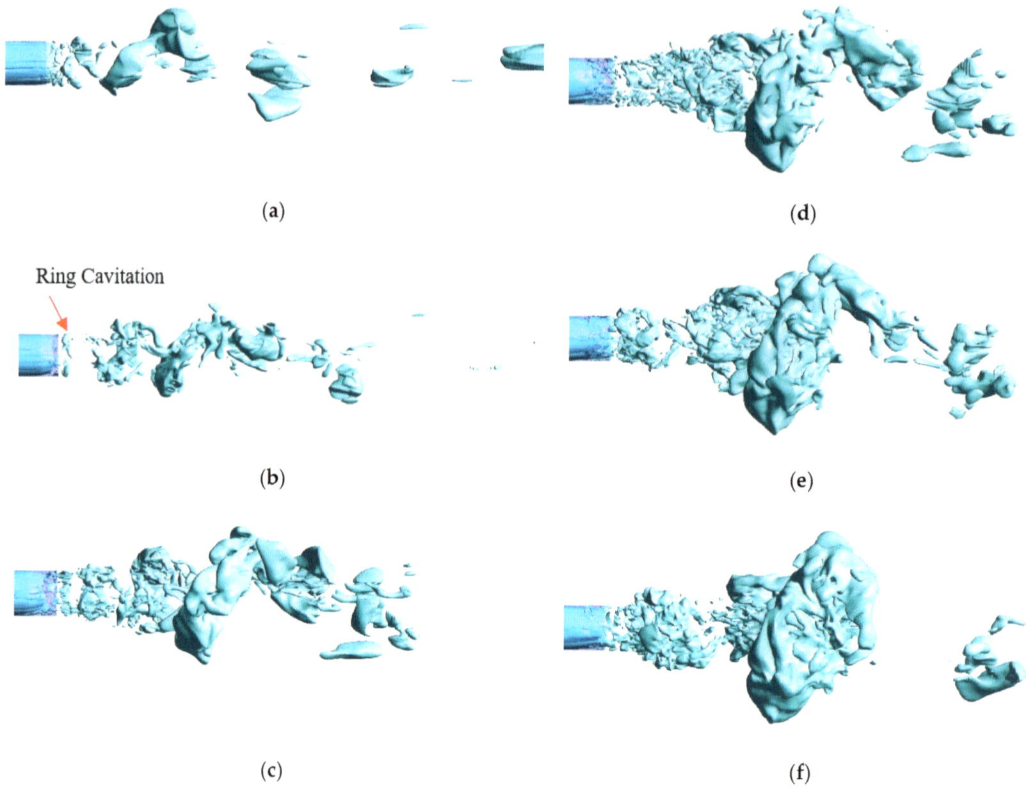

Figure 11. Cavitation evolution near the shear layer of the jet (σ = 0.05). (**a**) t = 0 s; (**b**) t = 60 μs; (**c**) t = 120 μs; (**d**) t = 180 μs; (**e**) t = 240 μs; (**f**) t = 300 μs.

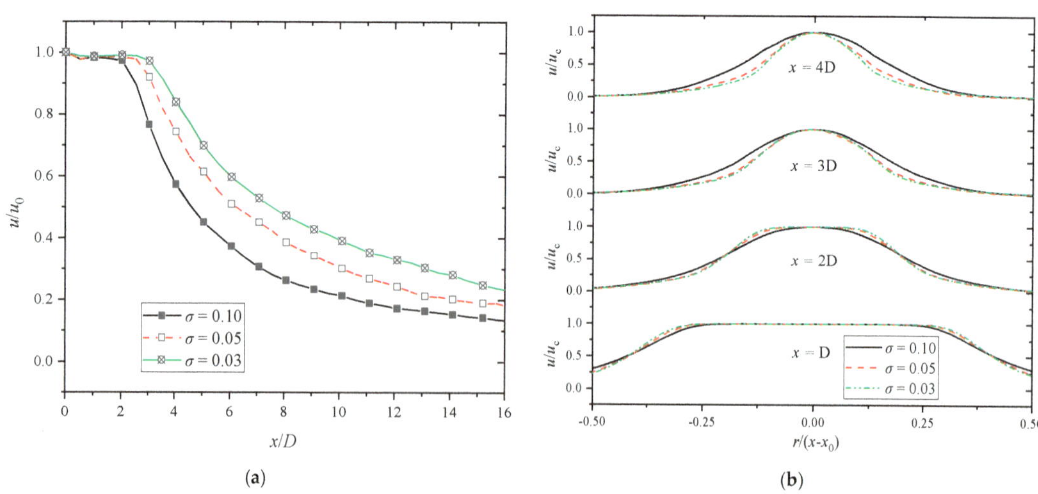

Figure 12. Axial velocity along the center line and radial direction at different x/D locations. (**a**) Streamwise variation; (**b**) Radial variation.

According to the research about jets, the spreading rate can be quantitatively analyzed using the jet's half-width and mixing layer thickness. The definition of the jet half-width is the radial length where the axial velocity decreases to half of the center value. For a typical shear flow, the mixing layer is defined as follows [30]:

$$\delta_w = \frac{\overline{u}_1 - \overline{u}_2}{(\partial \overline{u} / \partial y)_{max}} \tag{16}$$

where $\overline{u}_1$ and $\overline{u}_2$ are time-averaged streamwise velocities at different sides of the shear layer. Figure 13 illustrates the shear layer thickness and half-width variation along the streamwise direction of the jet. The mixing layer thickness and the half-width increases slightly within the potential core for all three cavitation numbers, and then increases fast further downstream. In general, the mixing layer and half-width of the jet are larger for the case with a higher cavitation number. Since the jet velocity and the Reynolds number are higher in the case with a low cavitation number, the variation of the mixing layer along the streamwise direction is different from that of research on non-cavitating jets [29]. This indicates that the occurrence of the cavitation has a great effect on the momentum exchange in the mixing layer.

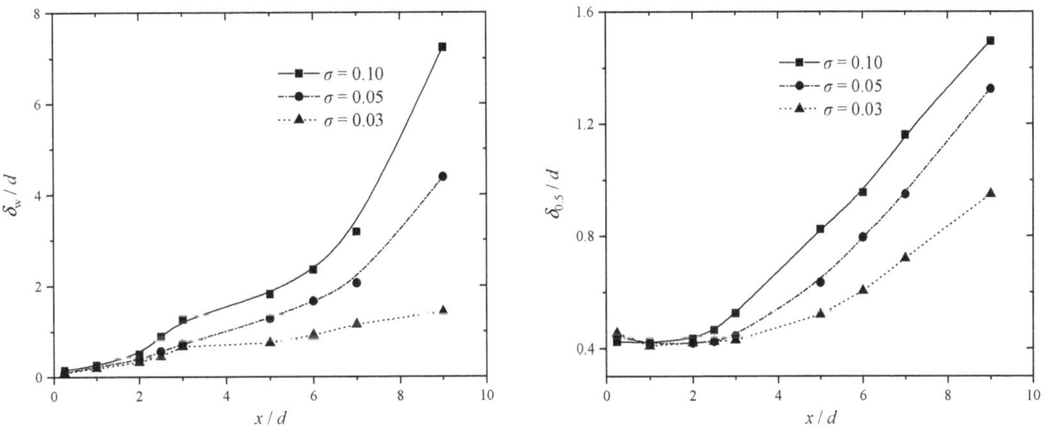

Figure 13. Variation of shear layer thickness and half-width variation along the streamwise direction. (**a**) Shear layer thickness; (**b**) Half-width variation.

4. Conclusions

The length of the cavitation clouds increases obviously with the decrease in the cavitation number; the scale of the jet for both radial and axial direction is similar for experimental and numerical results. The currently used numerical method can predict the evolution of the cavitation cloud in the jet accurately. The turbulence structures and the intensity are smaller in the region close to the nozzle, which increases gradually as the vortexes merge with each other and diffuse during motion. Comparing the monitored signals for the three different cavitation numbers, it can be seen that the velocity increases with the decrease in the cavitation number, since the pressure difference between the nozzle throat is increased. Under the effect of the viscose force of the fluid, the second instability happens on the ring-shaped vortexes. The vortex ring breaks into several parts while moving downstream, due to the entrainment of the low-momentum fluid. The streamwise vortex structures show up accompanying the second instability of the vortex ring. The core region of the jet is extended as the cavitation number is decreased, since the momentum and the initial velocity are higher for the condition with a lower cavitation number. The spreading rate is lower for the jet with a higher Reynolds number. This means that the momentum exchange between the jet and submerging water is reduced by the cavitation phenomenon.

Author Contributions: Data curation, W.S. and L.J.; Methodology, L.J.; Project administration, W.S.; Resources, W.S. and H.W.; Software, G.W.; Supervision, H.W.; Visualization, W.L.; Writing—original draft, Y.Y. and G.W.; Writing—review and editing, Y.Y. All authors have read and agreed to the published version of the manuscript.

Funding: This research was funded by the National Key Research and Development Project of China (No. 2019YFB 2005300), National High-Tech Ship Scientific Research Project of China (No. MIIT [2019] 360), National Natural Science Foundation of China (No. 51979138), National Natural Science Foundation of China (No. 273746), National Natural Science Foundation of China (No. 51979240), Jiangsu Natural Science Re-search Project (No. 19KJB470029), China Postdoctoral Science Foundation (No. 273746), and Natural Science Foundation of Jiangsu Province (No. BK20220609), Project funded by China Postdoctoral Science Foundation (No.2022TQ0127).

Institutional Review Board Statement: Not applicable.

Informed Consent Statement: Not applicable.

Data Availability Statement: Data on the analysis and reporting results during the study can be obtained by contacting the authors.

Conflicts of Interest: The authors declare no conflict of interest.

References

1. Soyama, H. Cavitating Jet: A Review. *Appl. Sci.* **2020**, *10*, 7280. [CrossRef]
2. Vasilakis, E.S.; Kyriazis, N.; Koukouvinis, P.; Farhat, M.; Gavaises, M. Cavitation induction by projectile impacting on a water jet. *Int. J. Multiph. Flow* **2019**, *114*, 128–139. [CrossRef]
3. Ijiri, M.; Shimonishi, D.; Nakagawa, D.; Yoshimura, T. New water jet cavitation technology to increase number and size of cavitation bubbles and its effect on pure Al surface. *Int. J. Lightwght Mater. Manuf.* **2018**, *1*, 12–20. [CrossRef]
4. Soyama, H. Cavitation Peening: A Review. *Metals* **2020**, *10*, 270. [CrossRef]
5. Soyama, H.; Takeo, F. Comparison between cavitation peening and shot peening for extending the fatigue life of a duralumin plate with a hole. *J. Mater. Process. Technol.* **2016**, *227*, 80–87. [CrossRef]
6. Zhang, Y.; Zang, W.; Zheng, J.; Cappietti, L.; Zhang, J.; Zheng, Y.; Fernandez-Rodriguez, E. The influence of waves propagating with the current on the wake of a tidal stream turbine. *Appl. Energy* **2021**, *290*, 116729. [CrossRef]
7. Zhang, Y.; Zhang, J.; Lin, X.; Wang, R.; Zhang, C.; Zhao, J. Experimental investigation into downstream field of a horizontal axis tidal stream turbine supported by a mono pile. *Appl. Ocean. Res.* **2020**, *101*, 102257. [CrossRef]
8. Zhang, Y.; Zhang, Z.; Zheng, J.; Zhang, J.; Zheng, Y.; Zang, W.; Lin, X.; Fernandez-Rodriguez, E. Experimental investigation into effects of boundary proximity and blockage on horizontal-axis tidal turbine wake. *Ocean. Eng.* **2021**, *225*, 108829. [CrossRef]
9. Saito, Y.; Sato, K. Instantaneous Behavior of Cavitation Clouds at Impingement of Cavitating Water-Jet. *Prog. Multiph. Flow Res.* **2007**, *2007*, 47–53. [CrossRef]
10. Sato, K.; Sugimoto, Y.; Ohjimi, S. *Pressure-Wave Formation and Collapses of Cavitation Clouds Impinging on Solid Wall in a Submerged Water Jet*; University of Michigan: Ann Arbor, MI, USA, 2011.
11. Yang, Y.; Li, W.; Shi, W.; Wang, C.; Zhang, W. Experimental Study on Submerged High-Pressure Jet and Parameter Optimization for Cavitation Peening. *Mechanika* **2020**, *26*, 346–353. [CrossRef]
12. Zhang, W.; Ma, N.; Ma, J.; Li, C.; Ren, J. Experimental Study on the Unsteady Characteristics and the Impact Performance of a High-Pressure Submerged Cavitation Jet. *Shock. Vib.* **2020**, *2020*, 8862849.
13. Nakano, K.; Hayakawa, M.; Fujikawa, S.; Yano, T. Cavitation Bubbles in a Starting Submerged Water Jet. *Nihon Kikai Gakkai Ronbunshu B Hen Trans. Jpn. Soc. Mech. Eng. Part B* **2001**, *69*, 5. [CrossRef]
14. Sawamura, T.; Koshika, N. Velocity Measurement of Cavitation in Cavitating Jet Flow. *Proc. JSME Annu. Meet.* **2003**, *2003*, 155–156. [CrossRef]
15. Gopalan, S.; Katz, J.; Knio, O. Near-field flow structure and cavitation inception in jets. In Proceedings of the 3rd ASME·JSME Joint Fluids Engineering Conference, San Francisco, CA, USA, 18–23 July 1999.
16. Gopalan, S.; Katz, J.; Knio, O. The flow structure in the near field of jets and its effect on cavitation inception. *J. Fluid Mech.* **1999**, *398*, 1–43. [CrossRef]
17. Mao, N.; Kang, C.; Zhou, M.M. POD Analysis of Submerged Jet Structure and Cavitation Cloud Evolution. *J. Eng. Thermophys.* **2019**, *40*, 2036–2042.
18. Miltner, M.; Jordan, C.; Harasek, M. CFD simulation of straight and slightly swirling turbulent free jets using different RANS-turbulence models. *Appl. Therm. Eng.* **2015**, *89*, 1117–1126. [CrossRef]

19. Yang, Y.; Shi, W.; Tan, L.; Li, W.; Chen, S.; Pan, B. Numerical Research of the Submerged High-Pressure Cavitation Water Jet Based on the RANS-LES Hybrid Model. *Shock. Vib.* **2021**, *2021*, 6616718. [CrossRef]
20. Yang, Y.; Li, W.; Shi, W.; Zhang, W.; El-Emam, M.A. Numerical Investigation of a High-Pressure Submerged Jet Using a Cavitation Model Considering Effects of Shear Stress. *Processes* **2019**, *7*, 541. [CrossRef]
21. Wang, Y.; Yuan, G.; Yoon, Y.-K.; Allen, M.G.; Bidstrup, S.A. Large eddy simulation (LES) for synthetic jet thermal management. *Int. J. Heat Mass Transf.* **2006**, *49*, 2173–2179. [CrossRef]
22. Wang, Y.C.; Tan, L.; Cao, S.L.; Gao, C.C. Numerical study on flow structures of high Reynolds number submerged jet in an axis-symmetric cavity. In Proceedings of the 2014 ISFMFE-6th International Symposium on Fluid Machinery and Fluid Engineering, Wuhan, China, 22 October 2014.
23. Wang, Y.C.; Gao, C.C.; Tan, L.; Cao, S.L. Hysteresis properties of submerged jet cavitation in axis-symmetrical cavity. *J. Vib. Shock.* **2015**, *34*, 118–122.
24. Manninen, M.; Taivassalo, V.; Kallio, S. *Mixture Model for Multiphase Flow*; U.S. Department of Energy, Office of Scientific and Technical Information (OSTI): Oak Ridge, TN, USA, 1996.
25. Nicoud, F.; Ducros, F. Subgrid-Scale Stress Modelling Based on the Square of the Velocity Gradient Tensor. *Flow Turbul. Combust.* **1999**, *62*, 183–200. [CrossRef]
26. Zwart, P.J.; Gerber, A.G.; Belamri, T. A two-phase flow model for predicting cavitation dynamics. In Proceedings of the 5th International Conference on Multiphase Flow, Yokohama, Japan, 30 May–4 June 2004.
27. Friedlander, S.; Lipton-Lifschitz, A. Chapter 8 Localized Instabilities in Fluids. *Handb. Math. Fluid Dyn.* **2003**, *2*, 289–354.
28. Gohil, T.B.; Saha, A.K.; Muralidhar, K. Large eddy simulation of a free circular jet. *J. Fluids Eng.* **2014**, *136*, 051205. [CrossRef]
29. Gohil, T.B.; Saha, A.K.; Muralidhar, K. Numerical study of instability mechanisms in a circular jet at low Reynolds numbers. *Comput. Fluids* **2012**, *64*, 1–18. [CrossRef]
30. Brown, G.L.; Roshko, A. On density effects and large structure in turbulent mixing layers. *J. Fluid Mech. Digit. Arch.* **1974**, *64*, 775–816. [CrossRef]

sustainability

MDPI

Article

Hydraulic Oscillation Analysis of the Hydropower Station with an Equivalent Circuit-Based Hydraulic Impedance Scheme

Shuangqing Yan [1], Yang Zheng [2,*], Jinbao Chen [2] and Yousong Shi [3]

1 Department of Information and Technology, Shanghai Pudong Development Bank, Wuhan 430030, China
2 School of Power and Mechanical Engineering, Wuhan University, Wuhan 430072, China
3 School of Civil and Hydraulic Engineering, Huazhong University of Science and Technology, Wuhan 430074, China
* Correspondence: zhengyang@whu.edu.cn; Tel.: +86-13545875030

Abstract: The concept of hydraulic impedance is widely used for periodic or oscillating flow scenarios in applied hydraulic transients. This paper proposes an equivalent circuit-based discrete hydraulic impedance model for the hydraulic system of the power station. The equivalent hydraulic circuit is established by an analogy between the water hammer wave propagation in pressurized pipes and the electromagnetic wave propagation in conductors. The discrete hydraulic impedance at a given location is obtained by calculating the total specific impedance of the series and parallel circuits from one end to the other. In addition, the numerical solution process to obtain the natural frequencies of the system via the proposed model is designed in detail. Furthermore, oscillation characteristics of the pipelines in the hydropower station are investigated. The variation trends of the decay coefficients of different orders of oscillation of the system and the influence of turbine impedance on the oscillation frequencies are discussed, respectively.

Keywords: hydraulic impedance; equivalent circuit; hydropower station; oscillation analysis

Citation: Yan, S.; Zheng, Y.; Chen, J.; Shi, Y. Hydraulic Oscillation Analysis of the Hydropower Station with an Equivalent Circuit-Based Hydraulic Impedance Scheme. *Sustainability* **2022**, *14*, 11410. https://doi.org/10.3390/su141811410

Academic Editors: Yuquan Zhang, Jianhua Zhang, Ling Zhou and Bin Huang

Received: 16 August 2022
Accepted: 9 September 2022
Published: 12 September 2022

Publisher's Note: MDPI stays neutral with regard to jurisdictional claims in published maps and institutional affiliations.

1. Introduction

Transient analyses of hydropower systems often are carried out via numerical simulation schemes based on time and space discretization, such as the method of characteristics (MOC) [1], the finite difference/volume method [2,3], and the transfer function/differential equations [4,5]. The basic idea of these time domain numerical schemes is to transform continuous systems into discrete systems [6], thus reducing the degree of freedom in the system model and calculating the time evolutions of system states with satisfactory computational cost and numerical precision. However, most time domain numerical schemes are under the constraint of the Courant condition [7], where the Courant number is required to be close to one to restrain the numerical dissipation. The strong constraint will bring a substantial computational burden to the complex high-order systems. In addition, due to the nonlinearities, such as the dissipative terms and the uncertain parameters in these models [8], it is difficult to be implemented in the system stability analysis. Therefore, frequency domain methods, as another alternative, were extensively explored for stability analysis [2,9].

The conventional frequency domain schemes, e.g., the hydraulic impedance method [10–14] and the transfer matrix method [2], are widely used in applied hydraulic transients. The hydraulic impedance, defined as the complex head deviation divided by the complex flow rate deviation, is always formulated as a function of the complex frequency s. The relationship between the piezometric head and the discharge at an arbitrary longitudinal position along a pipe or a channel can be explicitly described with the model. This method shows potential for the leakage calibration [15], blockage detection [16,17], and the resonance analysis [18]. Suo and Wylie [8] obtained the response of a steady oscillatory flow with the linearized Saint-Venant equations of a pressurized pipe. Kim [11] formulated an impedance matrix model for complicated pipe networks and discussed the superiority of the impulse response method over other time domain schemes. In addition, he also discussed the dynamic matrix computation method [12] and incorporated the genetic algorithm into the address-oriented impedance matrix for the generic calibration of the system parameters of the pipe networks with complex topology [13]. Zhou et al. [14] investigated the self-excited hydraulic and mechanical vibration characteristics of the pumped storage power station based on the hydraulic impedance method. The free and forced oscillation responses were obtained and analyzed through this approach. Yang [9] systematically revealed the mathematical expressions of the hydraulic impedances of different components in the hydropower station, including the hydraulic turbine, the pressurized pipe, and the reservoir. Moreover, the stability of the hydropower station under different operating conditions is investigated with the hydraulic impedance method.

The traditional continuous hydraulic impedance can handle complex frequency domain modeling of simple pipe networks, such as pipes in series, pipe networks with single branches, and single loops. However, it cannot address the modeling of high-dimensional hydraulic networks with multi-loops very well. To fully take advantage of the well-developed circuit theory, the methodology of the complex frequency domain equivalent circuit-based hydraulic impedance model is inspired to calculate the specific impedance of any given location along the complicated pipeline by considering the pipe network as a discrete system. The basic idea of this resolution, inspired by an analogy between the hydraulic circuit and the electrical conductor, was first mentioned by Paynter [19] and Jaeger [20]. Souza Jr. [21] performed the numerical analysis of hydraulic transients in a hydropower plant using this methodology. Dr. Nicolet [22] extended the equivalent circuit model (ECM) to the mathematical modeling of different hydraulic facilities and validated its effectiveness in the simulation of hydraulic transients in engineering practice. Over the past few years, Zhao et al. [23] applied ECM to the dynamic analysis of operating condition conversion processes in a pumped storage plant. Zheng et al. [24] expanded the ECM to describe hydraulic transients of a hydraulic system with pipe and open channel flows and achieved satisfactory simulation precision. To the best of the authors' knowledge, most of the existing literature about this approach only focused on its privilege in the time domain numerical simulation [19–24]. Seldom was the equivalent circuit-based complex frequency-domain analysis reported. Although Nicolet [25] discussed the application of the frequency domain ECM to obtain the frequency responses in the reservoir–pipe–valve system, the system discussed is too simple to fully reflect the complex frequency domain description capability of the ECM. Therefore, it is necessary to further explore the features of the complex frequency domain ECM in more complicated hydraulic systems.

This paper proposes an equivalent circuit-based discrete hydraulic impedance model for the power station system. The equivalent hydraulic circuit is established by an analogy between the water hammer wave propagation in pressurized pipes and the electromagnetic wave propagation in conductors. The discrete hydraulic impedance at any discrete node is obtained by calculating the total specific impedance from one boundary side to the other according to the circuit theory. Subsequently, the detailed procedure of the free oscillation response analysis of the system via the proposed discrete impedance model is introduced. According to the oscillation analysis process, the variation trends of the decay coefficients of different orders of oscillation for the diversion and tailrace pipelines are discussed. Furthermore, the influence of turbine impedance on the decay coefficients and the frequencies corresponding to different orders of oscillation are investigated, respectively.

The rest of this paper is organized as follows: First, the methodology of the equivalent circuit-based hydraulic impedance for different hydraulic components, such as the pressurized pipe and the hydraulic turbine is introduced in Section 2. In Section 3, the overall equivalent circuit topology for the hydraulic system of the studied hydropower station is presented, and the detailed numerical solution process of the free oscillation response is illustrated. Then, numerical simulation results of the free oscillation responses for the hydropower station system are presented, and the variation trends of the system oscillation responses are systematically analyzed in Section 4. Eventually, the conclusions are given in Section 5.

2. Methodology of the Discrete Hydraulic Impedance

2.1. Equivalent Circuit of a Pressurized Pipe

The continuity equation and the momentum equation are the basic physical rules to describe the movement of fluids, such as wave propagation [26,27] and water delivery [28]. To mathematically characterize the dynamic behavior of fluid transients in a pressurized pipe, the well-known one-dimensional Saint-Venant equations [2,9] are stated in Equation (1),

$$\begin{cases} \frac{\partial H_i}{\partial x} + \frac{1}{gA_i}\left[\frac{\partial Q_i}{\partial t} + v_i\frac{\partial Q_i}{\partial x}\right] + \frac{\lambda_i Q_i|Q_i|}{2gD_iA_i^2} = 0 \\ \left[\frac{\partial H_i}{\partial t} + v_i\frac{\partial H_i}{\partial x}\right] + \frac{a_i^2}{gA_i}\frac{\partial Q_i}{\partial x} = 0 \end{cases} \tag{1}$$

where the subscript i denotes the pipe number. Hydraulic phenomena are characterized by a high wave speed a and a relatively much lower flow velocity v, so the convective terms $v_i\frac{\partial}{\partial x}$ related to the transport characteristic can be neglected with respect to the propagative terms $\frac{\partial}{\partial t}$ in Equation (1). Thus, when the piezometric head H and discharge Q are taken as state variables, the Saint-Venant equation in Equation (1) has the same formulation as the telegraphist's equations [22,25]. As expressed in Equation (2),

$$\begin{cases} \frac{\partial H_i}{\partial x} + L_{ei}\frac{\partial Q_i}{\partial t} + R_{ei}(Q_i)Q_i = 0 \\ \frac{\partial Q_i}{\partial x} + C_{ei}\frac{\partial H_i}{\partial t} = 0 \end{cases} \tag{2}$$

where the values of the equivalent RLC parameters per unit length are stated in Equation (3),

$$\begin{cases} R_{ei} = \frac{\lambda_i|Q_i|}{2gD_iA_i^2}\ [\mathrm{s/m^3}] \\ C_{ei} = \frac{gA_i}{a_i^2}\ [\mathrm{m}] \\ L_{ei} = \frac{1}{gA_i}\ [\mathrm{s^2/m^3}] \end{cases} \tag{3}$$

The equivalent resistance R_{ei}, the equivalent inductance L_{ei}, and the equivalent capacitance C_{ei} are related to the head losses through the pipe, the inertia effect of the water, and the storage effect caused by pressure variation, respectively. These three different circuit components are essential to precisely model a pressurized pipe. The equivalent circuit of a pressurized pipe containing n segments is displayed in Figure 1. The hydraulic transients along a pipeline are characterized as a discrete RLC circuit in which the state variables, i.e., the piezometric head and the discharge, can be obtained only for given locations x and given times t, according to the sizes of the time step dt and the spatial mesh dx.

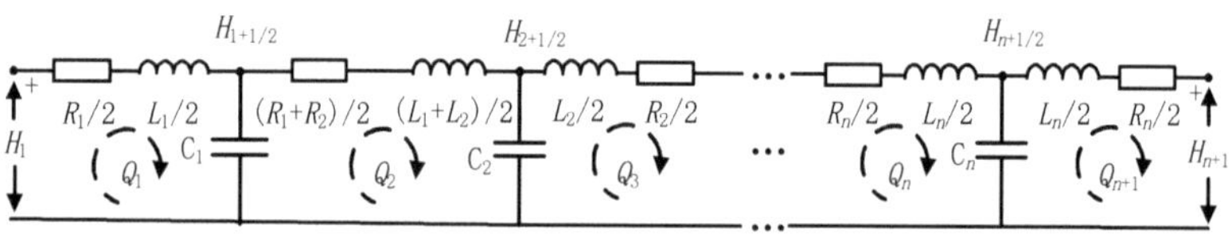

Figure 1. The equivalent circuit of a pressurized pipe.

Apart from the time domain differential equations for numerical simulation [19–25], the ECM can also be used to investigate the complex frequency-related features of the hydraulic system, such as the transfer matrix and the hydraulic impedance. The transfer matrix is often used to calculate the forced oscillation responses of the system, while the hydraulic impedance is more suitable for free oscillation analysis. In this study, we focus on the free oscillation analysis. If a pipe is modeled by n T-shape equivalent circuit segments, the equivalent impedance is calculated recursively from one end of the pipe according to the circuit theory. The equivalent impedance of the ith loop can be obtained with Equation (4),

$$Z_{equ_i} = (R_i + s \cdot L_i/2) + \frac{\left(R_i + s \cdot L_i/2 + Z_{equ_{i+1}}\right) \cdot \frac{1}{s \cdot C}}{R_i + s \cdot L_i/2 + Z_{equ_{i+1}} + \frac{1}{s \cdot C}} \tag{4}$$

where, $R_i = R_{ei} \cdot dx$, $L_i = L_{ei} \cdot dx$ and $C_i = C_{ei} \cdot dx$. The positive directions of the piezometric head and discharge are set as directions shown in Figure 1. When the impedance is calculated from the upstream side to the downstream side, the directions of the piezometric head and discharge follow the preset positive direction, otherwise, the direction of discharge follows the negative direction. It should be emphasized that Equation (4) works when discharges follow the directions marked in Figure 1. Otherwise, the hydraulic impedance $Z'_{equ_i} = -Z_{equ_i}$ when the discharges are in the opposite direction, where Z_{equ_i} is the impedance calculated with Equation (4).

The proposed complex frequency domain ECM transforms the piping system into a spatial discretization approximation of the continuous system, while the traditional impedance method computes the impedance of a certain location directly through the analytical solution. The pipe is divided into several pipe segments in the ECM, which brings inevitable numerical error led by spatial discretization. However, the well-developed circuit theory enables the ECM to be applied to systems with complex hydraulic layouts. For complete frictionless systems, the system impedances can be calculated at one end for a given range of frequencies and identify which satisfies the known boundary conditions. For dissipative systems, the problem then becomes searching for the complex frequencies satisfying all the hydraulic boundary conditions. It usually leads to an iterative numerical solving process searching for the minimum of a predefined objective function.

2.2. *Typical Hydraulic Boundary Treatments*

The treatment of different boundaries is also important for hydraulic system modeling [29].

2.2.1. Hydraulic Turbine

The hydraulic turbine is usually considered one of the most complicated hydraulic boundaries in the transient analysis of the hydropower station system. According to the theory proposed by Wylie and Streeter [30], the hydraulic impedance of the hydraulic turbine can be reflected by Equation (5),

$$Z_T = \frac{2n}{n_{11} D_1 \left(Q_{11} - \frac{\partial Q_{11}}{\partial n_{11}} \cdot n_{11}\right)} \tag{5}$$

where the impedance is highly related to the characteristic curves of the hydraulic turbine. n and D_1 are the rotational speed and runner's diameter, respectively. n_{11} and Q_{11} denote the unit rotational speed and unit discharge, respectively.

Suppose that $k_1 = Q_{11}/n_{11}$ and $k_2 = \partial Q_{11}/\partial n_{11}$, it is found that only if $k_1 < k_2$, will the hydraulic impedance of the turbine Z_T be negative, otherwise the value of Z_T is positive. Seen from Figure 2, the slope of Q_{11}/n_{11} is always positive and the slope of $\partial Q_{11}/\partial n_{11}$ is negative, which means that the turbine impedance $Z_T > 0$ when the power station is in steady turbine operation.

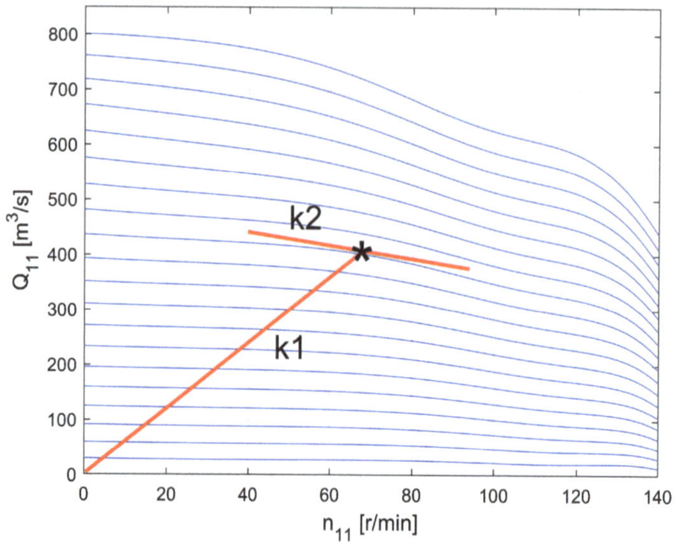

Figure 2. The discharge characteristic curve of a hydraulic turbine.

2.2.2. Reservoir

For most scenarios of applied hydraulic transients, the water level of a reservoir is often assumed to keep constant during the whole process of interest regardless of the possible variation in discharges, i.e., the piezometric head variation $h_r = 0$. Therefore, the reservoir boundary can be treated as an open end, i.e., the impedance $Z_r = 0$.

3. Oscillation Analysis of the Hydraulic System of a Hydropower Station

3.1. Introduction to the System Plant

A hydropower station system, firstly introduced in [9], is taken as the system plant in this study. The structural schematic of the system is depicted in Figure 3.

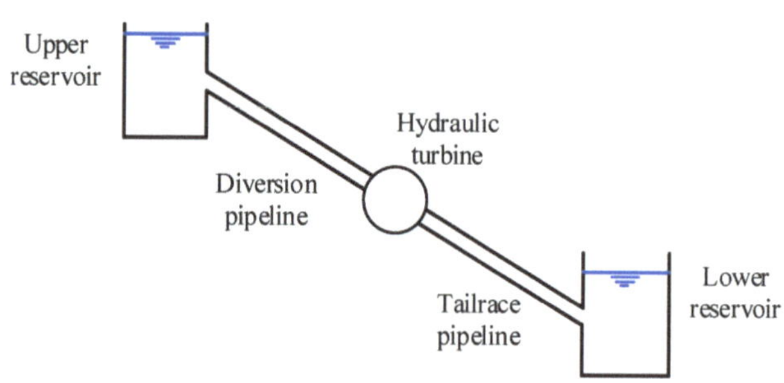

Figure 3. Structural schematic of the power station of interest.

The system is composed of an upstream reservoir, a diversion pipeline, a hydraulic turbine, a tailrace pipeline, and a downstream reservoir. The equivalent circuit of the hydropower station system is illustrated in Figure 4. The essential parameters of the pipeline system in the hydropower station are given in Table 1.

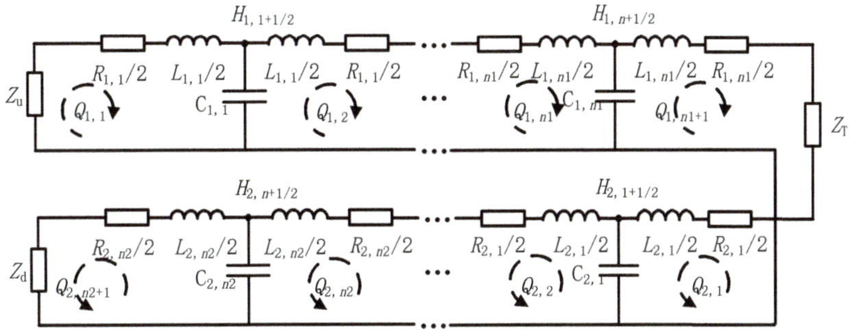

Figure 4. Equivalent circuit of the hydraulic system of a hydropower station.

Table 1. Parameters of the pipeline system in the hydropower station.

Pipe No.	*L* (m)	*A* (m^2)	*a* (m/s)	Z_c (s/m^2)	f_{10} (Hz)
#1	1000.0	50.24	1100.0	2.2308	0.2751
#2	100.0	78.50	1000.0	1.2979	2.5013

3.2. Frequency Response Validation of ECM

The frequency responses of the downstream end of the diversion pipe of the hydropower station in Figure 3, obtained with the proposed equivalent circuit-based impedance and the traditional continuous model [9] under different spatial steps, are displayed in Figure 5, respectively.

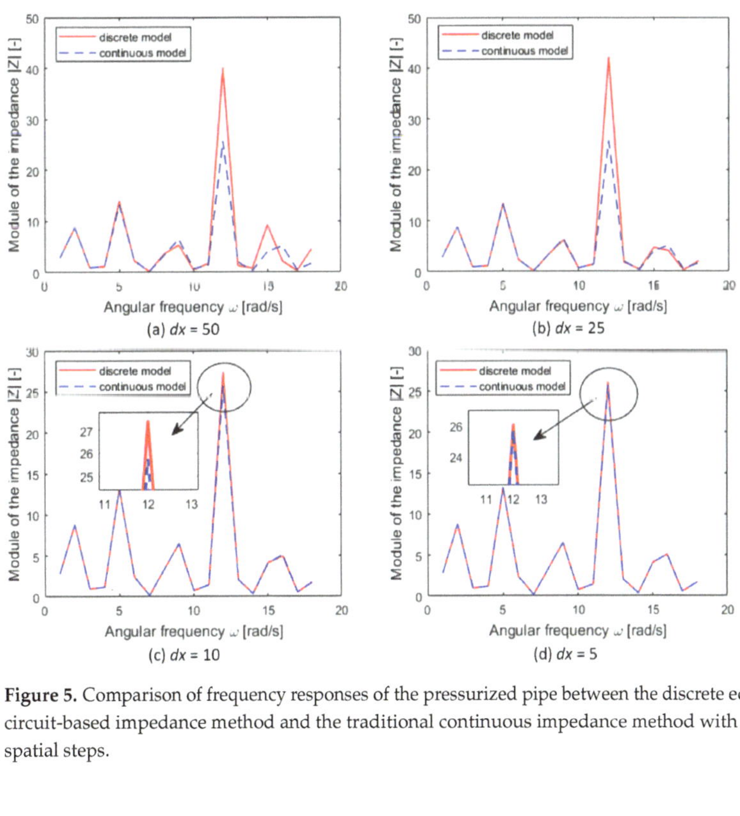

Figure 5. Comparison of frequency responses of the pressurized pipe between the discrete equivalent circuit-based impedance method and the traditional continuous impedance method with different spatial steps.

The modeling accuracy of the discrete system models is sensitive to the choice of spatial discretization. Compared with the frequency-related analytical solutions obtained from the continuous model, the discrete impedance model tends to have more precise responses for lower frequency regions. The modeling performances for higher-order oscillations are more prone to deteriorate if dx is not small enough, as shown in Figure 5. In addition, the modeling precision for higher frequency oscillations increases when the value of the spatial step further decreases.

3.3. Free Oscillation Response Analysis of the Hydropower Station System

From Figure 5, the impedance at the upstream reservoir Z_{equ} also can be obtained by calculating the total impedance of the series and parallel circuits using Equation (4) from the downstream side to the upstream side according to the circuit theory. Since the upstream and downstream reservoirs are considered to be two open ends, the impedances of the two reservoir boundaries are $Z_u = Z_d = 0$. The system stability is highly related to the oscillation characteristics of the hydraulic circuit in the power station. The oscillation patterns of different orders of the hydropower system are obtained by searching for the complex frequencies that satisfy the boundary condition $Z_{equ}(s) = Z_u = 0$ via the free oscillation response analysis. The real and imaginary parts of each complex frequency represent the decay coefficient and the angular frequency of the oscillation of the corresponding order, respectively.

It is not easy to obtain the analytical solutions to the equation $Z_{equ}(s) = Z_u = 0$, for the expression of the equation is quite complex. For this reason, the well-known Newton–Raphson method [2,9,25] is used here to calculate the numerical solutions. Due to the spatial discretization of the pipeline, the mathematical expression of the overall complex frequency domain ECM model is quite complicated, and the model's order is pretty high. Hence, it is usually rather difficult to deduce the derivative of the objective function of the optimization problem directly. For this reason, we use the first-order linear approximation of the Newton iterative method to avoid the derivation deduction process. The detailed process for this problem is shown in Figure 6.

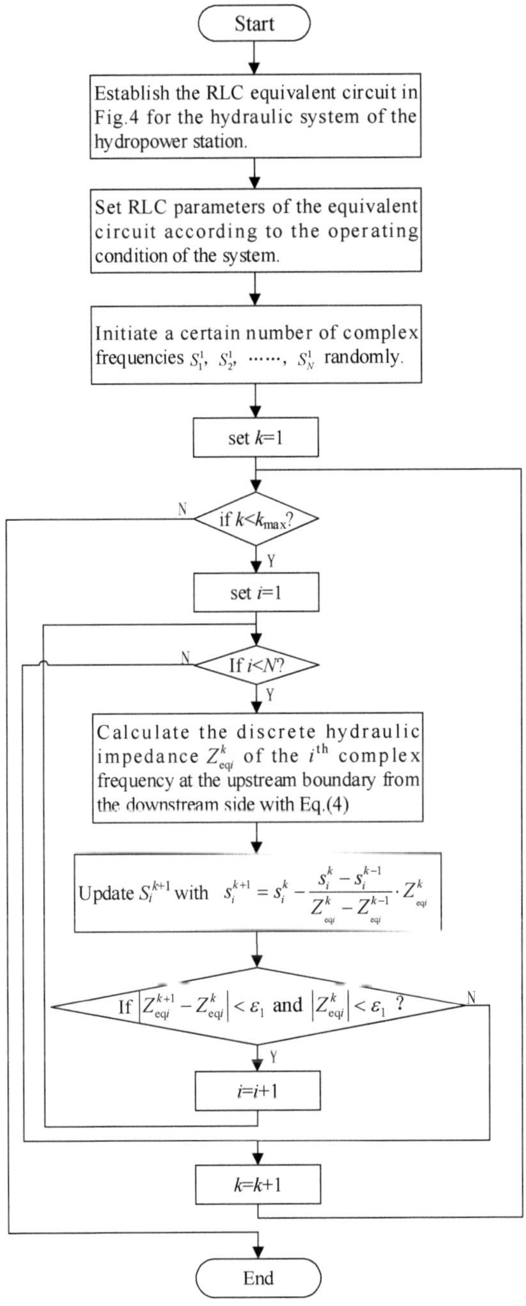

Figure 6. Flow chart of the free oscillation responses analysis.

4. Simulation Result Analyses

Simulation experiments were conducted to reveal the features of different orders of oscillation in the hydropower station system. Since there are two pipelines in the system plant, i.e., the diversion pipeline and the tailrace pipeline, the oscillation characteristics of either pipe may be influenced by the other. In addition, the influence of the hydraulic turbine impedance on the system's oscillation patterns needs further investigation. Note that the friction losses along the pipelines are neglected to explore the damping effect of the hydraulic turbine.

4.1. Oscillation Mode Characteristics of the Overall Hydropower Station System

It is known that the continuous model of the hydropower station system has infinite degrees of freedom. Therefore, the oscillation order for this system can be as high as infinite. However, as the proposed ECM is a discrete approximation to the system, only the characteristics of the lower orders of oscillation are precise in the modeling, as discussed in Section 2.1. Firstly, we investigate the decay coefficients of different orders of oscillation. It is found that the damping characteristics of the system oscillations when $Z_T \leq 3.5$ and $Z_T \geq 3.6$ are quite different from each other. The variation curves of the decay coefficients of the oscillations under different hydraulic turbine operating conditions with relatively small values of Z_T ($Z_T \leq 3.5$) are illustrated in Figure 7.

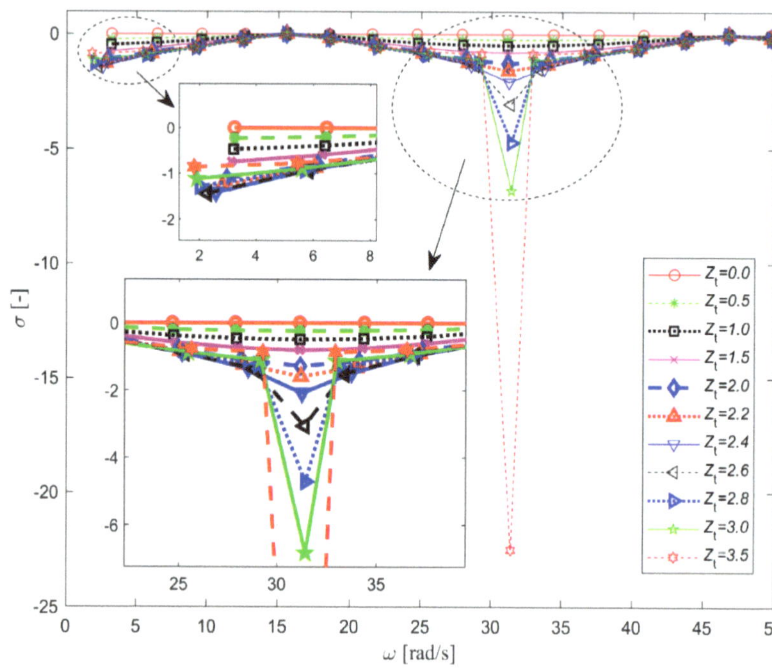

Figure 7. Decay coefficients of different orders of oscillations under different operating conditions of the hydraulic turbine when $Z_T \leq 3.5$.

Figure 7 shows that the all the decay coefficients gradually move towards zero when the angular frequency is less than 15 rad/s. At the 5th order of oscillation (about 15.58 rad/s), the decay coefficient reaches the crest and is quite close to zero (but is still negative). When the angular frequency is within 15 rad/s to 32 rad/s, the decay coefficient first decreases gradually as the angular frequency increases. At the 10th order of oscillation (about 31.16 rad/s), the decay coefficient reaches the nadir and its value is much smaller than those of the neighboring orders. Then, the decay coefficient gradually increases and reaches another crest at the 15th order of oscillation (about 46.73 rad/s), where the value at this crest is also negative and close to zero.

The decay coefficients of the oscillations of the 1st, 2nd, 3rd, 4th, 6th, 7th, 8th, 9th, 11th, 12th, 13th, 14th, and 15th orders are within the range of [−2, 0], but the value of the decay coefficient of the 10th-order oscillation varies from −22 to 0, which is quite different from the oscillations of other orders. The corresponding angular frequency ranges from 31.1 rad/s to 31.4 rad/s, which is very close to twice the base natural frequency of pipe #2 given in Table 1.

Different from that of the hydraulic turbine operating conditions with smaller values of Z_T ($Z_T \leq 3.5$), the variation curve of the decay coefficient of the system frequency responses under the hydraulic turbine operating conditions with larger values of Z_T ($Z_T \geq 3.6$) is illustrated in Figure 8.

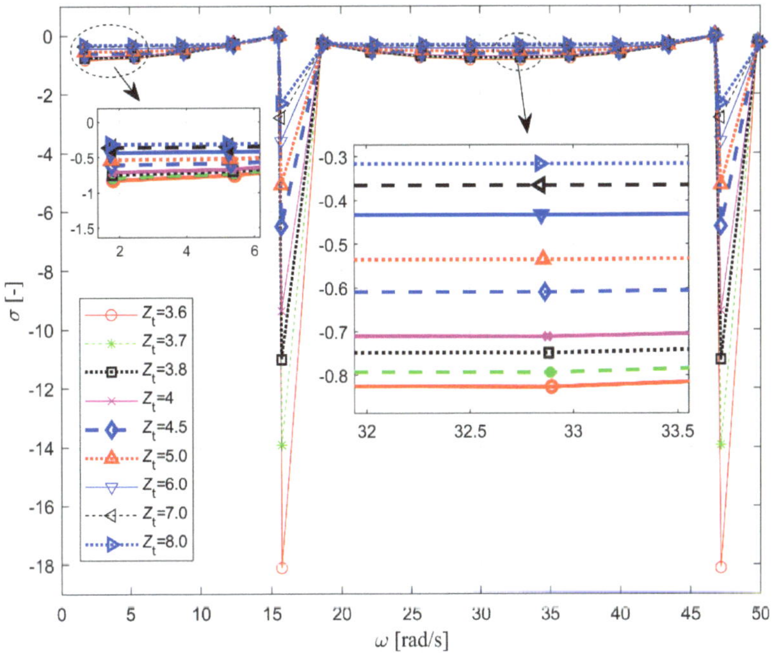

Figure 8. Decay coefficients of different orders of oscillations under different operating conditions of the hydraulic turbine when $Z_T \geq 3.6$.

Figure 8 shows that all the decay coefficients gradually increase to zero when the angular frequency is less than 15 rad/s, which is the same trend compared to that in Figure 7. At the 6th-order oscillation (about 15.70 rad/s), the decay coefficient suddenly decreases sharply to the nadir (which is much smaller than the value of neighboring orders) and then rapidly surges to a normal value (about σ = −0.28) at the 7th-order oscillation. The decay coefficient at the 6th-order oscillation ranges from −18.11 to −2.32 when Z_T decreases from 8 to 3.6. While the angular frequency is within 16 rad/s to 33 rad/s, the decay coefficient at first decreases gradually as the angular frequency increases. The decay coefficient reaches the nadir at the 11th-order oscillation (about 32.88 rad/s). Then, the decay coefficient gradually increases and reaches another crest at the 15th order of oscillation (about 46.72 rad/s), where its value is negative and close to zero. Afterward, the decay coefficient suddenly decreases sharply to the unusual nadir again at the 16th-order oscillation (about 47.11 rad/s) and rapidly rises to the normal crest (within the range of [−0.26, −0.21]) at the 17th-order oscillation (about 49.85 rad/s). From Figure 8, we can also observe the angular frequencies of the two unusual oscillations (the 6th-order and the 16th-order) are 15.70 rad/s and 47.11 rad/s, respectively. Their values are close to the 1st-order (15.71 rad/s) and 3rd-order (47.12) natural frequencies of pipe #2, respectively.

It is proven that in a simple reservoir–pipe–valve hydraulic system [25], the frequencies of the free oscillation responses at the valve inlet are even harmonics when the hydraulic impedance of the valve Z_v is smaller than the characteristic impedance of the pipe Z_c (i.e., $Z_v < Z_c$). In contrast, the frequencies of the free oscillation responses at the valve inlet are odd harmonics when $Z_v > Z_c$. This result is quite related to our aforementioned findings. In this study, the oscillation frequency distribution obtained in the hydropower station system also presents obvious odd harmonics performance when Z_T is relatively large and even harmonics performance when Z_T is relatively small. The hydraulic turbine acts as the valve in the reservoir–pipe–valve system. However, the system plant is more complex and consists of multiple pipelines, the coupling effect of the pipes greatly influences the frequency responses of the overall system. The critical hydraulic turbine impedance Z_T to determine the two different oscillation modes is located within the range of [3.5, 3.6]. According to the analysis above, the 10th-order oscillation when $Z_T \leq 3.5$ and the 6th-order and 16th-order oscillations when $Z_T \geq 3.6$ are considered to correspond to the natural frequencies of pipe #2.

4.2. Variation Trends of the Decay Coefficients of Oscillations of the Two Pipes

Apart from the three oscillations related to pipe #2, all other oscillations are considered related to pipe #1. The decay coefficients corresponding to different orders of oscillation related to pipe #1 under the different conditions of Z_T are depicted in Figure 9. As seen in Figure 9, the decay coefficients of the 5th-order oscillation and the 14th-order oscillation are kept close to zero for all values of Z_T. The angular frequencies of the two orders are 15.58 rad/s and 46.73 rad/s, respectively. These two frequencies are close to the 1st-order and 3rd-order oscillations of pipe #2. The 4th-order, 6th-order, 13th-order, and 15th-order oscillations of pipe #1 have relatively weak damping capability because the absolute values of their decay coefficients for all Z_T are relatively small. The reason why they have slow convergence characteristics is that the 4th-order and the 6th-order oscillations are next to the 5th-order oscillation, and the 13th-order and 15th-order oscillations are next to the 14th-order oscillation. On the contrary, the 1st-order, 9th-order, and 10th-order oscillations have a relatively strong damping capability because the absolute values of their decay coefficients for all Z_T are relatively large. The reason why they have fast convergence characteristics is that the angular frequencies of the 9th-order oscillation and the 10th-order oscillation are next to the 2nd-order oscillation of pipe #2, and the 1st-order oscillation is next to the 0th-order oscillation (i.e., the DC component).

In addition, the decay coefficients of the oscillation modes related to pipe #1 are within the range of $[-1.6, 0]$. For each order of oscillation, the decay coefficient value decreases to the corresponding nadir as Z_T increases when Z_T is small, then the decay coefficient increases as Z_T continues to increase. However, the critical hydraulic turbine impedance of the decay coefficient nadir for each oscillation order is different.

The decay coefficient variation curves of the first three orders of oscillation related to pipe #2 with different Z_T are shown in Figure 10. It is seen that the 1st-order oscillation and the 3rd-order oscillation can be observed when $Z_T \geq 3.6$ and their decay coefficients σ nearly have the same variation trend with each other. σ increases gradually as the value of Z_T increases. On the contrary, the 2nd-order oscillation is observed when $Z_T \leq 3.5$, and σ gradually decreases as the value of Z_T increases.

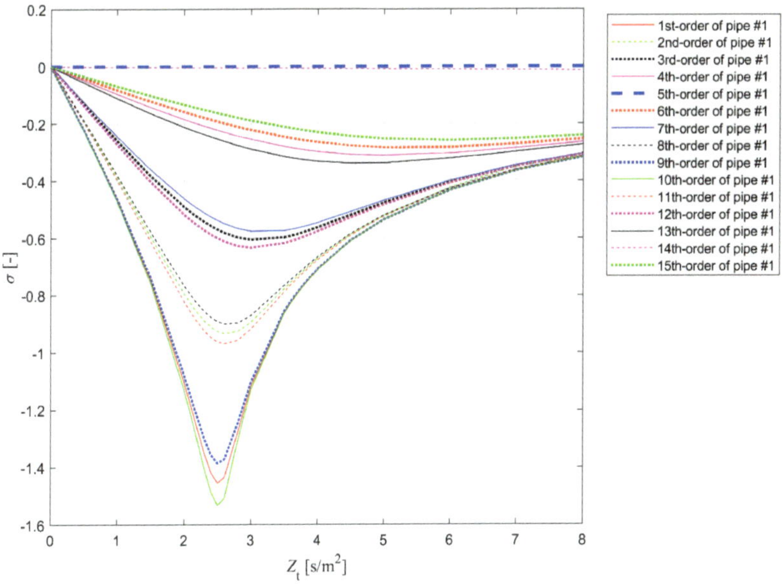

Figure 9. Damping coefficients correspond to different orders of oscillation related to pipe #1 under the conditions of different Z_T.

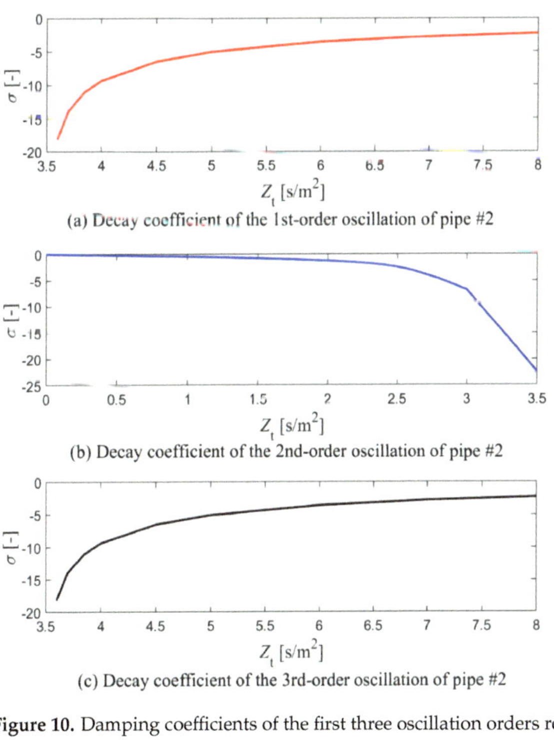

(a) Decay coefficient of the 1st-order oscillation of pipe #2

(b) Decay coefficient of the 2nd-order oscillation of pipe #2

(c) Decay coefficient of the 3rd-order oscillation of pipe #2

Figure 10. Damping coefficients of the first three oscillation orders related to pipe #2 with different Z_T.

4.3. Influence of the Turbine Impedance on the Oscillations Frequencies Distribution

In Sections 4.1 and 4.2, the decay coefficients corresponding to different orders of oscillation are investigated. In this subsection, we reveal the variation trend of the oscillation frequencies of the system. From Figures 7 and 8, the oscillation modes of the hydropower system are separated into two parts, i.e., the oscillations related to pipe #1 and the oscillations related to pipe #2. As discussed in Section 4.1, the oscillation frequencies' distribution of the hydropower station system tends to present odd harmonics performance when Z_T is relatively large and even harmonics performance when Z_T is relatively small. The critical turbine impedance of the two oscillation patterns is within the range of [3.5, 3.6]. Therefore, the variation trends of the frequencies of different orders of oscillation of the system with different Z_T are illustrated in Figure 11.

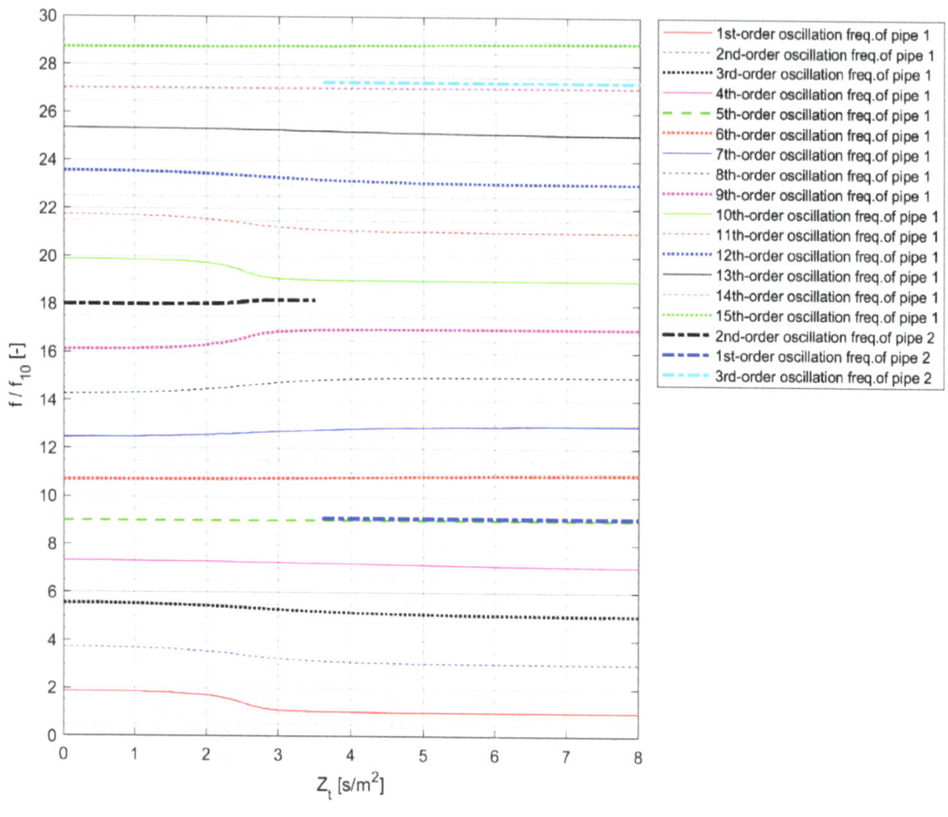

Figure 11. Variation trends of the frequencies of different orders of oscillation of the system with different Z_T.

Both frequency responses of pipe #1 and pipe #2 when $Z_T \leq 2$ can be considered even harmonics, although the frequencies of some oscillation orders of pipe #1 in this region are greatly affected by the coupling effect of the two pipes. On the contrary, the frequency responses of pipe #1 and pipe #2 when $Z_T \geq 5$ are odd harmonics for pipe #1 and pipe #2. Furthermore, an obvious frequency transition region can be observed between Z_T = 2 and Z_T = 5. The oscillation modes of pipe #1 gradually shift to the odd harmonics from the even harmonics as Z_T increases. During this transition, the oscillation modes of pipe #1 are between the two oscillation modes. However, the oscillation modes of pipe #2 suddenly change at a critical turbine impedance (within the region of $Z_T \in$ [3.5, 3.6]).

5. Conclusions

An equivalent circuit-based hydraulic impedance model for a hydropower station is proposed in this paper. The free oscillation analysis of the hydraulic system of the hydropower station with this method is carried out. Several conclusions are condensed below:

(1) Compared with the analytical frequency responses of the traditional continuous hydraulic impedance method, the frequency responses of the proposed equivalent circuit-based discrete impedance method for higher frequency impulses are more prone to being inaccurate. The overall modeling accuracy is sensitive to the precision of the spatial meshing.

(2) There exists a critical hydraulic turbine impedance Z_{TC} to determine the oscillation modes of the system. Performances of the system's oscillation pattern for a relatively small Z_T and a relatively large Z_T are different. When $Z_T < Z_{TC}$, the oscillation modes are even harmonics, and when $Z_T > Z_{TC}$, the oscillation modes are odd harmonics.

(3) An obvious transition region of the frequency response for pipe #1 can be observed around the critical turbine impedance. The oscillation frequency modes of pipe #1 gradually shift to the odd harmonics from the even harmonics as Z_T increases. However, the oscillation modes of pipe #2 suddenly change at a critical turbine impedance, and no transition process is observed.

Author Contributions: Conceptualization, S.Y. and Y.Z.; methodology, S.Y.; validation, S.Y., Y.Z. and J.C.; formal analysis, S.Y.; investigation, Y.S.; resources, Y.S.; data curation, S.Y.; writing—original draft preparation, S.Y. and Y.Z.; writing—review and editing, S.Y.; visualization, S.Y.; supervision, Y.Z.; project administration, Y.Z.; funding acquisition, Y.Z. All authors have read and agreed to the published version of the manuscript.

Funding: This research was funded by National Natural Science Foundation of China (grant number: 52009096), China Postdoctoral Science Foundation (grant number: 2022T150498) and the Fundamental Research Funds for the Central Universities (grant number: 2042022kf1022).

Institutional Review Board Statement: Not applicable.

Informed Consent Statement: Not applicable.

Data Availability Statement: Not applicable.

Conflicts of Interest: The authors declare no conflict of interest.

Nomenclature

v	Flow velocity
a	Wave propagation speed
t	Time
x	Horizontal location along the pipeline
H	Piezometric head
Q	Discharge
A	Sectional area of the pipe
D	Diameter of the pipe
g	Gravitational acceleration
λ	Friction loss coefficient of the pipe
R_{ei}	Equivalent resistance per unit length
L_{ei}	Equivalent inductance per unit length
C_{ei}	Equivalent capacitance per unit length
R	Equivalent resistance
L	Equivalent inductance
C	Equivalent capacitance
Z_{equ}	Equivalent hydraulic impedance
s	Laplace operator, also the complex frequency
Z_T	Turbine impedance
D_1	Diameter of the runner

n_{11}	Unit rotational speed
Q_{11}	Unit discharge
Z_r	Reservoir impedance
h_r	Piezometric head variation of the reservoir
Z_c	Characteristic impedance of the pipe
ω_0	Base frequency of the pipe

References

1. Zheng, Y.; Chen, Q.; Yan, D.; Liu, W. A two-stage numerical simulation framework for pumped-storage energy system. *Energy Conv. Manag.* **2020**, *210*, 112676. [CrossRef]
2. Chaudhry, M.H. *Applied Hydraulic Transients*; Springer: London, UK, 2014.
3. Alubokin, A.A.; Gao, B.; Ning, Z.; Yan, L.L.; Jiang, J.X.; Quaye, E.K.; Evans, K. Numerical simulation of complex flow structures and pressure fluctuation at rotating stall conditions within a centrifugal pump. *Energy Sci. Eng.* **2022**, *10*, 2146–2169. [CrossRef]
4. Chen, G.; Liu, C.; Fan, C.W.; Han, X.Y.; Shi, H.B.; Wang, G.H.; Ai, D.P. Research on damping control index of ultra-low-frequency oscillation in hydro-dominant power systems. *Sustainability* **2020**, *12*, 7316. [CrossRef]
5. Lu, X.; Li, C.; Liu, D.; Zhu, Z.; Tan, X. Influence of water diversion system topologies and operation scenarios on the damping characteristics of hydropower units under ultra-low frequency oscillations. *Energy* **2022**, *239*, 122679. [CrossRef]
6. Chen, J.; Xiao, Z.; Liu, D.; Hu, X.; Ren, G.; Zhang, H. Nonlinear modeling of hydroturbine dynamic characteristics using LSTM neural network with feedback. *Energy Sci. Eng.* **2021**, *9*, 1961–1972. [CrossRef]
7. Guo, W.; Zhu, D. Critical stable sectional area of downstream surge tank of hydropower plant with sloping ceiling tailrace tunnel. *Energy Sci. Eng.* **2021**, *9*, 1090–1102. [CrossRef]
8. Liu, D.; Li, C.; Malik, O.P. Operational characteristics and parameter sensitivity analysis of hydropower unit damping under ultra-low frequency oscillations. *Int. J. Electr. Power Energy Syst.* **2022**, *136*, 107689. [CrossRef]
9. Yang, J.D. *Applied Fluid Transients*; Science Press: Beijing, China, 2018.
10. Suo, L.; Wylie, E.B. Impulse response method for frequency-dependent pipeline transients. *J. Fluids. Eng. ASME* **1989**, *111*, 478–483. [CrossRef]
11. Kim, S.H. Impedance matrix method for transient analysis of complicated pipe networks. *J. Hydraul. Res.* **2007**, *45*, 818–828. [CrossRef]
12. Kim, S.H. Dynamic memory computation of impedance matrix method. *J. Hydraul. Eng.* **2011**, *137*, 122–128. [CrossRef]
13. Kim, S.H. Address-oriented impedance matrix method for generic calibration of heterogeneous pipe matrix network systems. *J. Hydraul. Eng.* **2008**, *134*, 66–75. [CrossRef]
14. Zhou, J.; Suo, L.; Hu, M. Study on self-excited vibration of hydro-mechanical system in pumped storage plants. *J. Hydraul. Eng.* **2007**, *38*, 1080–1084.
15. Gong, J.Z.; Lambert, M.F.; Simpson, A.R.; Zecchin, A.C. Single-event leak detection in pipeline using first three resonant responses. *J. Hydraul. Eng.* **2013**, *139*, 645–655. [CrossRef]
16. Duan, H.F.; Lee, P.J.; Ghidaoui, M.S.; Tung, Y.K. Extended blockage detection in pipelines by using the system frequency response analysis. *J. Water Res. Plan. Manag.* **2011**, *138*, 55–62. [CrossRef]
17. Zhang, C.; Wang, Y.T.; Liu, Y.; Ding, W. Vulnerability Analysis of Urban Drainage Systems: Tree vs. Loop Networks. *Sustainability* **2017**, *9*, 397. [CrossRef]
18. Kim, S.H.; Choi, D. Dimensionless Impedance Method for General Design of Surge Tank in Simple Pipeline Systems. *Energies* **2022**, *15*, 3603. [CrossRef]
19. Paynter, H.M. Surge and water hammer problems. *ASCE Trans.* **1953**, *146*, 962–1009. [CrossRef]
20. Jaeger, C. *Fluid Transients in Hydro-Electric Engineering Practice*; Blackie: Glasgow, UK, 1977.
21. Souza, O.H.; Barbieri, N.; Santos, A.H. Study of hydraulic transients in hydropower plants through simulation of nonlinear model of penstock and hydraulic turbine model. *IEEE Trans. Power Syst.* **1999**, *14*, 1269–1272. [CrossRef]
22. Nicolet, C.; Greiveldinger, B.; Hérou, J.J.; Kawkabani, B.; Allenbach, P.; Simond, J.J.; Avellan, F. High-order modeling of hydraulic power plant in islanded power network. *IEEE Trans. Power Syst.* **2007**, *22*, 1870–1880. [CrossRef]
23. Zhao, Z.; Yang, J.; Yang, W.; Hu, J.; Chen, M. A coordinated optimization framework for flexible operation of pumped storage hydropower system: Nonlinear modeling, strategy optimization and decision making. *Energy Convers. Manag.* **2019**, *194*, 75–93. [CrossRef]
24. Zheng, Y.; Chen, Q.; Yan, D.; Zhang, H. Equivalent circuit modeling of large hydropower plants with complex tailrace system for ultra-low frequency oscillation analysis. *Appl. Math. Model.* **2022**, *103*, 176–194. [CrossRef]
25. Nicolet, C. *Hydroacoustic Modelling and Numerical Simulation of Unsteady Operation of Hydroelectric Systems*; Ecole Polytechnique Federale de Lausanne: Lausanne, Switzerland, 2007.
26. Zhang, Y.; Zang, W.; Zheng, J.; Cappietti, L.; Zhang, J.; Zheng, Y.; Fernandez-Rodriguez, E. The influence of waves propagating with the current on the wake of a tidal stream turbine. *Appl. Energy* **2021**, *290*, 116729. [CrossRef]
27. Zhang, Y.; Zhang, J.; Lin, X.; Wang, R.; Zhang, C.; Zhao, J. Experimental investigation into downstream field of a horizontal axis tidal stream turbine supported by a mono pile. *Appl. Ocean. Res.* **2020**, *101*, 102257. [CrossRef]

28. Liu, D.; Li, C.; Malik, O.P. Nonlinear modeling and multi-scale damping characteristics of hydro-turbine regulation systems under complex variable hydraulic and electrical network structures. *Appl. Energy* **2021**, *293*, 116949. [CrossRef]
29. Zhang, Y.; Zhang, Z.; Zheng, J.; Zhang, J.; Zheng, Y.; Zang, W.; Lin, X.; Fernandez-Rodriguez, E. Experimental investigation into effects of boundary proximity and blockage on horizontal-axis tidal turbine wake. *Ocean. Eng.* **2021**, *225*, 108829. [CrossRef]
30. Wylie, E.B.; Streeter, V.L. *Fluid Transients in Systems*; Prentice-Hall: London, UK, 1993.

sustainability

MDPI

Article

Improved Neural Network Algorithm Based Flow Characteristic Curve Fitting for Hydraulic Turbines

Hong Pan [1], Chenyang Hang [1,*], Fang Feng [1], Yuan Zheng [1,2] and Fang Li [1]

1 College of Energy and Electrical Engineering, Hohai University, Nanjing 211100, China
2 College of Water Conservancy and Hydropower Engineering, Hohai University, Nanjing 210003, China
* Correspondence: 211606010028@hhu.edu.cn

Abstract: One of the most important characteristic curves in the economic operation of hydropower plants is the turbine flow characteristic curve, which illustrates the law of fluctuation between the characteristic parameters of the turbine under various operating situations. This article proposes an IPSO-LSTM-based refinement method for fitting the turbine flow characteristic curve using deep learning methods, and verifies the effectiveness of the method by comparison to solve the problem that traditional mathematical fitting methods are difficult to meet the requirements of today's many complex working conditions. Firstly, a deep LSTM network model is established based on the input and output quantities, and then the IPSO method is used to find the optimum number of neurons, the learning rate and the maximum number of iterations of the LSTM units in the network model and other key parameters to determine the relevant training parameters. The results show that the model can effectively improve the accuracy of fitting and predicting the turbine flow characteristics, which is of great significance to the study of the economic operation of hydropower plants and the non-linear characteristics of the turbine.

Keywords: LSTM; hydraulic turbines; flow characteristic curves; deep learning; IPSO

Citation: Pan, H.; Hang, C.; Feng, F.; Zheng, Y.; Li, F. Improved Neural Network Algorithm Based Flow Characteristic Curve Fitting for Hydraulic Turbines. *Sustainability* **2022**, *14*, 10757. https://doi.org/10.3390/su141710757

Academic Editors: Bin Huang, Ling Zhou, Yuquan Zhang and Jianhua Zhang

Received: 30 July 2022
Accepted: 24 August 2022
Published: 29 August 2022

1. Introduction

In-plant economic operation of hydropower plants is an important way to achieve reasonable load distribution and efficient economic operation of hydropower plant units. In the process of unit load allocation, the solution algorithm will obtain the relationship between hydraulic head, flow and output at the corresponding moment according to the flow characteristic curve of the unit, and then carry out the load allocation among units [1,2]. Therefore, the accuracy of the unit flow characteristic curve fitting has an important impact on the load distribution of hydropower plant units and the economic operation of the plant.

Due to site and cost constraints, the flow characteristic curve of a hydropower plant unit is usually first tested by the manufacturer in a model, then converted to obtain the characteristic curve of a prototype turbine, and finally, the flow rate of power consumption of the unit is calculated according to the turbine output formula [3,4]. In this process, due to the size effect, there is a certain gap between the model test and the prototype test, and the unit parameters of the model and the prototype cannot fully satisfy the similarity law, and there is a difference in the efficiency of the unit under the same operating conditions. At the same time, there is also a large difference between the unit flow consumption calculated according to the unit output formula and the actual unit flow consumption [5]. Therefore, using the large amount of real machine data generated by the actual operation of the unit to process the flow characteristic curve is the most realistic way to reflect the relationship between the characteristic parameters of the unit at each operating condition.

Traditional mathematical interpolation and neural network methods have been applied in the fitting of hydraulic turbine characteristic curves. Zhang Rongsheng et al. [6] used triangular dissection for interpolation of discrete hydraulic turbine comprehensive characteristic data to avoid the duality of data obtained by geometry-based methods; Liu Dong et al. [7] used power station monitoring data input to modify the hydraulic turbine characteristic curve model, which effectively improved the simulation accuracy of the model; Li Junyi [8] and Huang Xianrong [9] used BP neural network and radial basis function neural network to fit the hydraulic turbine comprehensive characteristic curve respectively.

From long time scales, the actual operation of hydropower units accumulates a large number of historical operating characteristics parameters, each set of parameters corresponds to an operating condition, and each condition reflects the current moment of the unit load allocation decisions. These historical data have a certain degree of repeatability and the accumulated historical data have a certain reference significance for future unit operating conditions. The traditional research approach ignores the significance of historical data as a guide, and once the curve fitting model and method have been established, the accuracy and speed of calculation will not change over time. The deep learning method is self-learning and self-evolving [10]. It can improve the accuracy of fitting and predicting the unit's flow characteristics curve through continuous learning of operational data during the long-term operation of the unit so that the coupling relationship between the unit's characteristics parameters is closer to the actual operating conditions of the unit.

To this end, this article is based on the real machine operation data of a domestic hydropower station. Firstly, a deep LSTM (Long Short-Term Memory) network model is established based on the input and output quantities. Then the IPSO (Improved Particle Swarm Optimization) method is used to find the optimum number of neurons, network learning rate and maximum number of iterations of the LSTM units in the network model, determine the relevant training parameters and propose an IPSO-LSTM-based method for fitting the flow characteristics curve of a hydraulic turbine. Finally, the validity of the method is verified through relevant example analysis.

2. Theoretical Foundation

For traditional mathematical fitting methods, a logical and clear correspondence between the input and output quantities needs to be established. Once this mapping relationship is established, it cannot be modified and self-learning. The curve fitting method based on traditional neural networks, although capable of self-learning using real machine operating data, is not capable of extracting data features and mining historical data, and is unable to fully explore the relationship between historical operating data and the connection between historical operating data and future operating conditions.

To overcome the above disadvantages, this article proposes an IPSO-LSTM-based method for fitting the flow characteristics curve of hydraulic turbines, using the advantages of deep learning to train the network with a large amount of historical data to construct a mapping relationship between the input and output quantities. The advantage of the deep learning method is that its computational accuracy improves with the increasing amount of training data, but the disadvantage is that if the amount of data is small, the accuracy of the computational results is difficult to guarantee.

The basic framework diagram is shown in Figure 1, where H in the input quantity represents the turbine hydraulic head, P denotes output force and Q in the output quantity denotes flow rate. The basic steps of the turbine flow characteristic curve fitting method used in this article are as follows:

(1) Data pre-processing: Normalization of the collected unit operation data to eliminate the influence of different characteristic parameters with different magnitudes on the learning ability of the network.
(2) Network parameter search for optimization: According to the established deep learning network model, using the improved particle swarm algorithm to optimize the important network parameters and determine the key parameters.
(3) Model training and prediction: After the network parameters are determined, the deep learning network is trained using the training set data, and its prediction performance is evaluated on the test set.

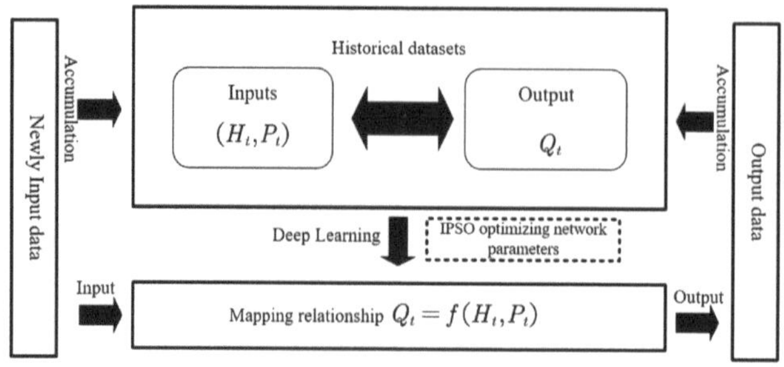

Figure 1. Basic framework chart.

2.1. Data Pre-Processing

The data in this article come from the leading hydropower station in Lancang River Basin. Many sensors in the hydropower plant can read the operation status data of hydropower units in real time. After reading the data, a round of data processing will be done in the hydropower plant. Except for normal data, this step greatly avoids abnormal values in the data set. In addition, when fitting the flow characteristic curve of a turbine, different variables have a wide range of values due to their different amplitudes. If the value is too large, considering the output of the non-linear activation function in the network model, the neuron may be saturated. Therefore, after obtaining the unit data, this research also needs to normalize the training data [11].

We normalize the collected unit hydraulic head, flow and output data and transform the values to the interval [0, 1]:

$$x' = \frac{x - x_{min}}{x_{max} - x_{min}}. \quad (1)$$

where $x = (H_t, P_t), y = Q_t$ represents the input and output of a data sample, and x_{min} and x_{max} denote the minimum and maximum values of each variable, respectively. When normalizing the training set data, both the input and output quantities need to be normalized, whereas the test set only needs to be normalized for the input quantities and the output quantities are inversely normalized.

2.2. LSTM Neural Network

As the operating conditions of the turbine vary with the external load and the characteristic data are collected on a time scale, this article uses Long Short-Term Memory (LSTM) to construct the model. Due to its unique design structure, LSTM is suitable for processing and predicting important events with very long intervals and delays in the time series. LSTM is an extension of the ordinary Recurrent Neural Networks (RNN), which solves the problem of gradient explosion or gradient disappearance during the training process of RNN neural networks, and it makes up for the disadvantage that it does not have a long-term memory function.

As shown in Figure 2, the LSTM network model contains three gates: an input gate, an output gate and a forget gate. Let the input sequence be $X_n = \{x_1, x_2, \ldots, x_n\}$ and the output sequence be $Y_n = \{y_1, y_2, \ldots, y_n\}$, where n is the length of the sequence; h_t is the output of the implied layer; c_t is the state cell of the implied layer, representing the state of the cell at the current moment; f_t is the output of the forget gate; i_t is the output of the input gate; o_t is the output of the output gate; σ and tanh are the activation functions [12]. W_{xf}, W_{xi}, W_{xc} and W_{xo} represent the corresponding weighting factor matrices respectively [13].

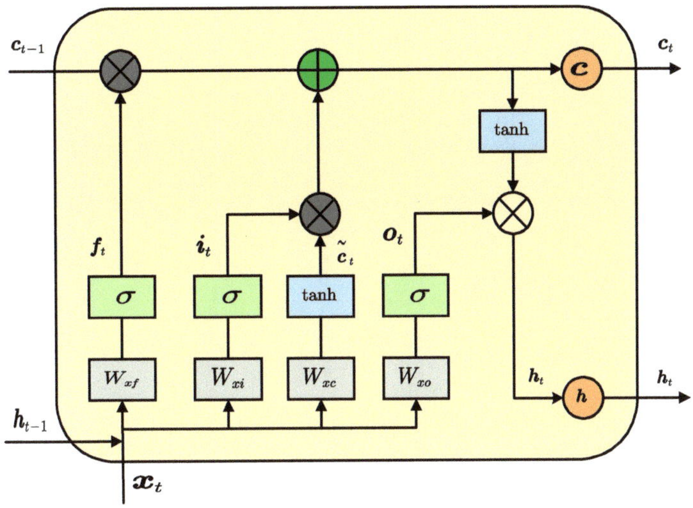

Figure 2. LSTM unit structure.

Each gate in the network is effectively a fully connected layer whose output is a vector with element size between [0, 1], which is used to control the transfer of information. The activation function σ(Sigmoid) of a gate can be expressed as:

$$\sigma(x) = \frac{1}{1 + e^{-x}} \tag{2}$$

The activation function tanh of the hidden layer in the LSTM unitary structure can be expressed as:

$$\tanh = \frac{e^{x} - e^{-x}}{e^{x} + e^{-x}} \tag{3}$$

2.3. Network Modeling

To better explore the hidden relationships between the characteristic parameters of the turbine, a deep neural network model is used for feature extraction of the data. At moment t, the LSTM network hidden layer structure established in this article is shown in Figure 3. The overall structure can be divided into three major parts: the input layer, the hidden layer and the output layer. As the input data has two features, hydraulic head and force (the output of hydraulic turbine), the input layer is set as a 2-dimensional full connection (FC) layer; the hidden layer has two LSTM layers, and each LSTM layer has multiple memory units inside. The output data is flow rate, so the output layer is set as a 1-dimensional full connection layer. Except for the data exchange between the neurons within the LSTM layer, the neurons in the remaining layers are independent of each other, and the only information is exchanged between the layers. The training set data $X_i(H_i, P_i)$ of length T is input to the hidden layer after normalization. To prevent overfitting of the model, a Dropout layer [14] is added after each LSTM layer. It is then back-normalized and output from the output layer is obtained to give the predicted value Y of the model.

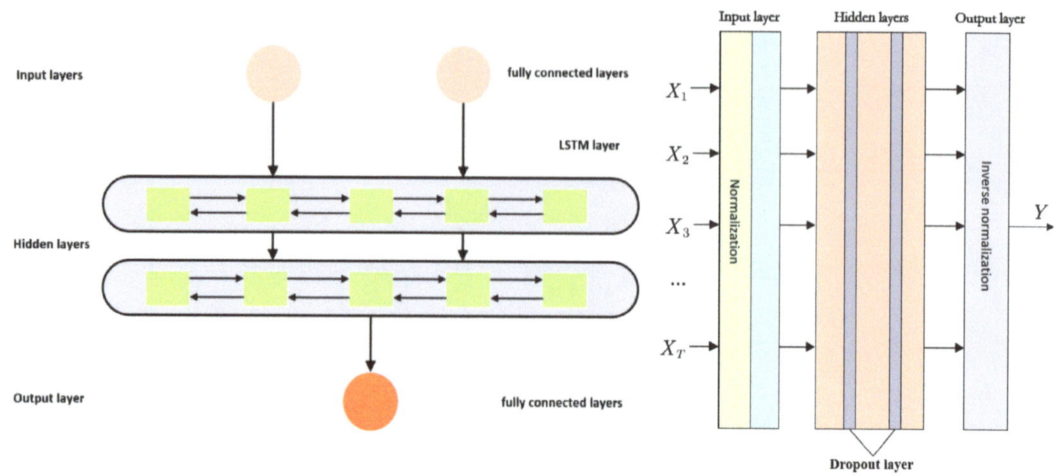

Figure 3. LSTM network structure.

The forward propagation process of LSTM networks can be summarized as [15]: At moment t, the input to the hidden layer of the LSTM network is x_t, the output is h_t and the memory cell is c_t. The input gate is used to control how much of the input data x_t flows into the memory cell at the current moment of the network, i.e., the amount saved into c_t, with the following values:

$$i_t = \sigma(W_{xi}x_t + W_{hi}h_{t-1} + b_i) \tag{4}$$

The forget gate is a key component of the LSTM unit, whose role is to determine the information to be retained or forgotten and to circumvent in some way the phenomenon of gradient disappearance or explosion that occurs when the gradient is propagated backward. The forget gate selects the information to be erased, i.e., it determines the magnitude of the effect of the information in the memory element c_{t-1} at the previous moment on the information in the memory element c_t at the current moment.

$$f_t = \sigma\left(W_{xf}x_t + W_{hf}h_{t-1} + b_f\right) \tag{5}$$

$$c_t = f_t \odot c_{t-1} + i_t \odot \tanh(W_{xc} + W_{hc}h_{t-1} + b_c) \tag{6}$$

The output gate determines the magnitude of the action of the memory cell c_t on the output value h_t at the current moment, i.e., the output value of the memory cell at moment t. The value of the output gate is shown in Equation (7) and the output h_t of the LSTM cell at moment t can be obtained by Equation (8).

$$o_t = \sigma(W_{xo}x_t + W_{ho}h_{t-1} + b_o) \tag{7}$$

$$h_t = o_t \odot \tanh(c_t) \tag{8}$$

In this article, the Adam algorithm [16] is used to train the LSTM network model. The product of the input quantities and the corresponding weight coefficients are first calculated, and then the sum of these products is passed through the action of the activation function. In backpropagation, the gradient value of the loss function for each parameter is calculated, and this is used to update the weight parameter W and the bias parameter b to complete the training process of the network model. Mean Square Error (*MSE*) was chosen as the loss function, and the formula was calculated as follows:

$$MSE = \frac{1}{n}\sum_{i=1}^{n}\left[\widetilde{Y}_i - Y_i\right]^2 \tag{9}$$

where n denotes the sequence length, $\widetilde{Y}_i$ denotes the label value, and Y_i denotes the predicted value of the network model.

3. IPSO-Based Neural Network Parameter Optimization

The number of neurons and layers in the hidden layer of a neural network directly determines the fitting ability of the model, while the learning rate and the number of iterations affect the training process and the effect of the model. To solve the problem that the number of hidden layer neurons, learning rate and the number of iterations is difficult to be determined, an Improved Particle Swarm Optimization (IPSO) algorithm is used to optimize the hyper-parameters of the established neural network model to determine the optimal parameter settings of the network model.

3.1. IPSO Algorithm

The basic principle of Particle Swarm Optimization (PSO) can be summarized as follows: in the D-dimensional search space, the position and velocity information of the particles are randomly initialized, and the $\boldsymbol{i}$-th particle is the velocity and position can be denoted as $z_i = (z_{i1}, z_{i2}, \ldots, z_{id}, \ldots, z_{iD})$ and $v_i = (v_{i1}, v_{i2}, \ldots, v_{id}, \ldots, v_{iD})$. The fitness value of each particle is obtained from the fitness function, from which two extremes of the particle population can be determined: the individual extremum $p_i = (p_{i1}, p_{i2}, \ldots, p_{id}, \ldots, p_{iD})$ and the global extremum $p_g = \left(p_{g1}, p_{g2}, \ldots, p_{gd}, \ldots, p_{gD}\right)$. The particle velocity and position information are updated as shown in Equations (10) and (11) [17].

$$v_{id}^{k+1} = w \cdot v_{id}^{k} + c_1 \cdot rand \cdot \left(p_{id} - z_{id}^{k}\right) + c_2 \cdot rand \cdot \left(p_{gd} - z_{id}^{k}\right) \tag{10}$$

$$z_{id}^{k+1} = z_{id}^{k} + v_{id}^{k+1} \tag{11}$$

where $I = 1, 2, \ldots, n$; $d = 1, 2, \ldots, D$; k is the number of iterations;

w is the inertia weight; c_1 and c_2 are the learning factors for the individual and the whole population respectively, generally taken as a random number between (0, 2);

v_{id}^{k} is the velocity of the $\boldsymbol{i}$-th particle in dimension d at the $\boldsymbol{k}$-th iteration;

z_{id}^{k} is the position of the $\boldsymbol{i}$-th particle in dimension d at the$\boldsymbol{k}$-th iteration;

p_{id} is the individual extreme value of the $\boldsymbol{i}$-th particle in the d dimension;

p_{gd} is the extreme value in d dimension in the current iteration of the population.

To make the particle swarm algorithm superiority-seeking ability and speed, non-linear variation of weights is used to improve the performance of PSO [18]. The form of w variation in this article is shown in Equation (12).

$$w = w_{max} - (w_{max} - w_{min}) \cdot \arcsin\left(\frac{t}{t_{max}} \cdot \frac{2}{\pi}\right) \tag{12}$$

where w_{max} and w_{min} are the maximum and minimum values of w, respectively; t is the current number of iterations; and t_{max} is the maximum number of iterations.

When the particle swarm first starts searching, t is small and w is close to its maximum value. Then, w decreases slowly, ensuring the algorithm's optimization capability. As t increases, w decreases non-linearly and the magnitude of the decrease increases, ensuring the algorithm's local optimization capability and thus making the algorithm more flexible in its search.

3.2. Network Parameter Optimization

For the key parameters in the established network model, such as the number of neurons in the first and second hidden layers (h_1, h_2), the model learning rate (ε) and the number of training sessions (ep), the dimension of the search solution space of the particles is set to 4 dimensions. The fitness value of the i-th particle is calculated as shown in Equation (13).

$$fit_i = \frac{1}{n}\sum_{j=1}^{n}\left(\hat{y}_t^j - y_t^j\right)^2 \quad (13)$$

where fit_i denotes the fitness value of the i-th particle; n is the number of samples in the test set; and $\hat{y}_t^j$ and y_t^j denote the predicted and labeled values of the network pa rameters for the j-th sample of the model corresponding to particle$i$. The predicted values are calculated based on the information of the parameters carried by the particle after training the network model.

As shown in Figure 4, the neural network parameters for IPSO optimization are calculated as follows:

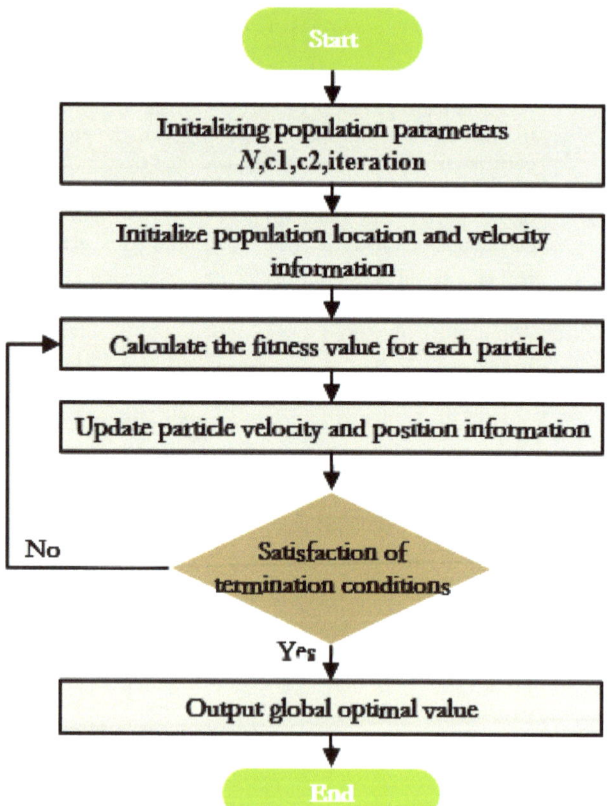

Figure 4. IPSO optimized network parameters flow chart.

(1) Initialize each individual's parameters. Determine the population size, the number of iterations, the learning factor and the range of intervals in which the velocity and position of the particles are taken.

(2) Initialize the position and velocity information of the particles. Randomly generate a certain number of population particles $X_{i,0}(h_1, h_2, \varepsilon, ep)$, $i = 1, 2, \ldots$, n, with the values of the particles in each dimension within a limited interval.

(3) Calculate the fitness value corresponding to each particle X_i according to the established objective function equation, determine the global extremes and individual extremes of the initial population, and use the best position of each particle as its historical best value.
(4) In each iteration, update the velocity and position information of the particles according to Equations (9) and (10), calculate the fitness value of each particle of the new population after the update, and determine the individual extreme value and global extreme value of the current population.
(5) Repeat steps (3) and (4) until the maximum number of iterations or accuracy is satisfied, output the optimal network parameters and train the network model.

3.3. Evaluation Indicators

To be able to evaluate the calculated results of the fitted models established, the mean square error, Mean Absolute Error (*MAE*), Root Mean Square Error (*RMSE*) and coefficient of determination (R^2) were selected as the basis for judging the effectiveness of the various fitting methods, which were calculated using the following formulae:

$$MAE = \frac{1}{n}\sum_{i=1}^{n}\left|\tilde{Y}_i - Y_i\right| \tag{14}$$

$$RMSE = \sqrt{\frac{\sum\limits_{i=1}^{n}\left(\tilde{Y}_i - Y_i\right)^2}{n}} \tag{15}$$

$$R^2 = 1 - \frac{\sum\limits_{i=1}^{n}\left(\tilde{Y}_i - Y_i\right)^2}{\sum\limits_{i=1}^{n}\left(\overline{Y}_i - Y_i\right)^2} \tag{16}$$

where n is the length of the sequence, $\tilde{Y}_i$ is the label value, Y_i is the predicted value of the network model and $\overline{Y}_i$ is the average of the label values. For unit flow characteristic curve fitting, the smaller the value of *MAE* and *RMSE* and the larger the value of R^2, the better the model is fitted and the higher the accuracy.

4. Case Analysis

4.1. Parametric Optimization

In this article, we take a leading hydropower station in a terrace basin as an example, and the experimental model data of the root unit and the real machine operation data are shown in Figure 5, which clearly shows that the real machine operation data are more concentrated, in line with the actual situation that the unit is mostly operated in the economic zone; the model data are full hydraulic head operating conditions data, which are more comprehensive. Using the comprehensiveness of the model data and the specificity of the real machine data, the training data set of the network model is combined, which effectively takes into account the full hydraulic head operating conditions of the unit, but also more comprehensively combines the efficient operating zone of the unit in the actual operation process, avoiding the prediction accuracy and precision problems of using a single data source for training the network.

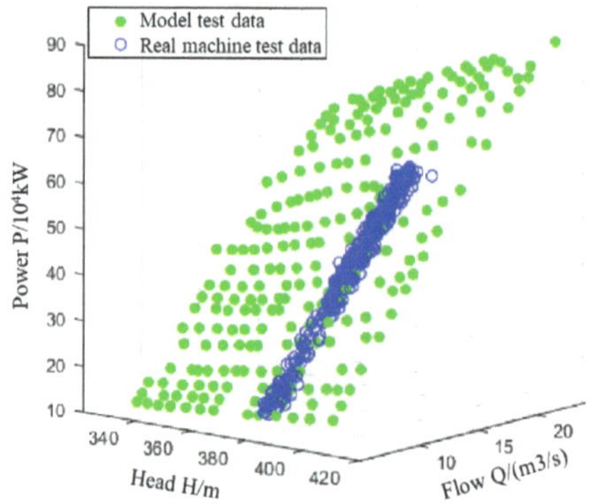

Figure 5. Hydraulic turbine model experiments and real machine operation data.

The operation data was collected in the form of 96 samples a day, collecting the characteristics data of the unit for one year of operation, with a total data volume of 365 × 96. The model test data was provided by the manufacturer, with a total of 1220 samples. The data set was divided into a training set, validation set and test set, where the training set accounted for 90% of the total quantity, the test set accounted for 10% and the validation set accounted for 20% of the training set. Afterwards the hyper-parameters of the established network model were optimized according to the divided data set.

The necessity of choosing a two-layer network is first verified. Separate single-layer and double-layer LSTM network models are built and validated using the processed data. The input dimension is 2-dimensional and the output dimension is 1-dimensional; the number of neurons in the hidden layer is 5; the maximum number of training sessions is 500; the learning rate is 0.01; and the activation function is a Sigmoid function. The prediction results of the single-layer and double-layer networks are shown in Table 1. As can be seen from the table, under the condition of the same network hyper-parameters, the MAE and MSE of the prediction results of the deep layer network are significantly smaller than those of the shallow layer network in multiple calculations, both of which are reduced by about 46%, and the accuracy and stability of the prediction results are significantly improved.

Table 1. Comparison of prediction accuracy between deep and shallow networks.

	MAE	MSE	RMSE
Single-layer	1.09440	0.00076	0.06373
Double-layer	0.59028	0.00041	0.02752

According to the above experimental results, the deep neural network model for the fitting of hydraulic turbine characteristic curve is constructed by using the advantage of deep network for data feature extraction, and the hyper-parameters of the network model are optimization-seeking selected. The optimized parameters are set as follows: the number of neurons in the input layer is 2; the number of neurons in the output layer is 1; the internal parameters of the LSTM unit are trained by Adam's algorithm; and the initialization method is Xavier. The parameters of the improved particle swarm algorithm are set as follows: the number of populations is 5; the number of evolution is 20; the learning factor of the particles is taken as $c_1 = c_2 = 2$, any particle $X_{i,0}(n_1, n_2, \varepsilon, h)$ for each dimension parameter in the range of [1, 100], [1, 100], [0.0001, 0.01] and [300, 600], and the velocity of the particle in the range of [−5, 5], [−5, 5], [−0.0005, 0.0005] and [−10, 10]. In IPSO, $w_{max} = 0.9$, $w_{min} = 0.1$; In PSO, $w = 0.5$.

In this article, the programming language is Python and the programming platform is PyCharm 2021.2. The version of TensorFlow and TensorFlow-GPU library is 2.4.0. The Keras module of TensorFlow framework is used for model building, and the hardware environment is as follows: GPU model is NVIDIA GeForce RTX 3070 and CPU model is Intel(R) Core(TM) i7-7800X CPU. The results are shown in Figure 6. Both PSO and IPSO results were averaged over 10 calculations. The IPSO method was then used to calculate the number of neurons in each hidden layer, the network learning rate and the number of training sessions, with the same algorithm settings as before. It can be clearly seen from the Figure 6 that the final fitness values of PSO and IPSO are similar, but the initial value of IPSO is significantly better than that of PSO. Moreover, PSO converged gradually at the 12th iteration, while IPSO converged at the 5th iteration, indicating that IPSO has better optimization ability. Therefore, IPSO has more advantages than PSO in overall performance.

The change in the number of neurons in each hidden layer of the LSTM network model is shown in Figure 7. As can be seen from the figure, the number of neurons in the first hidden layer of the network model, h_1, finally stabilizes at 40, and the number of neurons in the second hidden layer, h_2, finally stabilizes at 32 as the number of iterations increases. Figure 8 shows the variation of the network learning rate ε with the number of iterations, with the value eventually stabilizing at 0.0015 as the number of iterations increases. Figure 9 reflects the variation of the training number of the LSTM network with the number of IPSO iterations ep, with the final value stabilizing at 342. Based on the results of hyper-parameter search, i.e., $h_1 = 40, h_2 = 32, \varepsilon = 0.0015$ and ep = 342. After determining the network structure, the LSTM network was used to train the divided training set data.

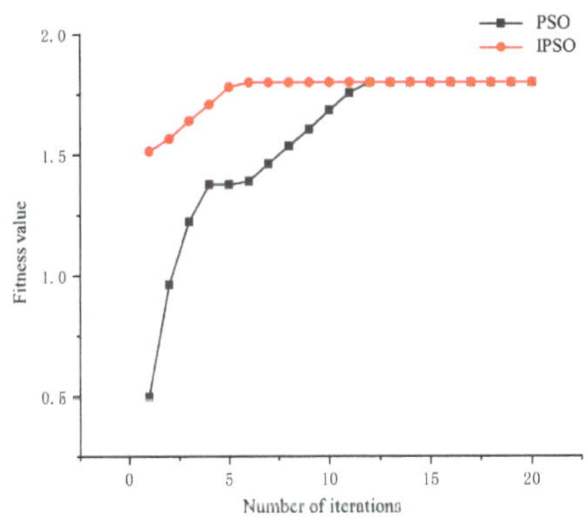

Figure 6. Change in the fitness of the objective function.

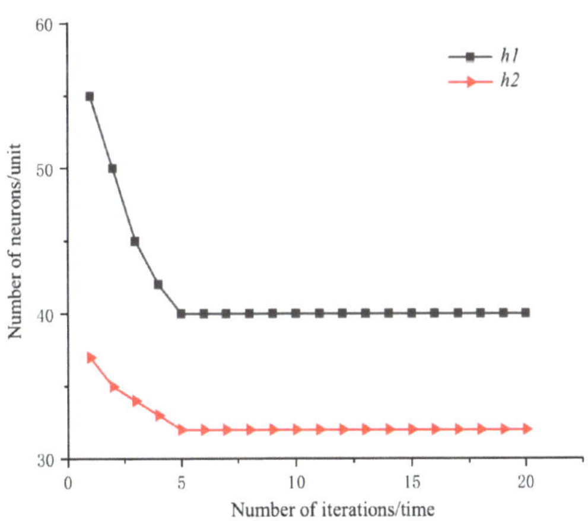

Figure 7. Change in neuron numbers.

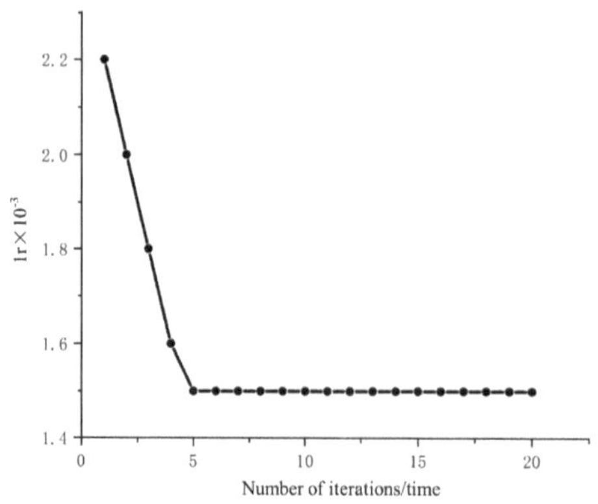

Figure 8. Change in online learning rates.

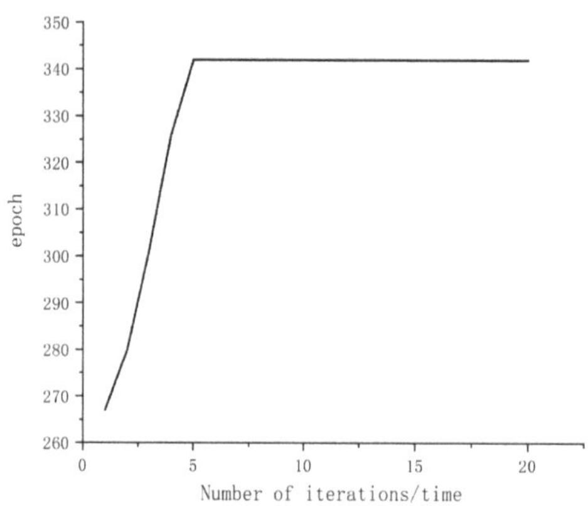

Figure 9. Change in number of network training sessions.

4.2. *Model Training and Prediction*

Once the important parameters of the network model have been determined using the IPSO method, the model can be trained, and when the set accuracy or the maximum number of iterations is reached, the training is completed, the trained network model is saved, and finally the performance of the model on the test set data is observed.

The network structure and the output dimension of each layer of the network established in the article are shown in Table 2. There are four layers in the network. In order to prevent the model from overfitting, the Dropout method is used to process between two LSTM network layers respectively, and the ratio is set to 0.2.

Table 2. Network structure.

Network Layer (type)	Output Dimension
dense1 (FC)	2
lstm1 (LSTM)	h_1
dropout1 (Dropout)	h_1
lstm2 (LSTM)	h_2
dropout2 (Dropout)	h_2
dense2 (FC)	1

To demonstrate the self-learning and self-evolving capability of the deep learning method, the training set sample data volumes were taken to be 3, 6 and 9 months, respectively, and the prediction results were then evaluated using three metrics: MSE, MAE and R^2. The training parameters were set to epochs = ep, bath size = 32, learning rate = ε; Adam's algorithm was chosen as the optimization algorithm and MSE was chosen as the loss value. The training was stopped when the number of epochs was reached and the trained model was used to predict the test set data and the results were compared to both Support Vector Machine (SVM) [19] and Extreme Learning Machine (ELM) [20] methods. The results are shown in Figures 10 and 11. It can be seen from the figures that the error value gradually decreases as the number of samples increases.

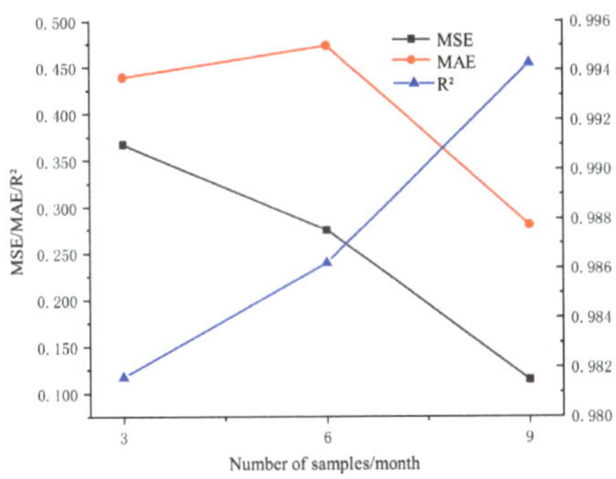

Figure 10. Self-learning process.

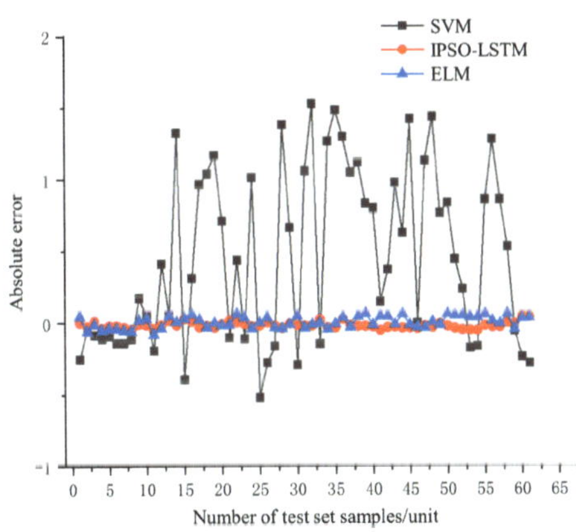

Figure 11. Comparison of predicted results.

As can be seen from Figure 10, as the number of samples in the training set keeps increasing, the value of R^2 increases from 98.16% at the beginning to 99.43%, MSE decreases from 36.71% to 11.43%, and MAE decreases from 43.90% to 28.03%. This is due to the fact that the LSTM-based deep learning model performs uninterrupted self-correction as training samples are accumulated, so its decision accuracy increases as the number of training samples increases, reflecting a strong self-learning capability.

For comparison, the number of neurons in the ELM network is set to h_1, and the maximum training number is ep = 342; the kernel function of SVM is set to Radial Basis Functions (RBF). The programming language for both models was still Python, and the running platform was PyCharm 2021.2. The prediction results were evaluated using three evaluation metrics, MSE, MAE and RMSE, as shown in Table 3.

Table 3. Comparison of predicted results.

Methods \ Indicators	MSE	MAE	RMSE
SVM	0.57100	0.5916	0.7556
ELM	0.00160	0.0347	0.0400
IPSO-LSTM	0.00086	0.0259	0.0293

As can be seen from Figure 11 and Table 3, the prediction results of IPSO-LSTM method are significantly better than those of SVM method and ELM method. Among them, SVM has a larger prediction error compared to the other two methods, while the MSE of the proposed method is reduced by about 46%, the MAE by about 25% and the RMSE by about 27% compared to the ELM method.

The final output of the IPSO-LSTM model is shown in Figure 12. The horizontal coordinate of the *XOY* plane is the hydraulic head *H*, the vertical coordinate is the outflow *P*, and the *Z*-direction represents the flow rate *Q* of the unit, which is also the output value of the network model. The blue dots in the figure indicate the predicted values and the red circles indicate the labeled values from the test set. As can be seen from the graph, the coordinates of the labeled and predicted values almost overlap, indicating that the proposed method has high prediction accuracy, reflecting the reliability of the method and providing an important guarantee for the economic and stable operation of the unit.

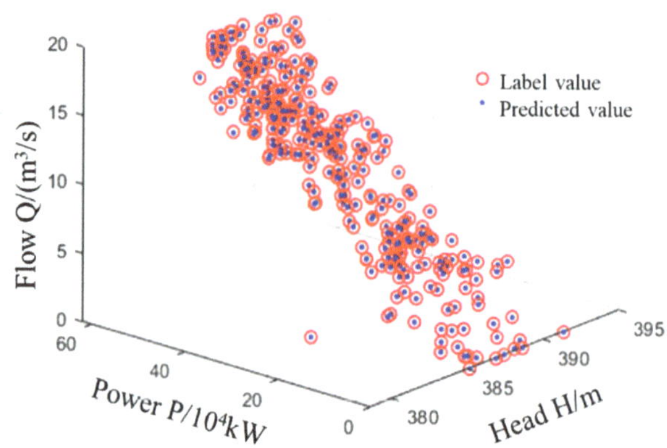

Figure 12. IPSO-LSTM fitting of turbine characteristic curves.

5. Conclusions

In this article, a method for fitting turbine flow characteristics curve with self-learning capability based on deep learning is proposed, and the conclusions obtained are as follows:

(1) The deep learning-based hydraulic turbine flow characteristic curve fitting method can continuously train and correct the model by continuously accumulating historical data, and its fitting accuracy will improve with the accumulation of the number of training sets, so the method has the characteristics of self-learning and self-correction.

(2) The training process of the IPSO-LSTM-based deep learning model for fitting the flow characteristics of hydraulic turbines is closely related to the number of training sets; the more samples in the training set, the faster the model training converges and the smaller the error.

(3) Since the deep learning-based hydraulic turbine flow characteristic curve fitting method uses a large amount of real machine operation data to train the model and construct a mapping relationship between input and output quantities, compared

with traditional mathematical fitting methods and machine learning methods, it can effectively discover the laws in a large amount of historical operation data, use the connection between historical data and future operating conditions, and make accurate calculations of future unit operating conditions characteristics and parameters. It can also make reasonable predictions for certain unknown operating conditions and has good adaptability.

(4) The hydraulic turbine flow characteristics curve fitting method combining hydraulic turbine model test data and real machine operation data proposed in this article, i.e., considering the full working condition operation characteristics of the hydraulic turbine in the model test and combining the vibration avoidance operation characteristics of the real machine data. Compared with the traditional fitting method, the fitting model learns the operating characteristics of the hydraulic turbine in each working condition and the actual operating characteristics at the same time. Therefore, the results of the fitting calculations using the method proposed in this article are accurate and economical at the same time.

Author Contributions: Software, F.L.; Writing—original draft, C.H. and F.F.; Writing—review & editing, H.P. and Y.Z. All authors have read and agreed to the published version of the manuscript.

Funding: This work was supported by the National Key R&D Program Key Project of Intergovernmental International Science and Technology Innovation Cooperation under Grant 2019YFE0105200, in part by the Science and Technology Project of Huaneng Group Headquarters under Grant HNKJ20-H20, and in part by the Multi-Dimensional Fault Diagnosis Method of Hydropower Unit Based on Improved Deep Learning Strategy under Grant 51809082.

Institutional Review Board Statement: Not applicable.

Informed Consent Statement: Not applicable.

Data Availability Statement: Not applicable.

Conflicts of Interest: The authors declare no conflict of interest.

References

1. Yang, Z.; Yang, K.; Su, L.; Hu, H. The short-term economical operation problem for hydropower station using chaotic normal cloud model based discrete shuffled frog leaping algorithm. *Water Resour. Manag.* **2020**, *34*, 905–927. [CrossRef]
2. Chen, S.; Liang, B.; Lu, H. Derivation of the aggregation-decomposition principle for optimal load distribution of hydropower station. *J. Hydraul. Eng.* **2017**, *48*, 1185–1194.
3. Zhao, W. *The Study of the Economic Operation of Hydropower Station Based on APSO-SA Algorithm*; Huazhong University of Science and Technology: Wuhan, China, 2016.
4. Zhang, Y.; Zang, W.; Zheng, J.; Cappietti, L.; Zhang, J.; Zheng, Y.; Fernandez-Rodriguez, E. The influence of waves propagating with the current on the wake of a tidal stream turbine. *J. Appl. Energy.* **2021**, *290*, 116729. [CrossRef]
5. Zheng, Y.; Chen, D. *Water Turbine*; China WaterPower Press: Beijing, China, 2011.
6. Zhang, R.; Liu, Z.; Wang, L.; Zhang, Y. Data interpolation by Delaunay triangulation for the combined characteristic curve of a turbine. *J. Hydroelectr. Eng.* **2011**, *30*, 197–201+221.
7. Liu, D.; Hu, X.; Zeng, Q.; Zhou, H.K.; Xiao, Z.H. Refined hydro-turbine characteristic curve model based on input-output correction. *J. Hydraul. Eng.* **2019**, *50*, 555–564.
8. Li, J.; Chen, Q.; Chen, G. Study on synthetic characteristic curve processing of Francis turbine conbined with BP neural network. *J. Hydroelectr. Eng.* **2015**, *34*, 182–188.
9. Huang, X.; Liu, D. The treatment of hydroturbine synthetic characteristic curve by radial basis networks. *J. Hydroelectr. Eng.* **2007**, *26*, 114–118.
10. Kilinc, H.C.; Yurtsever, A. Short-Term Streamflow Forecasting Using Hybrid Deep Learning Model Based on Grey Wolf Algorithm for Hydrological Time Series. *Sustainability* **2022**, *14*, 3352. [CrossRef]
11. Zhu, Q.; Li, H.; Wang, Z.; Chen, J. Short-Term Wind Power Forecasting Based on LSTM. *Power Syst. Technol.* **2017**, *41*, 3797–3802.
12. Dai, J.; Song, H.; Sheng, G. Prediction Method for Power Transformer Running State Based on LSTM Network. *High Volt. Eng.* **2018**, *44*, 1099–1106.
13. Zhou, D.; Zhuang, X.; Zuo, H. A hybrid deep neural network based on multi-time window convolutional bidirectional LSTM for civil aircraft APU hazard identification. *Chin. J. Aeronaut.* **2022**, *35*, 344–361. [CrossRef]
14. Nitish, S.; Geoffrey, E.; Hinton, A.K.; Ilya, S.; Ruslan, S. Dropout: A simple way to prevent neural networks from overfitting. *J. Mach. Learn. Res.* **2014**, *15*, 1929–1958.

15. Wang, X.; Wu, J.; Liu, C.; Yang, H.; Du, Y.; Niu, W. Exploring LSTM based recurrent neural network for failure time series prediction. *J. Beijing Univ. Aeronaut. Astronaut.* **2018**, *44*, 772–784.
16. Kingma, D.P.; Ba, J. Adam: A method for stochastic optimization. *arXiv* **2014**, arXiv:1412.6980.
17. Xue, L.; Dezhi, H.; Letian, D.; Qiuting, T. Intrusion detection of wireless sensor networks based on IPSO algorithm and BP neural network. *Int. J. Comput. Sci. Eng.* **2020**, *22*, 221–232.
18. Li, W.; Feng, F.; Jiang, Q. Prediction for railway passenger volume based on modified PSO optimized LSTM neural network. *J. Railw. Sci. Eng.* **2018**, *15*, 3274–3280.
19. Wang, Q.; Li, J.; Chen, S.; Wang, X. SVM-based Implicit Stochastic Scheduling Mode for Cascade Hydropower Stations. *MATEC Web Conf.* **2018**, *246*, 2046. [CrossRef]
20. Zhou, J.; Yu, X.; Jin, B. Short-Term Wind Power Forecasting: A New Hybrid Model Combined Extreme-Point Symmetric Mode Decomposition, Extreme Learning Machine and Particle Swarm Optimization. *Sustainability* **2018**, *10*, 3202. [CrossRef]

sustainability

MDPI

Article

Multi-Objective Optimization of Back-to-Back Starting for Pumped Storage Plants under Low Water Head Conditions Based on the Refined Model

Chen Feng [1], Guilin Li [1,*], Yuan Zheng [1], Daqing Zhou [1] and Zijun Mai [2]

[1] College of Energy and Electrical Engineering, Hohai University, Nanjing 211100, China
[2] Department of Rural Water Management, Nanjing Hydraulic Research Institute, Nanjing 210029, China
* Correspondence: windycity@hhu.edu.cn; Tel.: +86-150-7199-5381

Abstract: Pumped storage plants (PSP) must switch frequently between various working conditions. Moreover, PSPs can fall easily into an S-shaped zone under low water head conditions, especially during back-to-back starting (BTBS), which reduces the stability and safety of unit operation. In this paper, a nonlinear PSP model for BTBS is established by combining an electrical subsystem with a refined hydraulic-mechanical subsystem. The influences of the hydraulic, mechanical, and electrical factors on the BTBS process are investigated quantitatively. Choosing the speed overshoot and speed stable time as the optimization objectives, and considering a variety of constraints, the multi-objective particle swarm optimization (MOPSO) algorithm is introduced to study and optimize two typical startup strategies. The results show that: (1) The parameters of a hydraulic-mechanical-electrical system have a significant impact on BTBS process, and the most unfavorable working condition corresponds to the lowest water head; (2) In the control schemes, a novel constant excitation voltage strategy is proposed based on the multi-objective optimization scheme. Compared with the constant excitation current strategy or single-objective, the optimization strategy proposed can considerably improve the speed overshoot and the speed stable time by at least 68.27% and 3.22% under the worst working conditions. (3) It is further verified that the problem of trapping in the S-shaped region under various working conditions may be avoided by the obtained optimal control scheme. The results give prominence to the effectiveness of the proposed optimization strategy for maintaining the safety and stabilization of PSP operation.

Keywords: pumped storage plant; back-to-back starting; low water head condition; constant excitation voltage; multi-objective optimization; optimal control scheme

Citation: Feng, C.; Li, G.; Zheng, Y.; Zhou, D.; Mai, Z. Multi-Objective Optimization of Back-to-Back Starting for Pumped Storage Plants under Low Water Head Conditions Based on the Refined Model. *Sustainability* **2022**, *14*, 10289. https://doi.org/10.3390/su141610289

Academic Editor: Dino Musmarra

Received: 19 July 2022
Accepted: 15 August 2022
Published: 18 August 2022

1. Introduction

In order to cope with climate change, many renewable energy sources have been developed, such as tidal, wind, solar, and hydro energies [1]. As unstable renewable energy is connected to the grid in large quantities and the load demand is diversified, the power grid's supply and demand fluctuation is becoming more serious [2,3]. As an essential energy storage technology, a pumped storage plant (PSP) can effectively adjust the impact of unstable energy on the power system and enhance the consumption capacity of the power grid for unstable energy such as solar energy and wind power [4,5]. However, during the large fluctuation transition process of the pumped storage unit (PSU), due to the influence of the S-shaped curve of the pump turbine, the water hammer effect in the penstock of the diversion system causes an effect similar to the rapid closing of guide vanes, and the inflow decreases rapidly, resulting in the sharp rise of water pressure and speed, threatening the structural safety of unit and diversion pipelines [6,7]. The above operation characteristics and various potential operation accidents not only seriously endanger the safe and stable operation of PSPs, but also affect the safety and stability of the power supplies.

At present, the conventional startup mode of large PSPs at home and abroad is mainly via a static frequency converter starting and BTBS as standby [8–10]. During variable frequency starting, it is necessary to absorb electric energy from the power grid. If the power grid or the frequency converter fails, the unit cannot be started in this way, but the BTBS mode can meet the requirements well. Therefore, the experience and technology of BTBS should be explored in the context of building a resilient power grid. Among the numerous startup conditions of the PSU, the BTBS mode is the most complicated, requiring the mutual cooperation of monitoring, speed regulation, excitation, protection, and other systems of the two units, which presents an enormous challenge to the stable operation of PSPs. However, this issue has not been fully addressed until now. Previous studies have mainly focused on the transient electrical characteristics and electromechanical variation law. Relevant research findings are summarized as follows:

For BTBS modelling, previous studies focus on the electrical perspective by greatly simplifying the hydraulic and mechanical systems. The functions offered by turbine manufacturers and obtained by model tests, which are used to describe the dynamic behaviors among the mechanical torque, gate valve opening (GVO), and speed, have been widely adopted [11,12]. However, these models are mostly oversimplified based on strong assumptions in reflecting the coupling characteristics of the hydraulic-mechanical-electrical subsystems of a PSP under various working conditions, especially the S-shaped characteristics under low head conditions. Therefore, a nonlinear model with a detailed hydraulic-mechanical system, i.e., the logarithmic curve projection (LCP) method for pump turbine [13], and the method of characteristics (MOC) for conduit systems [14], has been adopted to improve the accuracy of the BTBS transient processes simulation in this work, which can overcome the shortcomings of the above-simplified models.

For analysis of influencing factors, Konidaris [12] systematically discussed the main parameters' impact on the performance of the BTBS based on the physical laws behind the transient process. Osburn and Atwater [15] described the design and testing of a BTBS system. The feasibility of this kind of startup is analyzed theoretically. Guidance was developed in order to correctly select the main parameters for a successful start [16]. By means of simulating the transient starting performance for a PSU, alternative designs and their effects on the startup ability of the PSU for reliable improvement of the starting system were studied [17]. Proposals were offered to achieve safe and reliable fault clearance on the basis of the fault characteristics and actual deexcitation test results of the PSP [18].

For control optimization, there are many applications of the MOPSO algorithm in energy systems; A. Beiranvand applied the MOPSO algorithm in the optimization analysis of thermal power generation [19], and M. Ghazvini proposed the Coupled Multi-Objective Evolutionary Approaches [20]. There are also multi-objective optimization cases in the field of materials and chemicals [21,22], which give some inspiration to this paper. However, the BTBS model is more complicated. There exists no report on the optimization of the BTBS control scheme, and only some research on the startup strategy of PSPs under turbine conditions has been involved. Xu et al. [23] studied the optimal control for the turbine startup process of PSPs under low head conditions. There are hydraulic, mechanical, and electrical interferences between two units, and more control factors must be taken into account during the BTBS process. Moreover, BTBS generally happens during the low power consumption period, and the upper-reservoir water storage has been exhausted during the peak power consumption period. Thus, the PSU can more easily be under low water head conditions and fall into the S-shaped zone.

Through reviews of relevant works, it was found that too little work has been devoted to BTBS in the previous research, which mainly focused on the BTBS workflow or the influence of electrical factors on the transient process of BTBS. Many other factors, such as effects from the S-shaped characteristics, are likely to affect the transient processes of BTBS and have not been studied. In addition, to our best knowledge, the control scheme optimization strategy for BTBS has not been proposed in previous works.

Accordingly, in this paper, the effect mechanisms of the hydraulic-mechanical-electrical coupling factors on BTBS are investigated, and the optimization strategy for BTBS control at low water head conditions is proposed. A nonlinear model for PSP by incorporating an electrical subsystem with a refined hydraulic-mechanical subsystem is established to conduct the study, and the effects of the electrical and hydraulic-mechanical subsystems on the process of BTBS are quantitatively investigated subsequently. Moreover, on the basis of the analysis of the influencing factors, a novel multiobjective optimization scheme for the BTBS control strategy is proposed. The innovations of this paper are as follows.

(1) It is the first to establish a nonlinear PSP model combining the electrical subsystem with the fine hydraulic and mechanical subsystems for BTBS;
(2) The effects of the hydraulic-mechanical-electrical parameters on the BTBS are comprehensively investigated on the basis of the model mentioned above;
(3) An innovative multi-objective optimization scheme is proposed for the control strategy of BTBS at low water head conditions for the first time, which is proven to be suitable for a variety of working conditions.

The remainder of the paper proceeds as follows: The nonlinear model of a PSP with a refined hydraulic-mechanical subsystem is constructed and validated in Section 2. In Section 3, the effects of the hydraulic-mechanical-electrical parameters on the BTBS processes are then studied. In Section 4, an innovative multi-objective optimization scheme for BTBS in the condition of low head is proposed on the basis of the analysis above. In Section 5, a case study is conducted and the optimization result is also authenticated in other working conditions. In Section 6, the conclusions of these analyses are brought forth. Figure 1 presents the flowchart of this research.

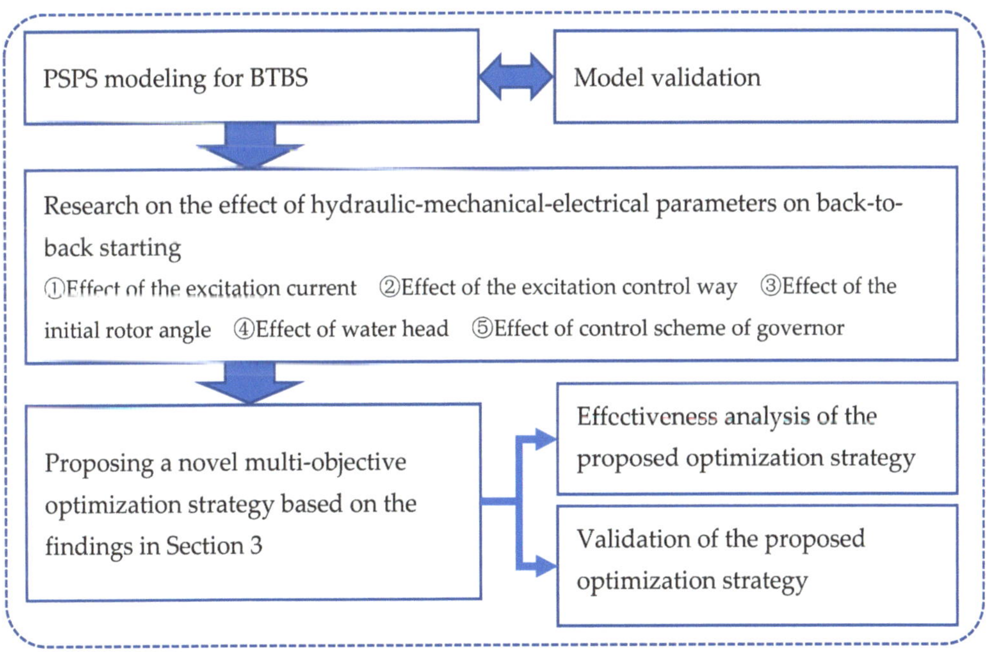

Figure 1. A flowchart of the research in this paper.

2. Refined Modelling of a Pumped Storage Plant for Back-to-Back Starting

The BTBS mode is also called synchronous starting mode, which uses one PSU as a generator to provide the current with increasing frequency. The other PSU to be started is used as a motor, and the current with variable-frequency output by the former is used to accelerate to the rated speed synchronously. The BTBS is affected by the coupling factors of the hydraulic-mechanical and electrical systems. The basic form of BTBS is shown in Figure 2.

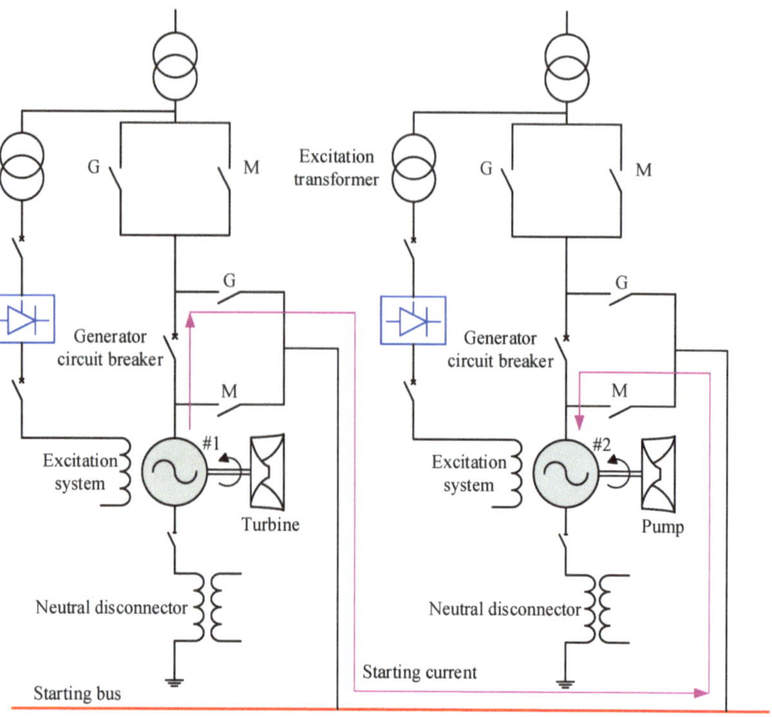

Figure 2. The basic form of back-to-back starting.

2.1. The Hydraulic-Mechanical Subsystem

(1) Conduit System

On the basis of hydromechanical theory, the momentum and continuity equations, which describe the unsteady pressure and flow in penstocks, can be formulated as follows [24]:

$$\frac{\partial Q}{\partial t} + gA\frac{\partial H}{\partial x} + \frac{f}{2dA}Q|Q| = 0 \tag{1}$$

$$a^2\frac{\partial Q}{\partial x} + gA\frac{\partial H}{\partial t} = 0 \tag{2}$$

The details of all variable symbols in the study are presented in the Nomenclature. The MOC method is used to solve the above equations, considering the elasticity of the water hammer in the penstocks [25].

(2) The Pump turbine Model

As the key equipment of energy conversion, it is crucial to model the pump turbine accurately. At present, the characteristic curves have been widely adopted in nonlinear modelling for pump turbines provided by manufacturers, as shown in Figure 3. The nonlinear model can be described by (3) and (4):

$$\begin{cases} M_{11} = f_M(Y, N_{11}) \\ Q_{11} = f_Q(Y, N_{11}) \end{cases} \tag{3}$$

$$\begin{cases} M_t = M_{11} D^3 H_t \\ Q_t = Q_{11} D^2 \sqrt{H_t} \\ N = N_{11} \sqrt{H_t} / D \end{cases} \quad (4)$$

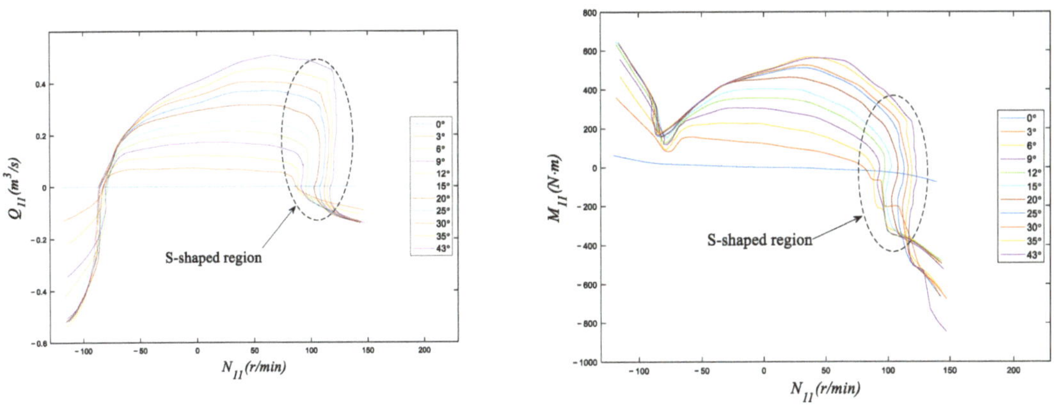

(a) **(b)**

Figure 3. Characteristic curves of a pump turbine. (**a**) Flow characteristic curve. (**b**) Moment characteristic curve.

As shown in Figure 3, the S-shaped area raises obstacles for the pump turbine modelling, which presents a multivalued characteristic. The Logarithmic-Curve-Projection (LCP) method [13] is proposed to convert the unit speed N_{11} in the way of logarithmic projection so as to overcome the difficulty. The equation is as follows:

$$\begin{cases} X = n_{11} / e^{q_{11}} \\ n_{11} = N_{11} / N_{11r} \\ q_{11} = Q_{11} / Q_{11r} \end{cases} \quad (5)$$

The results obtained by means of the LCP method are shown in Figure 4; the LCP curves reflect the change rule of Q_{11} and M_{11} with abscissa X. It immensely reduces the complexity of the calculation procedure problems whether compared with the Suter or improved Suter methods [26].

Thereafter, in a given moment, the unit moment and the unit flow are computed using interpolation in accordance with the LPC curves.

(3) The Turbine Governor System

The governor of the PSP is comprised of a controller and servomechanism. As the actuator of the governor, the servomechanism uses the output signal of the PID controller to drive the guide vanes. Considering the main nonlinear factors (dead zone, saturation, and output limiting), the whole model of the governor is shown in Figure 5.

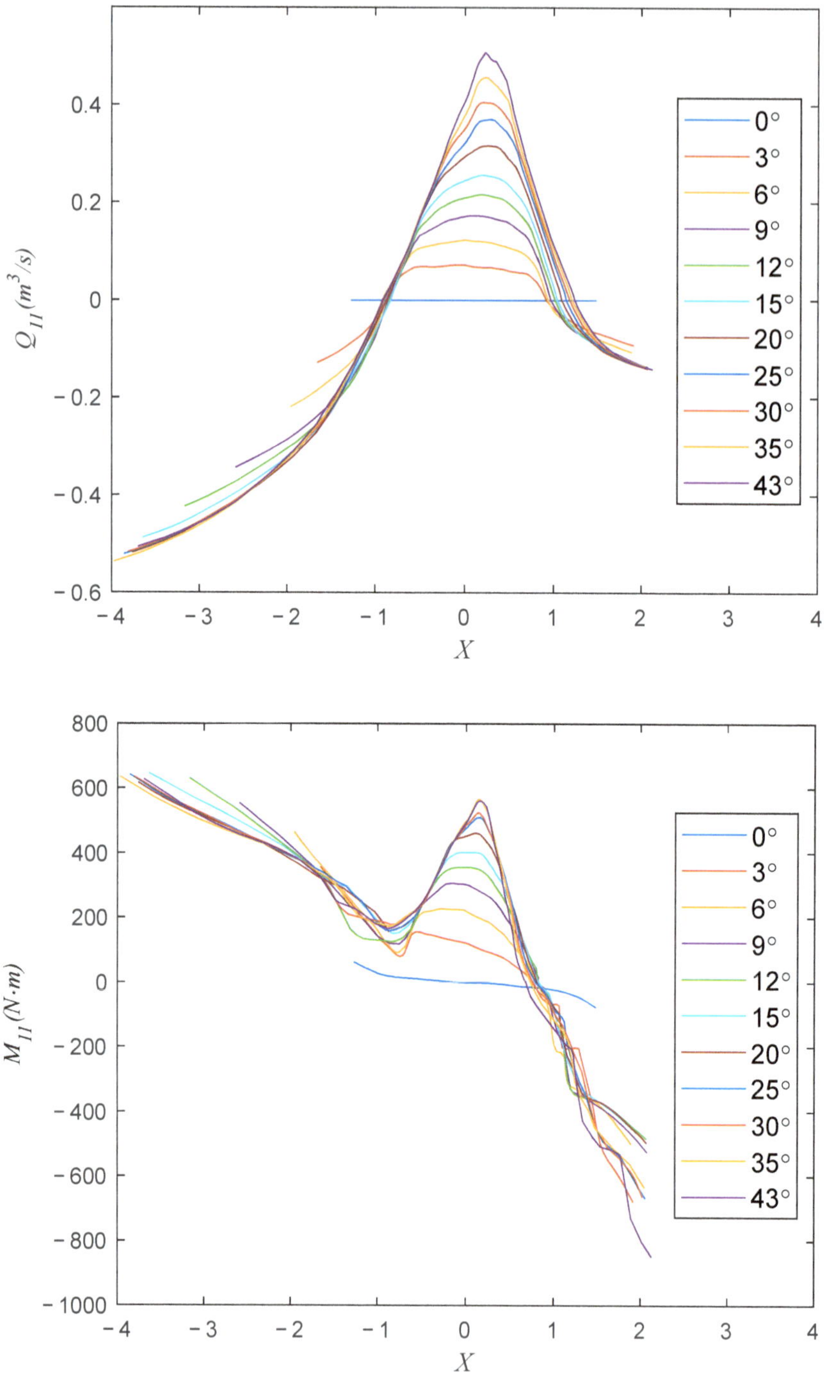

Figure 4. Logarithmic-Curve-Projection of the characteristic curves.

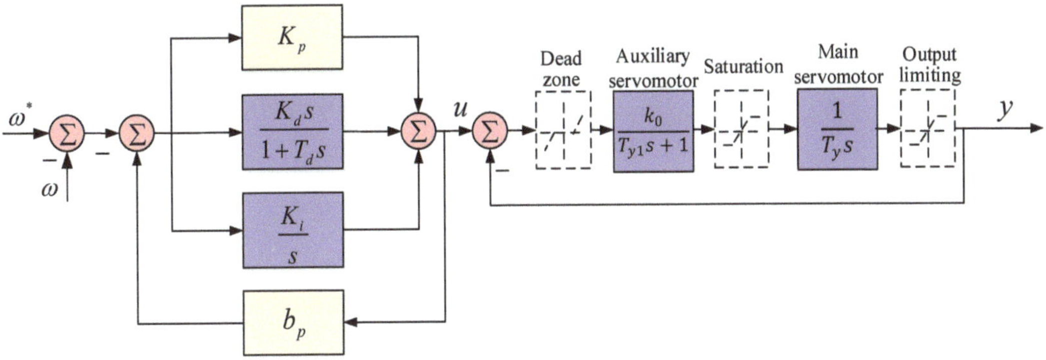

Figure 5. Block diagram of the governor system with nonlinear mechanical components.

2.2. The Electrical Subsystem

(1) Synchronous Machine

Since it is very difficult to gain an accurate value of the original parameters of the synchronous machine through analysis and calculation, 18 original parameters of a synchronous machine are usually converted into 11 machine parameters composed of steady-state, transient, and sub-transient parameters in engineering, which can be obtained directly through experiments. The mathematical model of a synchronous machine expressed by 11 machine parameters includes voltage balance equations of stator winding (8), excitation equations of stator winding (9), voltage balance equations of rotor winding (10), and a rotor motion Equation (11) [27].

$$\begin{cases} v_d = \dot{\varphi}_d - \omega\dot{\varphi}_q - R_a i_d \\ v_q = \dot{\varphi}_q + \omega\dot{\varphi}_d - R_a i_q \end{cases} \tag{6}$$

$$\begin{cases} \varphi_d = E_q'' - X_d'' i_d \\ \varphi_q = -E_d'' - X_q'' i_q \end{cases} \tag{7}$$

$$\begin{cases} T_{d0}'\dot{E}_q' = -\frac{X_d - X_d''}{X_d' - X_d''}E_q' + \frac{X_d - X_d'}{X_d' - X_d''}E_q'' + E_{fd} \\ T_{d0}''\dot{E}_q'' = E_q' - E_q'' - (X_d' - X_d'')i_d \\ T_{q0}'\dot{E}_d' = -\frac{X_q - X_q''}{X_q' - X_q''}E_d' + \frac{X_q - X_q'}{X_q' - X_q''}E_d'' \\ T_{q0}''\dot{E}_d'' = E_d' - E_d'' + \left(X_q' - X_q''\right)i_q \end{cases} \tag{8}$$

$$\begin{cases} \dot{\delta} = (\omega - 1)\omega_0 \\ T_j\dot{\omega} = M_t - M_e \end{cases} \tag{9}$$

(2) The Excitation System

A DC exciter is adopted, as shown in Figure 6 [28]. The primary elements of the excitation system block are the lead-lag compensator, main regulator, exciter, and damping block.

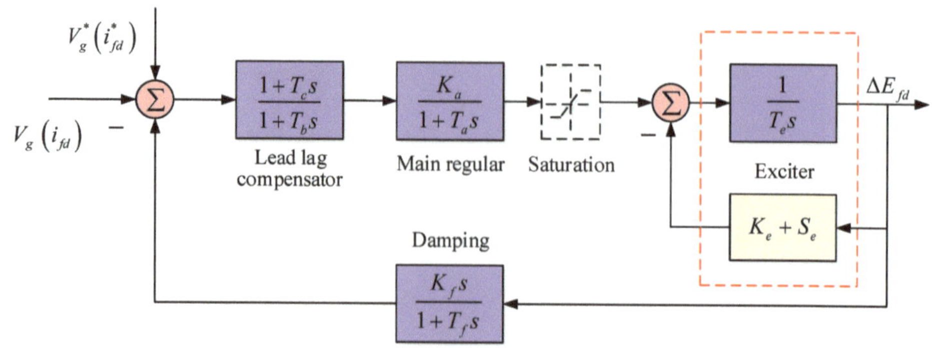

Figure 6. Block diagram of the excitation system.

2.3. Model Validation with On-Site Measurement

The proposed PSP model is validated by a case study in this subsection. In previous studies, in many cases the electrical subsystem model has been validated by measurement [29]. Therefore, the goal of this paper is to use the model to prove the hydraulic-mechanical subsystem.

The research data in this section are from a large PSP in Hubei Province of China, and the topology of its regulation system is shown in Figure 7. The parameters of the pump turbine and the generator and motor of the PSU are shown in Table A1, and the characteristic curves of the pump turbine are shown in Figure 3. The pipeline parameters of the diversion system that can be used for modelling after processing are shown in Table A2. The dimensional parameters of the surge chamber are shown in Table A3. In addition, the basic parameters of each module of the regulation system are shown in Table A4.

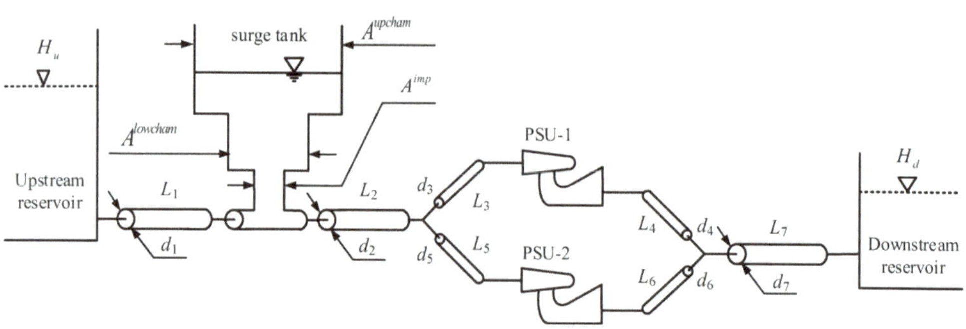

Figure 7. Layout diagram of the considered PSP.

In this case study, PSU-2 is offline when PSU-1 begins a single-unit load rejection. The water level of the upper reservoir is 306.98 m, and the downstream reservoir is 96.11 m, so the gross head is 210 m. When the PSU-1 operates normally with 300 MW load, the load is suddenly rejected. The guide vane is normally closed in 30 s. The variation waveform of various parameters in the simulation test is shown in Figure 8. Through analysis of the graph, the characteristic values of each parameter change are extracted and compared with the results of field measurements. Table 1 shows the comparison results.

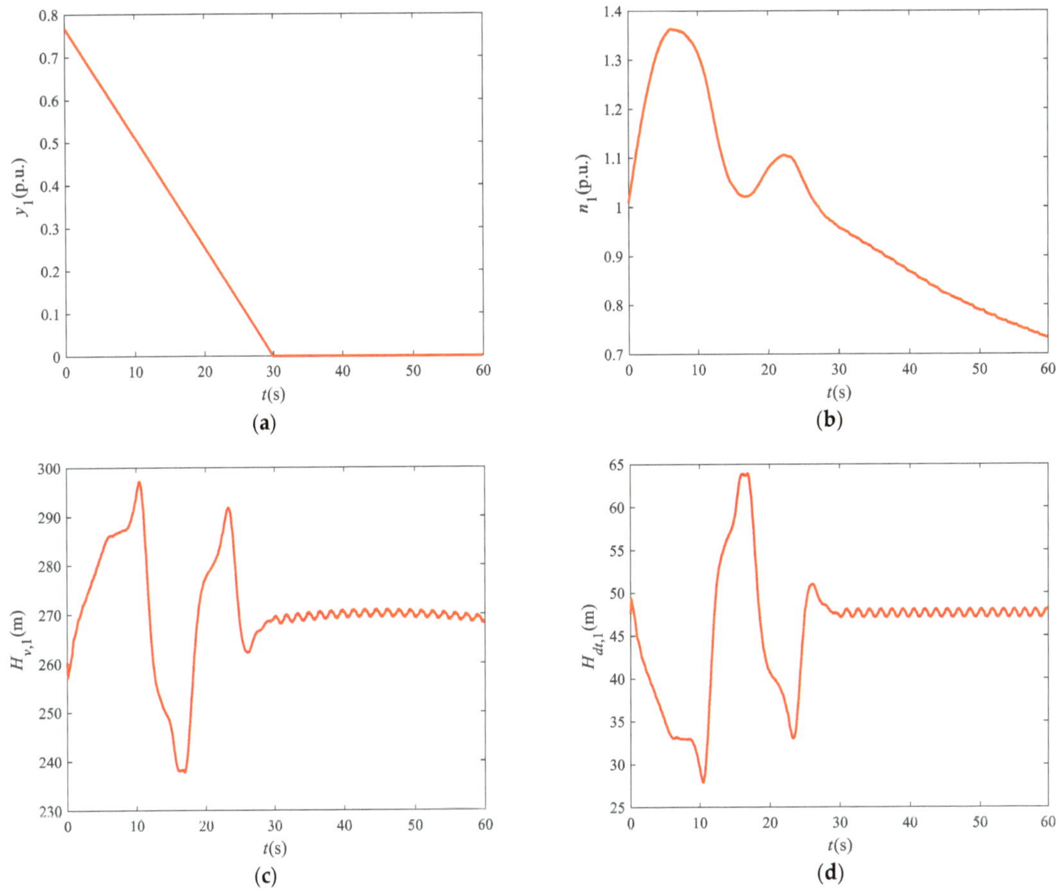

Figure 8. Simulation results of PSU-1 during a single-unit load rejection. (**a**) Guide vane opening. (**b**) Rotational speed. (**c**) Volute water pressure. (**d**) Draft tube water pressure.

Table 1. Full load rejection results of PSU-1 compared to on-site measurements.

Unit Number	Category	Maximum Pressure at Measuring Point of Volute Inlet	Minimum Water Pressure at MEASURING Point of Draft Tube	Maximum Speed
1#	Measurement	299.32 m	26.6 m	140%
	Refined model	297.05 m	27.9 m	136%
	Absolute error	−2.27 m	1.3 m	−4%
	Relative error	−0.75%	4.88%	2.85%

As revealed in Figure 8 and Table 1, the calculation results of the full load rejection transition process achieve consensus results with the on-site measurements, which effectively verifies the accuracy of the PSP mathematical model constructed in this paper.

3. Analysis of Factors Affecting Back-to-Back Starting

3.1. Excitation Current

The excitation system of the two units adopts separate excitation sources (grid-side power supply or auxiliary power supply), and the closed-loop control modes of the constant excitation current (CEC) are adopted for the two units. After the speeds of the two units are synchronously accelerated to 90% of the rated value, the excitation control modes are changed to the closed-loop working modes of the constant terminal voltage. Figure 9 is an example of successful BTBS (given excitation current $i_g^* = 1.2, i_m^* = 0.6$), while Figure 10 is an example of a failed start due to improper matching of excitation current (given that excitation current $i_g^* = 0.8, i_m^* = 0.5$).

To explore the impact of given excitation current (i_g^*, i_m^*) on the BTBS process, BTBS under different excitation currents (i_g^*, i_m^*) are simulated, as may be observed in Figure 11. The range of i_{fd}^* is [0, 2], the default values are kept for other parameters of the PSP.

In Figure 11, The blue circle within the green area means that the BTBS is successful when the excitation current is (i_m^*, i_g^*). It shows that the BTBS results are altered by the given excitation current. The excitation current of the generator and motor must maintain an appropriate proportion to ensure successful BTBS. In particular, the proportion is affected by the characteristics of the units, and the requirements of the excitation current for the

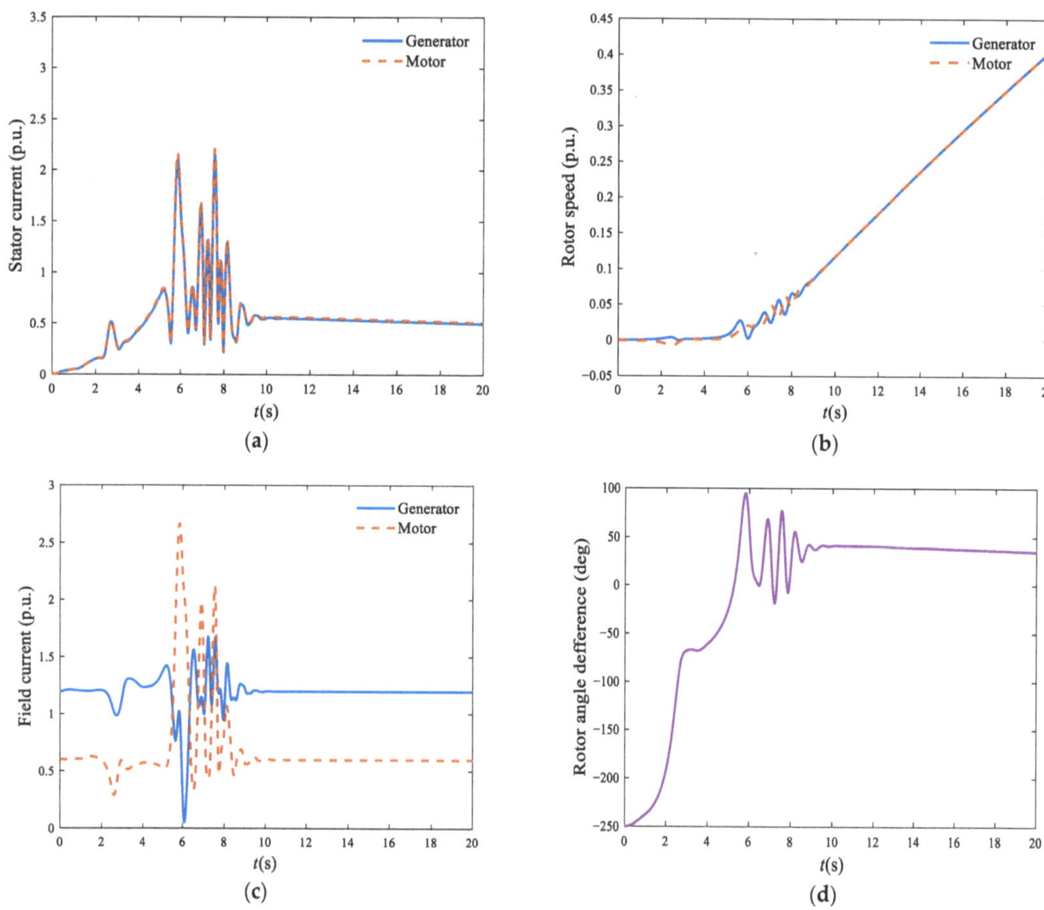

Figure 9. Successful BTBS of constant excitation current $i_{fd}^{g*} = 1.2, i_{fd}^{m*} = 0.6$. (**a**) Stator's current; (**b**) Rotor speed; (**c**) Field current; (**d**) Rotor angle difference.

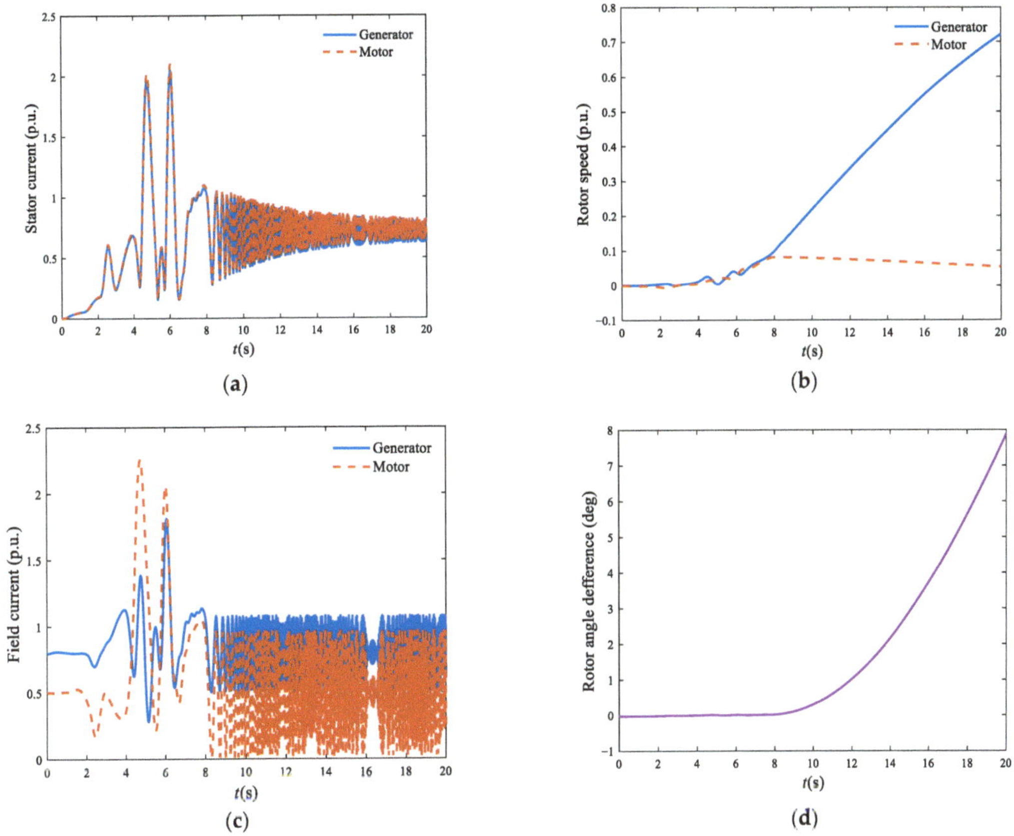

Figure 10. Unsuccessful BTBS of constant excitation current $i_{fd}^{g*} = 0.8, i_{fd}^{m*} = 0.5$. (**a**) Stator's current; (**b**) Rotor speed; (**c**) Field current; (**d**) Rotor angle difference.

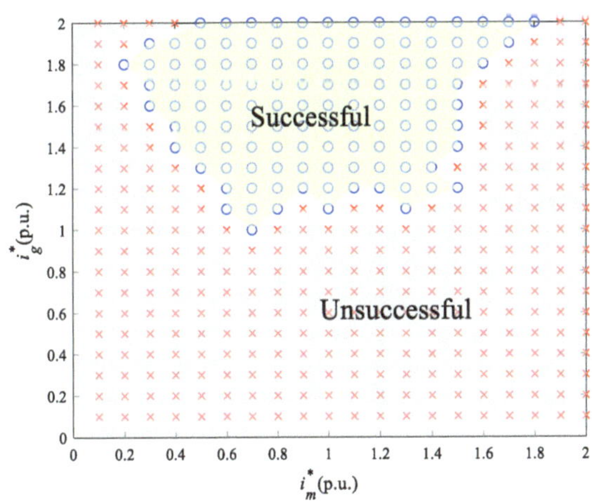

Figure 11. The influence of the given excitation current on the BTBS process.

3.2. Control Way of the Excitation System

There are usually four control modes of an excitation system: constant terminal voltage regulation, constant excitation current regulation, constant reactive power regulation, and constant power factor regulation [30]. However, in the BTBS process of a large PSU, the excitation mode is usually constant excitation current regulation or constant terminal voltage regulation.

One innovation of this paper is the discovery that the closed-loop working mode of constant excitation voltage (CEV) (Note: not constant terminal voltage) can also realize BTBS. Two control modes are simulated and compared. For the convenience of comparison, only the oscillation stages before the two machines run into synchronous acceleration are presented. The simulation results of BTBS with CEC ($i_g^* = 1.4, i_m^* = 1.2$) are shown in Figure 12. Figure 13 shows the simulation results of BTBS with CEV ($V_g^* = 1.4, V_m^* = 1.2$).

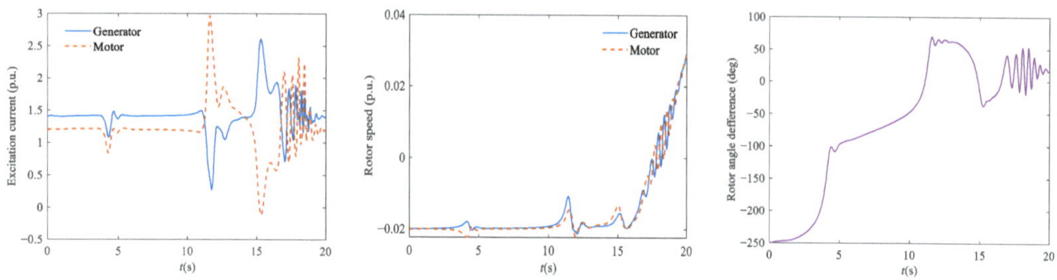

Figure 12. Simulation results of BTBS with CEC. (**a**) Excitation current; (**b**) Rotor speed; (**c**) Rotor angle difference.

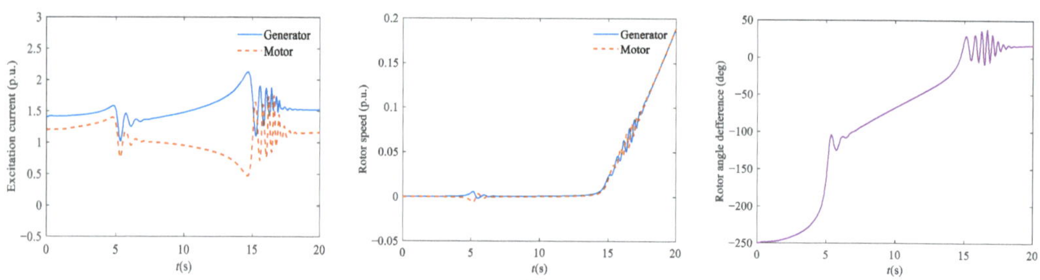

Figure 13. Simulation results of BTBS with CEV. (**a**) Excitation current; (**b**) Rotor speed; (**c**) Rotor angle difference.

By studying Figures 12 and 13, it is revealed that under the same initial conditions, the oscillation in the starting process of CEC is much larger than in that of CEV. The above physical process shows that there must be an oscillation process when the two machines enter the oscillation stage before synchronous acceleration. The rotor circuit passively plays a specific damping role in this oscillation process, and the excitation current will oscillate accordingly. If the CEC closed-loop mode is adopted, the excitation current of the two machines will be forced to remain constant, which will weaken the damping effect of the rotor circuit. When the closed-loop mode of CEV is adopted, the excitation current of the two machines is allowed to oscillate passively to a certain extent, which is equivalent to increasing the damping of the rotor circuit. In this way, the excitation circuit should adopt the closed-loop mode of CEV.

3.3. Initial Difference between Rotor Positions

Due to the randomness of the initial difference between rotor angles of two units, there must be an adjustment from the initial angle difference to the target angle difference, which begins after the water pump is started. Table 2 shows the BTBS results of the two machines under different initial angles. The variation of rotor angle difference under different initial conditions is shown in Figure 14.

Table 2. BTBS results of the two machines with different initial rotor angles.

Initial Rotor Angel Difference (°)	Start Time of Speed Rise (s)	Rotor Angle Difference at Steady State (°)	Description of BTBS
0	3	17.66	Successful start; Slight oscillation
−90	8	19.24	Successful start; Slight oscillation
−180	10	20.89	Successful start; Moderate oscillation
−270	10	21.19	Successful start; Moderate oscillation

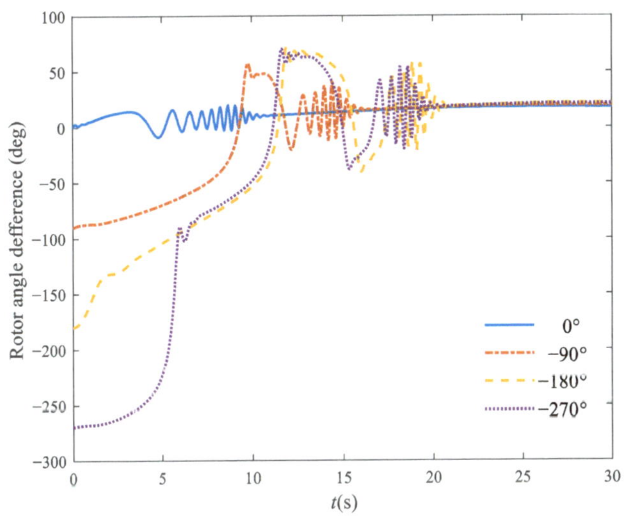

Figure 14. The influence of initial rotor angle differences between two machines.

By observing the simulation results, it can be observed that different initial rotor angles have a certain impact on the startup time of the unit, but have little impact on the target angle difference when the rotor angle difference is stable. With the change of initial rotor angle difference between units, the oscillation in the startup synchronization will change to varying degrees, but it will not affect the successful startup of the unit. Therefore, the influence of the initial difference between rotor positions on BTBS may be ignored.

3.4. Water Head

As an important working condition parameter, the working head is also one of the pivotal influencing factors of the hydraulic machinery transition process, which can be roughly estimated by the water level difference between the upper reservoir and lower reservoir. Here, the changes in water levels above and downstream represent the changes in the working head of the pump turbine, and explore the influence of the working head on BTBS. Therefore, based on the actual unit data of a large PSP in China and the refined model of the PSU, the BTBS process under different working heads is simulated by numerical simulation in this section. The actual data of the unit is as follows: the maximum working head is 217 m, the rated working head is 195 m, and the minimum working head is 178.1 m.

Figure 15 shows the dynamic process trajectory curve of unit flow and unit torque of the tractor under the working head of 200 m, 195 m, 190 m, and 185 m during the BTBS process. For the convenience of comparison, the critical multivalue points of each opening line on the flow characteristic curve and torque characteristic curve are connected to obtain the critical flow line and torque critical line dividing the S-shaped area, as shown in Figure 16. The critical boundary divides the closed area consisting of 0 unit velocity line, 6.9% opening line and 100% opening line into two parts. On the right is the S-shaped area and on the left is the area that can work steadily.

As observed in Figure 15, with the decrease of the working head, the opening of the no-load guide vane of the corresponding back-to-back starting tractor gradually increases. At this time, the unit can easily fall into the S-shaped area at the working point of rated speed. The strong nonlinearity of the pumped storage system in the S-shaped area and the instability of the control system act upon each other, especially when the PSU operates at 185 m head; the flow and torque in the S-shaped characteristic area fall into chaotic operation. This dynamic characteristic will lead the unit speed to swing to and fro, leading to a repeated adjustment of the guide vane opening. If there is no appropriate control strategy, the BTBS process of the unit under a low water head will fail. In summary, under the back-to-back starting condition, the lower the working head of the pump turbine, the more unfavorable it is to the hydraulic mechanical transition process of the unit.

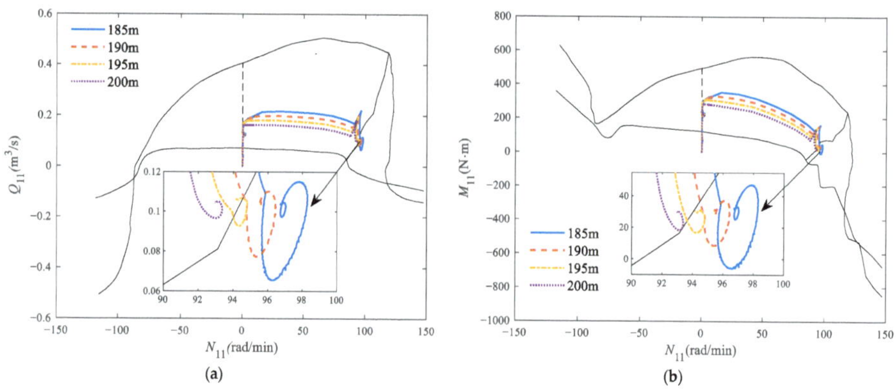

Figure 15. Dynamic process trajectory of moment and torque under different water heads. (**a**) Dynamic trajectory of flow; (**b**) Dynamic trajectory of the moment.

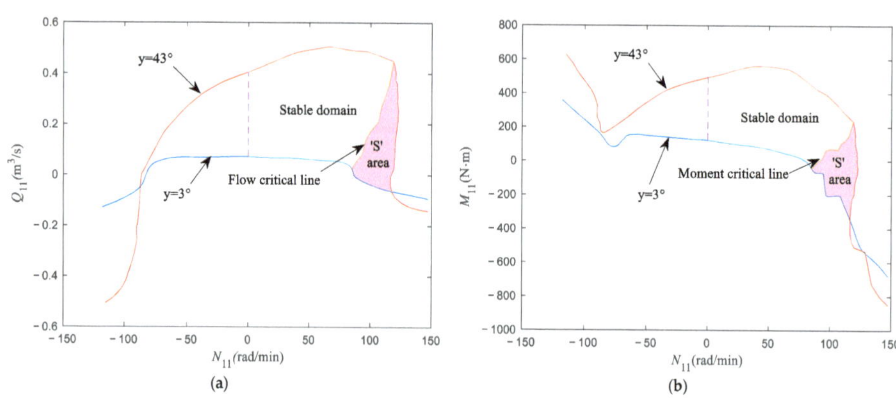

Figure 16. The critical curves of the stable domain and S-shaped region. (**a**) Flow critical curve; (**b**) Moment critical curve.

3.5. Control Scheme of the Governor

The BTBS process consists of both open-loop and closed-loop control. The strategy of slowly opening of the guide vane to a certain set point (related to the current water head) is adopted before the rotor speed of the driving machine reaches a certain value ω_S. After accelerating to 0.9, the PID controller is put into operation in closed-loop regulation according to the rated speed, as shown in Figure 17. The rotational speed transition curves of BTBS are plotted at different points of starting time T_s in Figure 18. Generally, the faster the guide vanes are opened, the faster the speeds of the units rise, but this is not absolute. As shown in Figure 18, when the starting time T_s =15 s, the rising speed of the driven

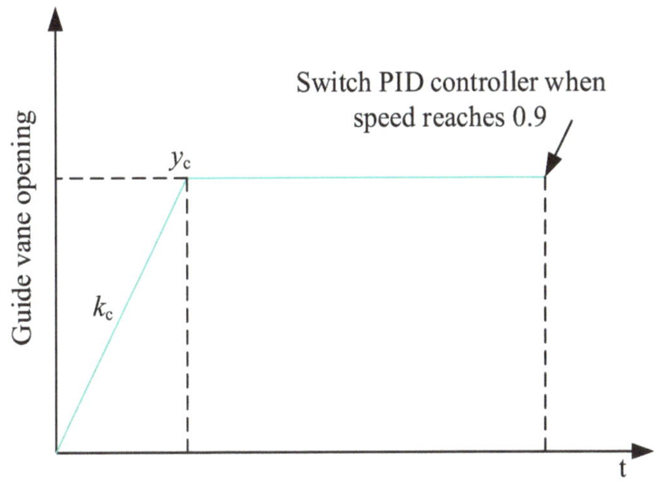

Figure 17. The one-stage direct control law.

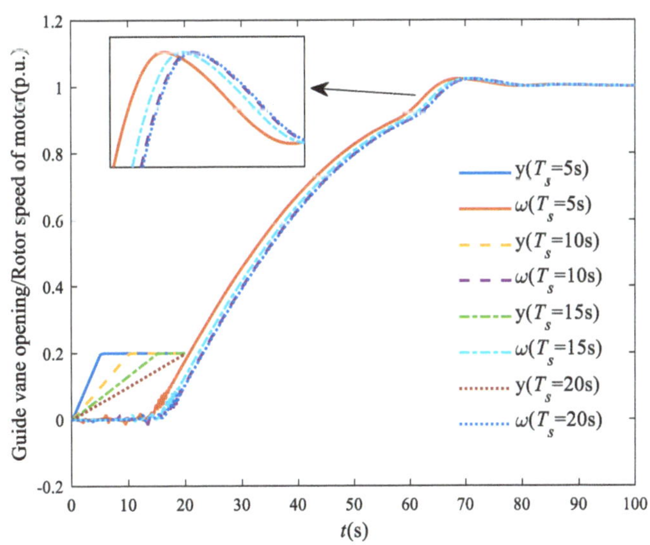

Figure 18. The results of different control schemes.

4. Optimization of BTBS Strategy Based on Multi-Objective Control

It is evident from Section 3 that the excitation current, control way of the excitation system, and control scheme of the governor have a marked impact on the BTBS process. It may even be unsuccessful under the low water head combination of worse control parameters, which is defined as the worst operating condition of BTBS. Therefore, this paper selects the working condition under a low water head as the research object.

The MOPSO algorithm [31,32] is introduced to complete the multi-objective optimization of the BTBS strategy. MOPSO is extensively used in many industries and has outstanding performers in the control optimization of PSU due to its superiority of easy implementation, high precision, and fast convergence [33,34].

4.1. Objective Function

In the BTBS process, the overshoot of rotational speed and the speed stable time can be chosen to respectively represent the stationarity and the rapidity on which we focus. Thus, the overshoot and stable time of rotational speed are included as the evaluation indexes to optimize the BTBS strategy on the basis of multi-objective control. While both evaluation indexes are substitutes, the velocity is more critical, so the speed overshoot of PSU-1 and speed rise time of PSU-2 are chosen as two optimization indexes.

The two objectives are defined as follows:

$$\begin{cases} \min F_1 = \omega_{1\max} - 1 \\ \min F_2 = t_{s2} \end{cases} \tag{10}$$

where $\omega_{1\max}$ is the relative value of the peak speed of driving machine PSU-1, and t_{s2} is the speed stable time of driven machine PSU-2. The description of the objective function is shown in Figure 19.

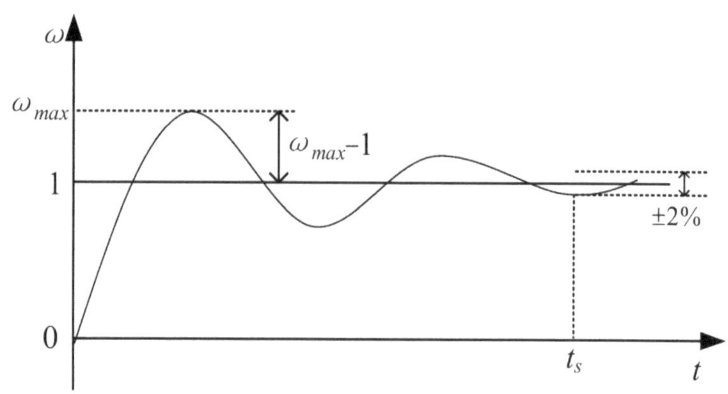

Figure 19. Diagram of speed transition process curve.

4.2. Decision Variables

The decision variable is the parameter of the BTBS strategy. For convenience of analysis, the speed regulation strategy adopts "one-stage guide vane direct opening + PID control", as shown in Figure 17. This paper studies two excitation strategies: constant excitation current (CEC) and constant excitation voltage (CEV). Their decision variables are summarized as follows:

- Scheme 1: The traditional CEC mode with one-stage DGVC+PID control. In one-stage DGVC, the guide vane of the driving machine is first opened at the rate k_c, and remains unchanged when the guide vane opening reaches y_c. When the speed reaches 90% of the rated speed, the PID controller will be put into operation. The excitation system of the two units shall first operate with the given excitation current, and switch to the constant terminal voltage mode when the speed reaches 90% of the rated speed. k_c generally takes the maximum rate, which is a known parameter. Therefore, the given excitation current i_g^*, i_m^*, and given opening y_c are selected as the decision variables in the open-loop stage. Parameters of PID controller, K_p, K_i, and K_d, are adopted as the decision variables in the closed-loop stage.

$$X_1 = \left[i_g^*, i_m^*, y_c, K_p, K_i, K_d\right] \tag{11}$$

- Scheme 2: The proposed CEV mode with one-stage DGVC+PID control. Similarly, the given excitation voltageV_g^*, V_m^*, given opening y_c, and the three parameters of PID controller, K_p, K_i, and K_d, are chosen as the decision variables.

$$X_2 = \left[V_g^*, V_m^*, y_c, K_p, K_i, K_d\right] \tag{12}$$

4.3. Constraint Conditions

There are three types of constraints extracted from engineering limitations according to our experiments in this study.

(1) Operation Time Constraint

$$0 < t_{s2} < t_{\max} \tag{13}$$

where t_{s2} is the stability time of PSU-2. $t_{\max}$ is the maximum operation time.

(2) The Boundary of Decision Variables

$$X_i \in [L, U] \tag{14}$$

where X_i is the decision variables of $i-$th scheme; L and U indicate the lower limits and the upper limits of the boundary.

(3) Rotor Speed Difference Between Two Machines According to the requirements of the BTBS, the rotational speed difference between the two units shall not exceed a certain limit value; otherwise, it is considered a startup failure:

$$|\omega_1 - \omega_2| \leq \omega_{\max}^d \tag{15}$$

where ω_1 is the rotor speed of the driving machine and ω_2 is the rotor speed of the driven machine. $\omega_{\max}^d$ is the maximum allowable rotor speed difference between the two units

4.4. Optimization Procedures

Based on the nonlinear simulation platform of the PSP system, this study puts forward two multi-objective optimization schemes of BTBS strategy, which are multi-objective constant excitation current (MOCEC) and multi-objective constant excitation voltage (MO-CEV). In this subsection, the MOPSO algorithm is introduced to optimize the proposed optimization scheme under the low head condition of BTBS, because once the optimized control schemes can be applied to poor conditions, it is able to adapt to those under normal circumstances. Figure 20 shows the flowchart of the optimization for BTBS with main optimized procedures.

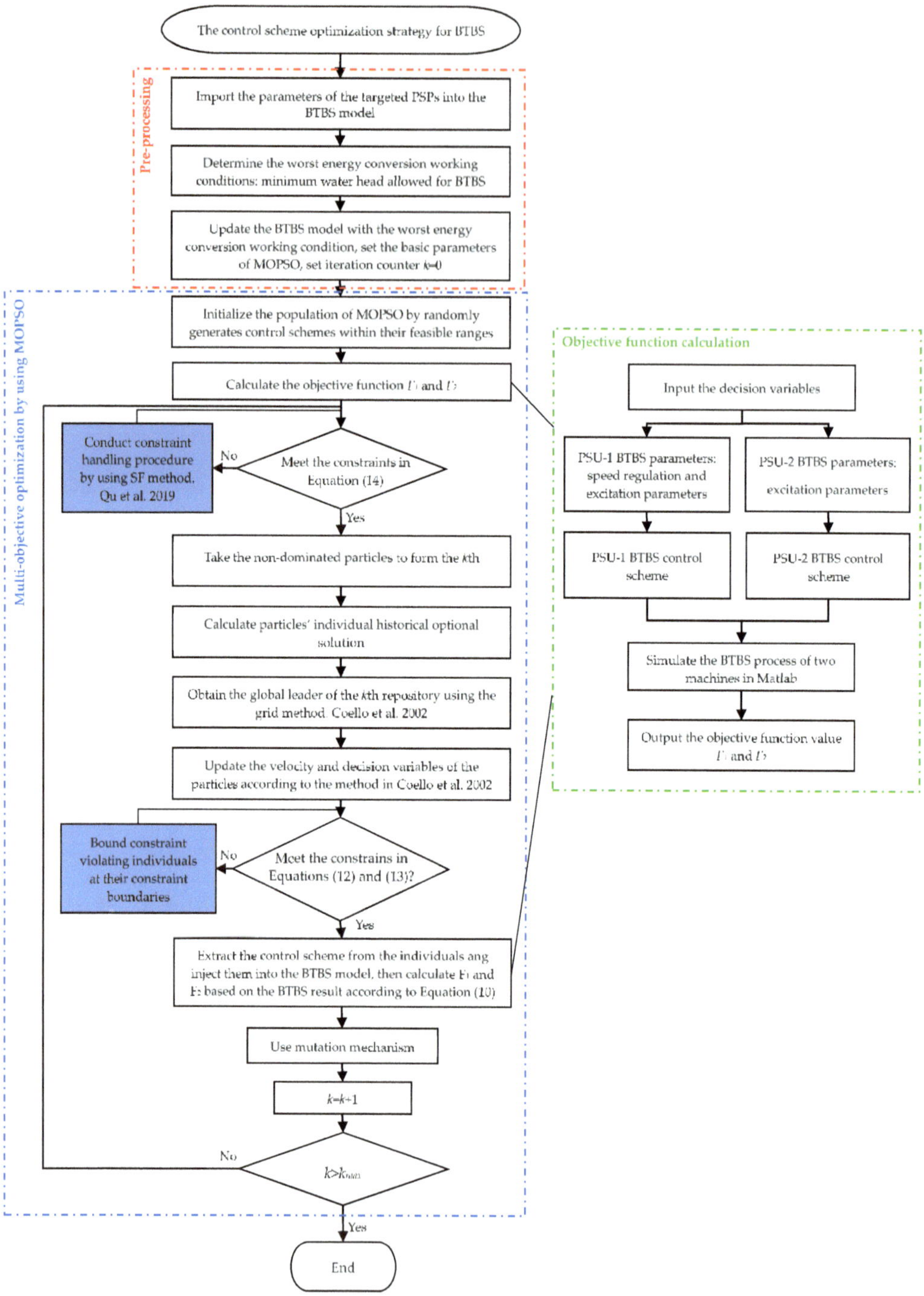

Figure 20. Multi–objective control scheme optimization for BTBS [31,34].

5. Case Study and Analysis

A real PSP in China is taken as the experimental object for the simulation analysis so as to certify the efficacy of the proposed scheme. The layout of the power station is shown in Figure 7. Both units participate in operation; that is, one pipe and two units operate under BTBS conditions. Among them, 1# machine operates as a tractor under the power generation condition of water turbine, and 2# unit operates as a driven machine under the pumping condition of the motor.

5.1. Model Parameters Setting

As shown in Figure 7, the capacities of the two units are the same. The basic parameters of the water diversion system, pump turbine, synchronous machine, governor, and excitation system are the same as those in Section 2.3, as shown in Tables A1–A4. All experiments in this section are carried out under the identical water head; that is, the water level of the upstream reservoir is 291.45 m, and that of the downstream reservoir is 106.11 m. The parameters of the multi-objective optimization algorithm are set as follows: the number of iterations is 200, the population size is 50, the initial value of the inertia weight is 0.5, the size of the external archive is 25, the inertia weight damping is 0.99, and the acceleration constants c_1 and c_2 are both 1.6. According to the analysis of experimental experience and results, the value ranges of decision variables X_1 and X_2 are obtained, as shown in Table 3.

Table 3. Boundaries of decision variables.

Decision Variables	Boundaries	Values					
X_1	L_1	0.1	0.1	0.1	1	0.1	0
	U_1	2.0	2.0	0.4	6	1	6
X_2	L_2	0.1	0.1	0.1	1	0.1	0
	U_2	2.0	2.0	0.4	6	1	6

The multi-objective optimization results named Pareto Set need further ordering to acquire the optimal solution that satisfies the needs of the decision-maker. Therefore, in this section, the multi-objective decision-making method based on relative target proximity is also used to sort the Pareto optimal solution set of MOCEC and MOCEV [35], and the weight information consisting of subjective weight α, objective weight β, and comprehensive weight γ required by the decision-making method are given in Table 4.

Table 4. The weight value of MOCEC and MOCEV schemes.

Schemes	α	β	γ
MOCEC	(0.90, 0.10)	(0.53, 0.47)	(0.91, 0.09)
MOCEV	(0.90, 0.10)	(0.63, 0.36)	(0.94, 0.06)

5.2. Introduction to Comparative Experiments

In order to demonstrate the superiority of the schemes mentioned above, single-objective optimization schemes based on the PSO algorithm are designed as comparative experiments, which are abbreviated as single objective constant excitation current (SOCEC) and single objective constant excitation voltage (SOCEV). The integrated time and absolute error (ITAE) index can be obtained for each unit, as shown in Equations (15) and (16):

$$F_{ITAE1} = \sum_{k=1}^{N_s} t(k)|1 - \omega_1(k)| \tag{16}$$

$$F_{ITAE2} = \sum_{k=1}^{N_s} t(k)|1 - \omega_2(k)| \tag{17}$$

where subscript 1 represents the variable of 1# unit, subscript 2 represents the variable of 2# unit. In order to consider the dynamic process performance indexes of the two units at the same time, the ITAE indexes of the two units are integrated with certain weights w_1 and w_2 as an objective function, as shown in Equation (17):

$$F_{ITAE} = w_1 F_{ITAE1} + w_2 F_{ITAE2} \tag{18}$$

where F_{ITAE} is the objective function of single objective scheme, and w_1 and w_2 are the weight of ITAE index of 1# unit and 2# unit respectively.

5.3. *Effectiveness Analysis*

For the MOCEC and MOCEV, the optimization result is a set of Pareto optimal solution sets. The particles in the Pareto optimal solution set are non-dominated solutions to each other. The particles in the Pareto optimal solution set are listed in Tables A5 and A6. To make the optimal solution satisfy the needs of the decision-maker, the decision-making method based on relative target proximity is used to sort the particles in the Pareto optimal solution set. The weight value of the decision-making method is shown in Table 4, so the particle ranking is selected as the optimal solution for the multi-objective optimization scheme. Table 5 shows the optimal solutions of the four schemes.

Table 5. Optimal decision variables of four schemes.

Variables \ Schemes	SOCEC	SOCEV	MOCEC	MOCEV
$i_g^*(V_g^*)$	1.2331	0.6599	0.9881	0.6049
$i_m^*(V_m^*)$	0.7422	0.5413	0.6272	0.4437
y_c	0.3985	0.3976	0.1924	0.1929
K_p	3.9083	4.6298	3.5060	3.5941
K_i	0.3630	0.3900	0.1000	0.1000
K_d	0.3190	0.7031	4.9311	5.0779

The optimal decision variables of the four schemes correspond to the four optimal BTBS strategies. The transition process curves of 1# and 2# units under BTBS conditions can be obtained by inputting the four optimal startup strategies into the nonlinear simulation platform of the PSP, as shown in Figure 21. Due to the short adjustment time of field current and rotor angle difference, these two parameters only show the transition process of the first 10 s. The performance indexes of rotor speed are shown in Table 6; $\Delta\omega_{max}$ represents the speed overshoot, t_r represents the speed rise time, and t_s represents the stability time.

In Figure 21a, it can be found that the speed overshoot of single-objective schemes is large, while the speed overshoot of multi-objective schemes is small; the speed fluctuation of multi-objective schemes is also small and the curves are more stable. Figure 21b shows the PSU-2′s speed transition processes of the four schemes. It can be observed that the speed curves of single-objective schemes fluctuate more, and the speed curves of the multi-objective schemes are more stable. Figure 21c–h shows the transition curves of other physical quantities. In Figure 21c–d, the field current regulation processes of the four schemes have a certain fluctuation, but the current transition process curve of the MOCEV scheme can reach the steady state fastest and the fluctuation is the smallest. In Figure 21e, the GVO adjustment process of the MOCEV scheme is faster and the oscillation times are the shortest. As seen in Figure 21f, the peak value of rotor angle difference of the four schemes is essentially the same, but the rotor angle fluctuation range of the MOCEV scheme is the smallest. As observed in Figure 21g, the terminal voltage rise processes of the four schemes are slightly different. The terminal voltage rise of the single-objective optimization methods is faster than those of the multi-objective optimization methods, and the impulse generated during voltage control switching of the CEC control scheme is smaller than that of the CEV control scheme. By observing Figure 21h, it can be concluded that during the BTBS under low water head conditions, the multi-objective optimization method can help the PSU avoid going deep into the S-shape zone, so as to avoid the repeated oscillation of active power and falling into chaos. Therefore, from the analysis of other working indexes of the two units, the multi-objective optimization schemes are better than the single-objective optimization schemes.

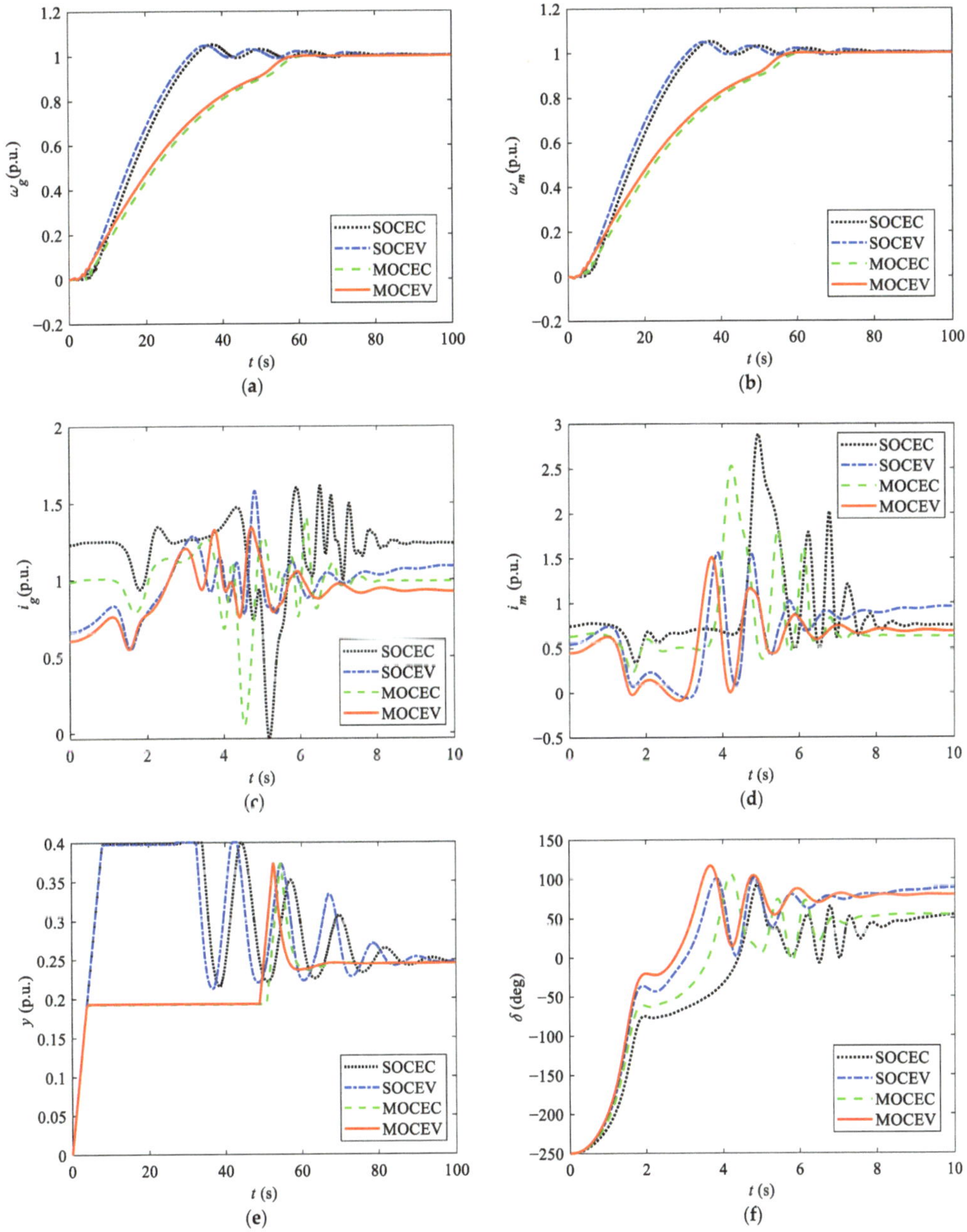

Figure 21. *Cont.*

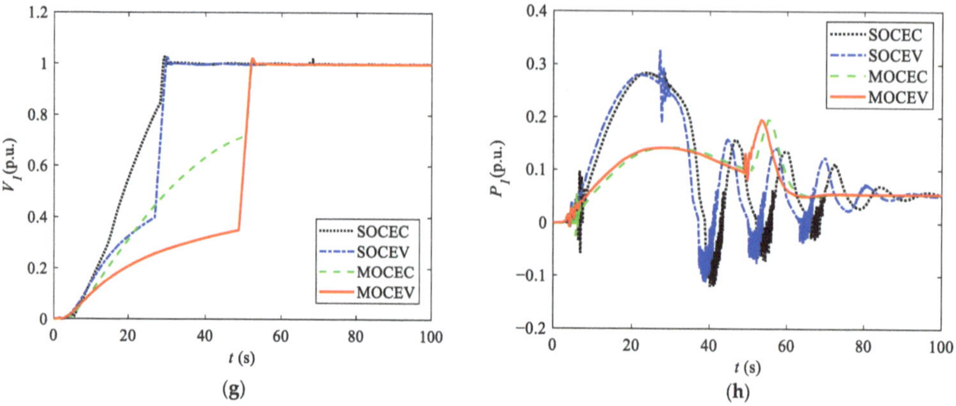

Figure 21. BTBS process of four schemes. (**a**) Rotor speed of PSU-1; (**b**) Rotor speed of PSU-2; (**c**) Field current of PSU-1; (**d**) Field current of PSU-2; (**e**) Guide vane opening; (**f**) Rotor angle difference; (**g**) Terminal voltage; (**h**) Active power.

Table 6. Speed performance indexes of four schemes.

Schemes / Indexes	SOCEC		SOCEV		MOCEC		MOCEV	
	PSU-1	**PSU-2**	**PSU-1**	**PSU-2**	**PSU-1**	**PSU-2**	**PSU-1**	**PSU-2**
$\Delta\omega_{max}$	0.0515	0.0518	0.0468	0.0472	2.7796×10^{-5}	4.1466×10^{-5}	8.7973×10^{-6}	2.2633×10^{-5}
t_r (s)	28.20	28.18	26.84	26.82	50.74	50.74	48.86	48.84
t_s (s)	90.68	90.68	99.30	99.30	60.98	60.94	59.00	58.98

For the single-objective schemes, the objective function is the weighted sum of the ITAE indexes of two units. It can be obtained from Table 6 that the speed overshoot of 1# unit is 4.98% and the speed rise time of 2# unit is 26.82 s for SOCEV, which is less than the value of SOCEC, but the speed stable time of 1# and 2# unit of SOCEC are slightly less than the value of SOCEV. For the multi-objective schemes, the objective function is speed overshoot of PSU-1 and speed stable time of PSU-2. The speed overshoot of 1# unit is 0.00088% and the speed stable time of 2# unit is 58.98 s for MOCEV, which is less than the value of MOCEC. Furthermore, comparing different schemes, the speed overshoot of multi-objective optimization schemes is much smaller than those of single-objective optimization schemes, and the speed stable time of multi-objective optimization schemes is also smaller. However, the speed rise time of multi-objective optimization schemes is greater than those of single-objective optimization schemes. In addition, among the four schemes, the speed overshoot of PSU-1 and speed stable time of PSU-2 for MOCEV are the smallest, so it is considered that the optimization effect of the MOCEV scheme is better.

To further reflect the advantages of the multi-objective scheme MOCEV, Figure 22 shows the Pareto fronts of MOCEC and MOCEV schemes.

The Pareto front are even-distributed shown in Figure 22, and the first 25 particles of the MOCEV scheme are non-dominated solutions. If one selects three particles from the far left, middle, and far right for specific analysis, one could simulate their speed transition process curve, as shown in Figure 23. In Figure 23, the speed overshoot of 1# unit of three particles increases in turn, while the speed stable time of 2# unit decreases in turn. Additionally, these three typical particles can help the unit avoid going deep into the S-shaped zone, which ensures the safety and stability of the BTBS process of the PSU under low head conditions. These three particles represent three kinds of particles. Decision-makers may choose one of them as the optimal solution for power plant operation according to actual needs. There are also 25 particles in front of the MOCEC scheme, and most of them are dominated by the particles at the front of the MOCEV. This phenomenon also shows that the MOCEV scheme is better than the MOCEC.

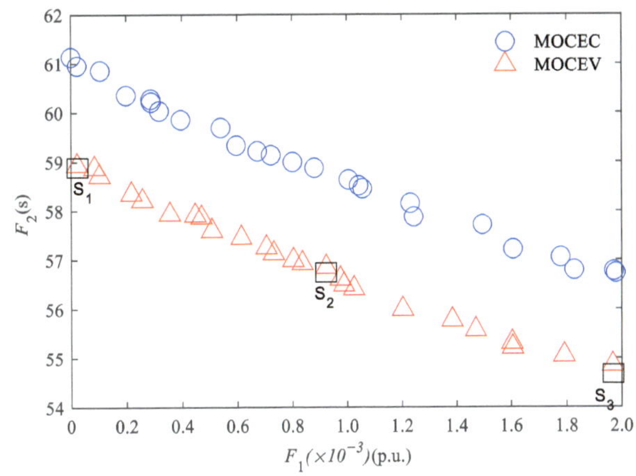

Figure 22. The Pareto fronts obtained by the multi-objective optimization strategy.

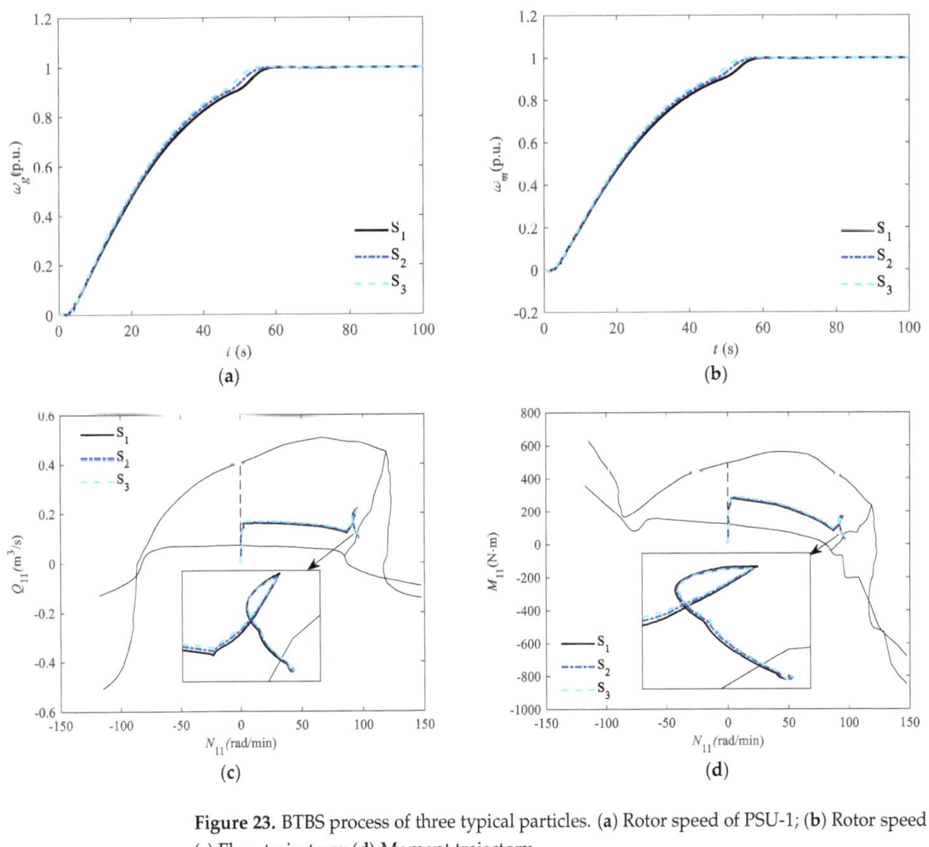

Figure 23. BTBS process of three typical particles. (**a**) Rotor speed of PSU-1; (**b**) Rotor speed of PSU-2; (**c**) Flow trajectory; (**d**) Moment trajectory.

5.4. Validation of the Proposed Optimization Strategy

This section verifies the effectiveness of the multi-objective optimization method MOCEV for BTBS conditions proposed in this paper under other conditions, and selects different upstream and downstream water levels to form the verification conditions; the corresponding data are listed in Table 7. The parameters of the optimization results in Section 5.4 are respectively substituted into the refined mathematical model of the regulation system of the PSU for the transition process simulation, and the key indicators of the transition process are obtained, as shown in Figure 24.

Table 7. Parameters and MOCEV results under various working conditions of BTBS.

Working Conditions	Water Level		Water Head (m)
	Upstream (m)	Downstream (m)	
N0	291	106	185
N1	295	105	190
N2	298	103	195
N3	303	103	200
N4	303	98	205
N5	308	98	210

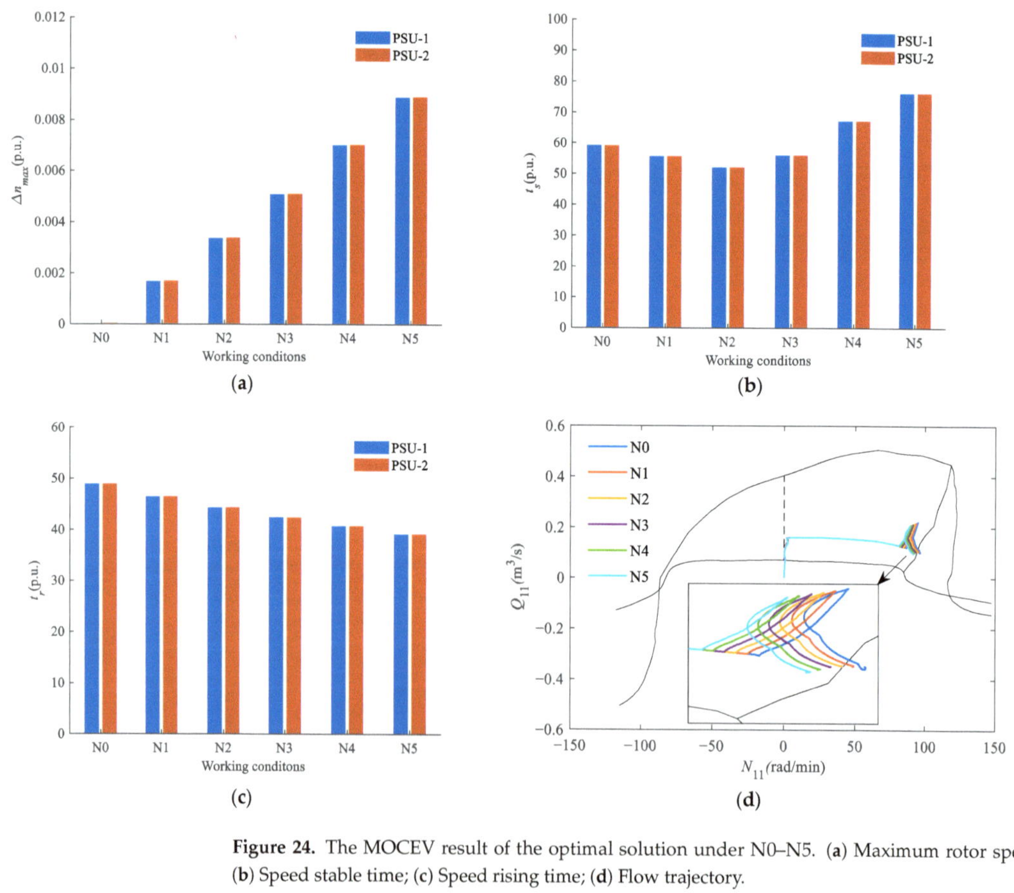

Figure 24. The MOCEV result of the optimal solution under N0–N5. (**a**) Maximum rotor speed; (**b**) Speed stable time; (**c**) Speed rising time; (**d**) Flow trajectory.

Figure 24a–c shows that, when the multi-objective optimization result is used as the control parameter for BTBS, with the water head increases, the velocity overshoot increases, the velocity stable time decreases first and then increases, and the speed rising time decreases. This is because the no-load GVO under a low water head is larger. When used for open-loop control under other water heads, the speed rise will be faster and the overshoot will be increased. However, in order to ensure that the unit does not fall into the S-shaped area, the integral parameter value under the low head is small, which will lead to a longer time for the speed to reach the steady state under other conditions. Therefore, with the increase of water head, the integral parameters can be appropriately increased to improve the regulation quality under other working conditions. In addition, by observing Figure 24d, it can be concluded that the control parameters optimized under a low water head can prevent the BTBS process of the unit from falling into the S-shaped zone within the full working condition.

6. Conclusions

A refined nonlinear model for BTBS of PSP is developed in this paper, the impacts from the electrical and hydraulic-mechanical subsystem on the operation of PSP during BTBS are investigated subsequently, and a novel MOCEV control strategy is proposed accordingly. The following conclusions are drawn from the simulation results.

1. The given value and control way of excitation current, the control scheme of the governor, and the water head have great influence on the transient process of BTBS. The control scheme of excitation current and guide vane should be selected as the decision variables in the BTBS optimization; the worst BTBS condition can be identified by the lowest water head.
2. The overshoot and stable time of the speed are contradictory. The traditional single-objective optimization scheme merely considers the single objective, which can very easily cause the unit to fall into the S-shaped area, resulting in severe fluctuations in speed and power.
3. Compared with the single-objective, the optimization strategy proposed can considerably improve the speed overshoot and the speed stable time by at least 68.27% and by 3.22% under the worst working condition. The optimization results show that the multi-objective scheme is a better choice than the single-objective scheme.
4. Compared with the MOCEC scheme, when the MOCEV scheme is adopted, the overshoot, rise time, and stable time are improved by 68.35%, 3.7%, and 3.2% in PSU-1, and 45.4%, 3.7%, and 3.2% in PSU-2. Thus, the MOCEV scheme is superior.
5. The proposed MOCEV optimal control scheme can effectively keep away from the S-shaped area and is verified by a real PSU.

Author Contributions: Conceptualization, methodology, software, data curation, writing, original draft preparation, C.F.; writing, editing, corresponding, G.L.; reviewing and editing, supervision, Y.Z.; supervision, D.Z.; visualization, investigation, Z.M. All authors have read and agreed to the published version of the manuscript.

Funding: The Fundamental Research Funds for the Central Universities (B220202005) and the National Natural Science Foundation of China (No. 51979086).

Institutional Review Board Statement: Approval for the study was not required in accordance with local/national legislation.

Informed Consent Statement: Not applicable.

Data Availability Statement: Not applicable.

Conflicts of Interest: The authors declare that they have no known competing financial interest or personal relationships that could have appeared to influence the work reported in this paper.

Nomenclature

Abbreviations	
PSP	pumped storage plant
BTBS	back-to-back starting
PSU	pump storage unit
CEV	constant excitation voltage
CEC	constant excitation current
LCP	logarithmic curve projection
MOC	method of characteristics
GVO	gate valve opening
MOCEC	multi-objective CEC
MOCEV	multi-objective CEV
SOCEC	single-objective CEC
SOCEV	single-objective CEV
ITAE	integrated time and absolute error
pu	per-unit value
Symbols	
Parameters	
A	cross sectional area of pipeline (m^2)
a	wave velocity (m/s)
b_p	permanent difference coefficient (pu)
D	the diameter of the turbine runner (m)
d	pipeline diameter (m)
f	friction coefficient (pu)
g	acceleration of gravity (m/s^2)
K_d	differential gain (pu)
K_e	self-excitation coefficient of exciter (pu)
K_f	damping coefficient (pu)
k_0	forward amplification factor (pu)
K_i	integral gain (pu)
K_p	proportional gain (pu)
R_a	resistance of stator winding (pu)
S_e	exciter saturation factor (pu)
T_a	amplifier time constant (s)
T_b	lead lag time constant (s)
T_c	lead lag time constant (s)
x	distance calculated from upstream (m)
T_d	differential time constant (s)
T_e	exciter time constant (s)
T_f	damping time constant (s)
T_j	mechanical time constant (s)
T_y	main servomotor response time (s)
T_{y1}	assistant servomotor response time (s)
T'_{d0}, T''_{d0}	transient and sub-transient time constants of open-circuit d-axis (s)
T'_{q0}, T''_{q0}	transient and sub-transient time constants of open-circuit q-axis (s)
X_q, X'_q, X''_q	synchronous, transient and sub-transient reactance of q-axis (pu)
X_q, X'_q, X''_q	synchronous, transient and sub-transient reactance of q-axis (pu)
Variables	
E'_d, E''_d	the transient and sub-transient internal EMF of d-axis (pu)
E_{fd}	excitation EMF (pu)
E'_q, E''_q	the transient and sub-transient internal EMF of q-axis (pu)
H	piezometric head (m)
H_t	the working head of pump turbine (m)
i_d, i_q	the current of d- and q-axis (pu)
i_{fd}	excitation current (pu)
i^*_g, i^*_m	excitation current setting value of generator and motor (pu)

K_a	amplifier coefficient (pu)
M_e	electromagnetic torque (pu)
M_t	the moment of pump turbine (pu)
M_{11}	unit torque (N/m^3)
N	the rotational speed of the turbine (r/min)
N_{11}	unit speed (m$^{1/2}$/s)
N_{11r}	rated unit speed (m$^{1/2}$/s)
Q	the water flow rate (m^3/s)
Q_t	the flow of pump turbine (m^3/s)
Q_{11}	unit flow (m$^{1/2}$/s)
Q_{11r}	rated unit flow (m$^{1/2}$/s)
u	controller output signal (pu)
V_g^*, V_m^*	excitation voltage setting value of generator and motor (pu)
v_d, v_q	d- and q-axis component of the voltage (pu)
Y	guide vane opening (deg)
y	main servomotor output signal (pu)
ω, ω^*	relative and given value angular shaft velocity (m)
φ_d, φ_q	the internal EMF of d- and q-axis (pu)
δ	rotor angle (rad)

Appendix A

Table A1. PSP design parameters.

Parameters	Values	Parameters	Values
Rated speed (r/min)	250	Rated capacity (MVA)	334
Rated water-head (m)	195	Rated voltage (kV)	15.75
Rated water flow (m^3/s)	176.1	Rated current (A)	12,244
Power rating (MW)	306	Rated frequency (Hz)	50
Turbine runner diameter (m)	5.26	Power factor	0.90
100% guide-vane opening (°)	43.01	Moment of inertia (ton/m^2)	19,300

Table A2. Basic parameters of equivalent pipeline of diversion system.

Number	Length (m)	Diameter (m)	Wave Velocity (m/s)	Roughness
L_1	1113.94	9.21	1100	0.014
L_2	206.71	8.97	1120	0.014
L_3	250.77	4.77	1204	0.011
L_4	173.52	6.90	1161	0.010
L_5	260.30	4.80	1204	0.011
L_6	173.52	6.89	1160	0.012
L_7	295.27	10.82	1050	0.014

Table A3. Basic parameters of surge chamber.

Sectional Area of the Impedance Hole (m^2)	Inflow Loss Coefficient	Outflow Loss Coefficient	Sectional Area (m^2)	Altitude (m)
19.63	0.0009217	0.0006767	19.63	231.70~268.30
			380.13	268.30~310.00
			530.93	310.00~320.00

Table A4. Basic parameters of each module of the regulation system model of pumped storage unit.

Excitation System		Synchronous Machine			
T_a	0.001	R_a	0.00125	X''_d	0.2
T_b	0	X_d	1.015	X''_q	0.195
T_c	0	X_q	0.627	T'_{d0}	12.6
T_e	0	X'_d	0.253	T''_{d0}	0.189
T_f	0.1	T_j	10.8	T''_{q0}	0.519
K_a	300	Speed Regulation System			
K_e	1	b_p	0.01	k_0	1
K_f	0.001	T_{y1}	0.02	T_y	0.2

Table A5. The detailed parameters corresponding to the Pareto front solutions of constant excitation current.

Solutions	F_1	F_2(s)	i_g^*	i_m^*	y_c	K_p	K_i	K_d
1	0.0000	61.1400	0.9805	0.6135	0.1923	3.4725	0.1000	5.1410
2	0.0000	60.9600	0.9881	0.6272	0.1924	3.5060	0.1000	4.9311
3	0.0001	60.8600	1.0098	0.6266	0.1932	3.4515	0.1000	5.0734
4	0.0002	60.3600	1.0188	0.5930	0.1938	3.6432	0.1000	5.2704
5	0.0003	60.2800	1.0329	0.6127	0.1942	3.5734	0.1000	5.1753
6	0.0003	60.2200	1.0125	0.6208	0.1945	3.5620	0.1000	5.3608
7	0.0003	60.0400	0.9989	0.6175	0.1944	3.5887	0.1000	5.1362
8	0.0004	59.8600	1.0071	0.5973	0.1949	3.5649	0.1000	5.1444
9	0.0005	59.7000	1.0080	0.6061	0.1954	3.3926	0.1000	4.9045
10	0.0006	59.3400	0.9943	0.6156	0.1964	3.5366	0.1000	5.3577
11	0.0007	59.2200	0.9977	0.6136	0.1962	3.4684	0.1000	4.9254
12	0.0007	59.1400	0.9970	0.6167	0.1970	3.4364	0.1000	5.1984
13	0.0008	59.0000	1.0160	0.6366	0.1971	3.5003	0.1000	5.0073
14	0.0009	58.8800	1.0035	0.6084	0.1968	3.3919	0.1000	4.6497
15	0.0010	58.6400	1.0143	0.6353	0.1984	3.3848	0.1000	5.1553
16	0.0010	58.5200	1.0118	0.6244	0.1980	3.3945	0.1000	4.7860
17	0.0011	58.4600	1.0115	0.6289	0.1988	3.3993	0.1000	5.1672
18	0.0012	58.1600	1.0112	0.6271	0.1997	3.3300	0.1000	5.1576
19	0.0012	57.8800	1.0027	0.6149	0.1999	3.4411	0.1000	5.2033
20	0.0015	57.7200	1.0255	0.6064	0.2001	3.2400	0.1000	4.5951
21	0.0016	57.2200	0.9945	0.6315	0.2014	3.3322	0.1000	4.9067
22	0.0018	57.0600	1.0057	0.6203	0.2022	3.2075	0.1000	4.8830
23	0.0018	56.8000	1.0073	0.6276	0.2028	3.3360	0.1000	5.0474
24	0.0020	56.7800	0.9945	0.6364	0.2038	3.3257	0.1000	5.1467
25	0.0020	56.7400	1.0045	0.6287	0.2029	3.1622	0.1000	4.6525

Table A6. The detailed parameters corresponding to the Pareto front solutions of constant excitation voltage.

Solutions	F_1	F_2(s)	V_g^*	V_m^*	y_c	K_p	K_i	K_d
1	0.0000	58.9400	0.6049	0.4437	0.1929	3.5941	0.1000	5.0779
2	0.0001	58.8800	0.5815	0.4172	0.1928	3.4206	0.1000	4.6254
3	0.0001	58.7200	0.5967	0.4302	0.1940	3.5515	0.1000	5.3448
4	0.0002	58.3600	0.6055	0.4150	0.1949	3.6035	0.1000	5.4766
5	0.0003	58.2200	0.5762	0.4380	0.1946	3.5755	0.1000	5.1809
6	0.0004	57.9400	0.5310	0.4862	0.1955	3.6367	0.1000	5.3869
7	0.0004	57.9200	0.5631	0.4490	0.1956	3.3992	0.1000	4.9775
8	0.0005	57.8800	0.5369	0.4775	0.1963	3.4578	0.1000	5.4568
9	0.0005	57.6000	0.5851	0.4409	0.1962	3.5772	0.1000	5.2510
10	0.0006	57.4600	0.5933	0.4596	0.1962	3.5212	0.1000	4.9430
11	0.0007	57.2600	0.6137	0.4421	0.1975	3.5201	0.1000	5.4057
12	0.0007	57.1400	0.5369	0.4708	0.1974	3.4673	0.1000	5.1295
13	0.0008	57.0000	0.5759	0.4408	0.1976	3.4549	0.1000	5.0178
14	0.0008	56.9400	0.5776	0.4511	0.1979	3.4562	0.1000	5.1085
15	0.0009	56.8600	0.5853	0.4415	0.1986	3.3694	0.1000	5.2359
16	0.0010	56.6200	0.5782	0.5092	0.1990	3.5058	0.1000	5.3278
17	0.0010	56.5000	0.5681	0.4310	0.1991	3.4715	0.1000	5.2946
18	0.0010	56.4200	0.5768	0.4593	0.1992	3.5151	0.1000	5.2587
19	0.0012	56.0000	0.5805	0.4346	0.2003	3.4980	0.1000	5.2770
20	0.0014	55.7800	0.5440	0.4646	0.2010	3.3650	0.1000	5.1138
21	0.0015	55.5800	0.5772	0.4403	0.2013	3.3715	0.1000	5.0147
22	0.0016	55.3200	0.5720	0.4422	0.2023	3.3496	0.1000	5.1389
23	0.0016	55.2200	0.5543	0.4567	0.2024	3.3941	0.1000	5.1392
24	0.0018	55.0600	0.5587	0.4434	0.2027	3.2578	0.1000	4.8341
25	0.0020	54.8600	0.5710	0.4508	0.2037	3.1818	0.1000	4.8972

References

1. Zhang, Y.; Zang, W.; Zheng, J.; Cappietti, L.; Zhang, J.; Zheng, Y.; Fernandez-Rodriguez, E. The influence of waves propagating with the current on the wake of a tidal stream turbine. *Appl. Energy* **2021**, *290*, 116729. [CrossRef]
2. Zhang, Y.; Zhang, Z.; Zheng, J.; Zhang, J.; Zheng, Y.; Zang, W.; Lin, X.; Fernandez-Rodriguez, E. Experimental investigation into effects of boundary proximity and blockage on horizontal-axis tidal turbine wake. *Ocean Eng.* **2021**, *225*, 108829. [CrossRef]
3. Zhang, Y.; Zhang, J.; Lin, X.; Wang, R.; Zhang, C.; Zhao, J. Experimental investigation into downstream field of a horizontal axis tidal stream turbine supported by a mono pile. *Appl. Ocean Res.* **2020**, *101*, 102257. [CrossRef]
4. Zheng, Y.; Chen, Q.; Yan, D.; Liu, W. A two-stage numerical simulation framework for pumped-storage energy system. *Energy Convers Manag.* **2020**, *210*, 112676. [CrossRef]
5. Zheng, Y.; Chen, Q.; Yan, D.; Zhang, H. Equivalent circuit modelling of large hydropower plants with complex tailrace system for ultra-low frequency oscillation analysis. *Appl. Math. Model.* **2021**, *103*, 176–194. [CrossRef]
6. Feng, C.; Li, C.; Chang, L.; Ding, T.; Mai, Z. Advantage analysis of variable-speed pumped storage units in renewable energy power grid: Mechanism of avoiding S-shaped region. *Int. J. Electr. Power Energy Syst.* **2020**, *120*, 105976. [CrossRef]
7. Zeng, W.; Yang, J.; Hu, J. Pumped storage system model and experimental investigations on S-induced issues during transients. *Mech. Syst. Signal Process.* **2017**, *90*, 350–364. [CrossRef]
8. Wang, X.; Na, R. Simulation of asynchronous starting process of synchronous motors based on Matlab/Simulink. In Proceedings of the Sixth International Conference on Electrical Machines and Systems, ICEMS 2003 IEEE, Beijing, China, 9–11 November 2003; Volume 2, pp. 684–687.
9. Wang, F.; Jiang, J. A novel static frequency converter based on multilevel cascaded H-bridge used for the startup of synchronous motor in pumped-storage power station. *Energy Convers Manag.* **2011**, *52*, 2085–2091. [CrossRef]
10. Magsaysay, G.; Schuette, T.; Fostiak, R.J. Use of a static frequency converter for rapid load response in pumped-storage plants. *IEEE Trans. Energy Convers.* **1995**, *10*, 694–699. [CrossRef]
11. Li, B.; Duan, Z.; Wu, J. Short-circuit analysis of pumped storage unit during back-to-back starting. *IET Renew. Power Gener.* **2015**, *9*, 99–108. [CrossRef]
12. Konidaris, D.N. Investigation of back-to-back starting of pumped storage hydraulic generating units. *IEEE Trans. Energy Convers.* **2002**, *17*, 273–278. [CrossRef]
13. Xu, Y.; Zhou, J.; Xue, X.; Fu, W.; Zhu, W.; Li, C. An adaptively fast fuzzy fractional order PID control for pumped storage hydro unit using improved gravitational search algorithm. *Energy Convers. Manag.* **2016**, *111*, 67–78. [CrossRef]
14. Chaudhry, H. *Applied Hydraulic Transients*, 3rd ed.; Springer: Berlin/Heidelberg, Germany, 2014.

15. Osburn, G.D.; Atwater, P.L. Design and testing of a back-to-back starting system for pumped storage hydrogenerators at Mt. Elbert Powerplant. *IEEE Trans. Energy Convers.* **1992**, *7*, 280–288. [CrossRef]
16. Rajamani, K.; Hariharan, M.V. Synchronous starting of synchronous machine. *Electr. Mach. Power Syst.* **1993**, *21*, 647–661. [CrossRef]
17. Karmaker, H.; Mi, C. Improving the starting performance of large salient-pole synchronous machines. *IEEE Trans. Magn.* **2004**, *40*, 1920–1928. [CrossRef]
18. Xiao, S.; Liu, J.; Kang, X. The coordination of Back-to-back starting and protections for pumped storage units. In Proceedings of the Asia-Pacific Power and Energy Engineering Conference, Shanghai, China, 27–29 March 2012; pp. 1–4.
19. Ghazvini, M.; Pourkiaei, S.M.; Pourfayaz, F. Thermo-economic assessment and optimization of actual heat engine performance by implemention of NSGA II. *Renew. Energy Res. Appl.* **2020**, *1*, 235–245.
20. Beiranvand, A.; Ehyaei, M.A.; Ahmadi, A.; Silvaria, J.L. Energy, exergy, and economic analyses and optimization of solar organic Rankine cycle with multi-objective particle swarm algorithm. *Renew. Energy Res. Appl.* **2021**, *2*, 9–23.
21. Sabbagh, O.; Fanaei, M.A.; Arjomand, A.; Ahmadi, M.H. Multi-objective optimization assessment of a new integrated scheme for co-production of natural gas liquids and liquefied natural gas. *Sustain. Energy Technol. Assess.* **2021**, *47*, 101493. [CrossRef]
22. Pourkiaei, S.M.; Pourfayaz, F.; Mehrpooya, M.; Ahmadi, M.H. Multi-objective optimization of tubular solid oxide fuel cells fed by natural gas: An energetic and exergetic simultaneous optimization. *J. Therm. Anal. Calorim.* **2021**, *145*, 1575–1583. [CrossRef]
23. Xu, Y.; Zheng, Y.; Du, Y.; Yang, W.; Peng, X.; Li, C. Adaptive condition predictive-fuzzy PID optimal control of start-up process for pumped storage unit at low head area. *Energy Convers. Manag.* **2018**, *177*, 592–604. [CrossRef]
24. Wylie, E.B.; Streeter, V.L.; Suo, L. *Fluid transients in systems*; Prentice Hall: Englewood Cliffs, NJ, USA, 1993.
25. Chaudhry, M.H. Characteristics and Finite-Difference Methods. In *Applied Hydraulic Transients*; Springer: New York, NY, USA, 2014; pp. 65–113.
26. Lai, X.; Li, C.; Zhou, J.; Zhang, N. Multi-objective optimization of the closure law of guide vanes for pumped storage units. *Renew. Energy* **2019**, *139*, 302–312. [CrossRef]
27. Machowski, J.; Lubosny, Z.; Bialek, J.W. *Power System Dynamics: Stability and Control*; John Wiley & Sons: Hoboken, NJ, USA, 2020.
28. Grigsby, L.L. *Power System Stability and Control*; CRC press: Boca Raton, FL, USA, 2007.
29. Yang, W.; Norrlund, P.; Bladh, J.; Yang, J.; Lundin, U. Hydraulic damping mechanism of low frequency oscillations in power systems: Quantitative analysis using a nonlinear model of hydropower plants. *Appl. Energy.* **2018**, *212*, 1138–1152. [CrossRef]
30. Zhao, L.; Gong, Z.S. Application of ABB F series excitation regulator in synchronous motor. *Electr. World* **2006**, *47*, 1–5. (In Chinese)
31. Coello, C.A.C.; Lechuga, M.S. MOPSO: A proposal for multiple objective particle swarm optimization. In Proceedings of the 2002 Congress on Evolutionary Computation, Honolulu, HI, USA, 12–17 May 2002; Volume 2, pp. 1051–1056.
32. Sheikholeslami, F.; Navimipour, N.J. Service allocation in the cloud environments using multi-objective particle swarm optimization algorithm based on crowding distance. *Swarm Evol. Comput.* **2017**, *35*, 53–64. [CrossRef]
33. Hou, J.; Li, C.; Guo, W.; Fu, W. Optimal successive start-up strategy of two hydraulic coupling pumped storage units based on multi-objective control. *Int. J. Electr. Power Energy Syst.* **2019**, *111*, 398–410. [CrossRef]
34. Qu, B.Y.; Suganthan, P.N. Constrained multi-objective optimization algorithm with an ensemble of constraint handling methods. *Eng. Optim.* **2011**, *43*, 403–416. [CrossRef]
35. Song, H.Z. A method based on objective adjacent scale for multi-objective decision-making and its application. *Math. Pract. Theory* **2004**, *5*, 30–36.

sustainability

Article

Optimal Configuration of Electrochemical Energy Storage for Renewable Energy Accommodation Based on Operation Strategy of Pumped Storage Hydro

Linjun Shi [1,*], Fan Yang [1,*], Yang Li [1], Tao Zheng [2], Feng Wu [1] and Kwang Y. Lee [3]

1 College of Energy and Electrical Engineering, Hohai University, Nanjing 210098, China
2 State Key Laboratory of Smart Grid Protection and Control, Nari Group Corporation, Nanjing 211106, China
3 Department of Electrical & Computer Engineering, Baylor University, Waco, TX 76798, USA
* Correspondence: 19990041@hhu.edu.cn (L.S.); yf199808@163.com (F.Y.)

Abstract: Due to the volatility of renewable energy resources (RES) and the lag of power grid construction, grid integration of large-scale RES will lead to the curtailment of wind and photovoltaic power. Pumped storage hydro (PSH) and electrochemical energy storage (EES), as common energy storage, have unique advantages in accommodating renewable energy. This paper studies the optimal configuration of EES considering the optimal operation strategy of PSH, reducing the curtailment of wind and photovoltaic power in the power grid through the cooperative work of PSH and EES. First, based on the curtailment of RES, with the goal of improving the accommodation of RES, a combined operation optimization model of PSH and EES is proposed. Then, an optimal configuration method of EES capacity is proposed to meet the power curtailment rate in the power grid. Finally, the simulation is carried out in the actual power grid and the CPLEX solver is used to solve the optimization, and the rationality and economy of the optimization are analyzed and discussed. The simulation results show that, based on the combined operation of PSH and EES, by rationally configuring the capacity of EES, the desired power curtailment rate of the power grid can be achieved, and the necessity of configuring variable speed units is verified.

Keywords: pumped storage hydro (PSH); electrochemical energy storage (EES); renewable energy accommodation; power curtailment; operation strategy; capacity configuration; variable speed unit

Citation: Shi, L.; Yang, F.; Li, Y.; Zheng, T.; Wu, F.; Lee, K.Y. Optimal Configuration of Electrochemical Energy Storage for Renewable Energy Accommodation Based on Operation Strategy of Pumped Storage Hydro. *Sustainability* **2022**, *14*, 9713. https://doi.org/10.3390/su14159713

Academic Editor: Pablo García Triviño

Received: 13 July 2022
Accepted: 5 August 2022
Published: 7 August 2022

1. Introduction

In recent years, in response to increasingly serious environmental problems, many countries have proposed renewable energy development goals to promote the transformation of the energy consumption structure. Wind power generation and photovoltaic (PV) power generation, as relatively mature renewable energy power generation technologies, have received extensive attention [1,2]. However, when wind power and PV power generation replace fossil-fuel power, due to their randomness, volatility and intermittency, they will have a greater impact, which may open spatial and temporal gaps between the availability of the energy and its consumption by the end-users [3]. Prioritizing the accommodation of wind power and PV power reduces the participation of conventional power sources (such as thermal power) in the power grid and, moreover, increases the output volatility of conventional power sources [4–6]. When these outputs cannot be coordinated and balanced with each other, it will lead to the curtailment of wind power and PV power. How to ensure the safe and stable operation of the power grid while accommodating wind, PV and other clean energy sources, has become a major challenge. In order to address these issues, it is necessary to develop suitable energy storage systems for the power grid.

As a means of transferring electrical energy in real time, energy storage devices can compensate for the volatility and randomness, and accommodate large-scale renewable energy, thereby reducing wind and PV power curtailment [7–9]. The pumped storage hydro (PSH) plant is currently a relatively mature large-scale energy storage device, which has various functions such as peak shaving, frequency modulation, phase modulation and spinning reserve [10,11]. The coordinated operation of PSH and renewable energy can improve the peak shaving capability of the system, which plays an important role in reducing generation curtailment and improving the utilization rate of renewable energy [12,13]. However, PSH is limited by the pump turbines, which must frequently change operating conditions to maintain the stability of the power grid. As the conversion process is a transient process, during this period, high-amplitude pressure pulsation, which threatens the safe and stable operation, will appear as well [14,15].

As PSH is limited by geographical location and pump turbines, electrochemical energy storage (EES) is an alternative large-scale storage device, which is more flexible and can also transfer wind and PV power [16,17]. Moreover, EES has a fast response and flexible charging and discharging characteristics, which can make up for the shortcomings of the slow response of traditional PSH units [18–20]. Suitable technologies of EES can be sodium-ion batteries [21], potassium-ion batteries [22], flow batteries [23] and special/safe types of lithium-ion batteries [24]. Therefore, the development of synergistic services between PSH and EES is of great significance for further improving the accommodation rate of renewable energy and ensuring the safe and stable operation of the power grid.

Under the background of renewable energy accommodation, there have been increasing studies on the optimal scheduling and related configuration issues of PSH and EES. When it comes to the study on energy storage for renewable energy accommodation, the existing literature mostly starts with the renewable energy accommodation by either PSH or EES alone. Referring to PSH for renewable energy accommodation, Ma et al. [25] introduced PSH into a microgrid containing wind and PV power and used PSH to compensate for the intermittent nature of renewable energy, thus verifying that PSH can achieve energy autonomy in remote communities. Considering the uncertainty of wind energy in time and space, Xia et al. [26] established a unit combination model with PSH to coordinate the short-term dispatch of wind power, and the simulation verified that the method can be applied to large-scale power systems. Lin et al. [27] considered the benefits of PSH and the economics of thermal power deep peak shaving, proposing an optimal scheduling model for combined PSH and thermal power peak shaving, and verified that the scheme can greatly improve the economy.

In addition, when it comes to renewable energy accommodation of EES, Dui et al. [28] introduced EES into wind farms and used its peak shaving and wind power forecast error compensation to reduce wind power curtailment. They also considered comprehensively the operational constraints of the power system and the uncertainty of wind power to optimize the capacity of EES installed in wind farms. Zhang et al. [29] combined the reactive power compensation of the PV inverter with the active power curtailment of EES and realized the improvement of the voltage curve and the reduction of the power curtailment while meeting the necessary load demand. Ye et al. [30] built a joint system of thermal power units and EES to participate in power grid peak regulation and renewable energy accommodation, considering the difficulty of grid peak regulation and the overall economy. They then proposed a hierarchical optimal dispatch model based on the optimal energy abandonment rate. It can be seen from these investigations that few study the synergistic accommodation of renewable energy by PSH and EES and discuss the need of configuring variable speed for PSH power units.

In order to improve the accommodation of renewable energy, this paper studies the synergistic operation of PSH and EES and develops the optimal configuration of EES, aiming at the shortcomings of the single energy storage form. Firstly, a combined operation optimization model of PSH and EES is formulated with the goal of improving renewable energy accommodation. Then, based on rationally arranging the operation of PSH, capacity optimization of EES is performed to meet the requirements of the power curtailment rate in the grid. Finally, based on the operation strategies of PSH in different scenarios, considering the participation of variable speed units, its influence on the optimization effect and operation economy is explored. Through the case study of the actual power grid, the dispatching strategies of PSH and EES are obtained based on the optimization model, and the configuration recommendations of EES are obtained based on the curtailment rate, thus verifying the effectiveness of the method. In addition, the discussion on the variable speed unit verifies its advantages in accurate compensation and provides new possibilities for the construction of pumped storage power stations in the future.

2. Operation Optimization Model of PSH

2.1. Power Curtailment and Objective Function after PSH Optimization

During the actual operation of the power grid, the balance between the generated power and the load demand should be maintained at every moment. Since the power generation of renewable energy is affected by external weather conditions, its output fluctuates greatly. Therefore, under normal circumstances, the dispatching center of the power grid will formulate corresponding operation strategies to increase the renewable energy power generation as much as possible under the premise of ensuring the safe operation of the power grid. However, the phenomenon of the power curtailment of wind and PV power will inevitably occur in the actual operation.

Considering factors such as the output characteristics of renewable energy and the transmission capacity of the power grid in the actual power grid, with the real-time power generation of wind and PV power as input, the power curtailment in each period of the power grid is optimized in the simulation. The initial total amount of curtailed power in the power grid is shown in the following formula:

$$W_{\mathrm{C,A}} = \sum_{t=1}^{T} P_{\mathrm{c},t} \Delta t \tag{1}$$

where $W_{\mathrm{C,A}}$ is the total amount of curtailed power before optimization, $P_{\mathrm{c},t}$ is the curtailed power in the t-th period, T is the total number of periods in the simulation cycle, and Δt is the length of the period.

By adding the PSH power station, renewable energy accommodation can be effectively realized, which is embodied in pumping water to store electric energy during the period of power curtailment, and electricity can be generated to release electric energy during the remaining periods when there is room for accommodation. It should be noted that if there is already power curtailment during this period, the power generation of PSH will exacerbate the power curtailment, and the power curtailment caused needs to be enforced. Therefore, the total amount of curtailed power after the participation of PSH is as follows:

$$W_{\mathrm{C,B}}' = \sum_{t=1}^{T} \left(P_{\mathrm{c},t} - \sum_{k=1}^{K} P_{k,t}^{\mathrm{hp}} + Y_{\mathrm{P}} \sum_{k=1}^{K} P_{k,t}^{\mathrm{hg}} \right) \Delta t \tag{2}$$

where $W_{\mathrm{C,B}}'$ is the preliminary total amount of curtailed power after the PSH optimization; $P_{k,t}^{\mathrm{hp}}$ and $P_{k,t}^{\mathrm{hg}}$ are the power of the k-th unit under pumping and generating conditions in the t-th period, respectively; K is the total number of units; Y_{P} is a Boolean variable indicating whether there is power curtailment in this period, taking 1 to indicate yes, and taking 0 to indicate no.

Since the curtailed power after optimization will not be less than 0, the part of the curtailed power that is less than 0 after the PSH compensation is set to 0, and the final total amount of curtailed power after PSH optimization is:

$$W_{\mathrm{C,B}} = \max\left(W_{\mathrm{C,B}}', 0\right) \tag{3}$$

where $W_{\mathrm{C,B}}$ is the final total amount of curtailed power after PSH optimization.

Therefore, knowing the initial power curtailment and parameters of PSH units, by reasonably arranging the output of PSH units, the total amount of curtailed power is reduced as much as possible, and the goal is to minimize the total amount of curtailed power after adding PSH optimization. The objective function is as follows:

$$\min F_1 = W_{\mathrm{C,B}} \tag{4}$$

where F_1 is the final total amount of curtailed power after PSH optimization.

2.2. *Constraints of PSH*

2.2.1. Power Constraints of PSH Units

Under pumping and generating conditions, the operating power of each PSH unit is between certain upper and lower limits:

$$y_{k,t}^{\mathrm{p}} P_{\mathrm{hp,min}} \leq P_{k,t}^{\mathrm{hp}} \leq y_{k,t}^{\mathrm{p}} P_{\mathrm{hp,max}} \tag{5}$$

$$y_{k,t}^{\mathrm{g}} P_{\mathrm{hg,min}} \leq P_{k,t}^{\mathrm{hg}} \leq y_{k,t}^{\mathrm{g}} P_{\mathrm{hg,max}} \tag{6}$$

where $P_{\mathrm{hp,max}}$ and $P_{\mathrm{hp,min}}$ are, respectively, the upper and lower limits of the pumping power of the unit; $P_{\mathrm{hg,max}}$ and $P_{\mathrm{hg,min}}$ are, respectively, the upper and lower limits of the generating power of the unit; $y_{k,t}^{\mathrm{p}}$ and $y_{k,t}^{\mathrm{g}}$ are Boolean variables that, respectively, represent whether the units are in the pumping condition and the generating condition, taking 1 to indicate yes, and taking 0 to indicate no.

2.2.2. Single Working Condition Constraints of PSH Units and Stations

In the case of the same unit or the same power station, the Boolean variable representing the pumping or generating condition is set to ensure the single working condition of the unit or power station. The sum of the two variables is less than or equal to 1, indicating that the working conditions of the unit or power station are consistent at the same time:

$$y_{k,t}^{\mathrm{p}} + y_{k,t}^{\mathrm{g}} \leq 1 \tag{7}$$

$$Y_{\mathrm{P}} + Y_{\mathrm{G}} \leq 1 \tag{8}$$

where Y_{P} and Y_{G} are Boolean variables that, respectively, represent whether the power station is in the pumping condition and the generating condition, taking 1 to indicate yes, and taking 0 to indicate no.

2.2.3. Constraint on the Number of Working Units

Whether in the pumping condition or in the generating condition, the maximum number of working units in the PSH station is K, as shown in the following:

$$\sum_{k=1}^{K} y_{k,t}^{\mathrm{p}} \leq K Y_{\mathrm{P}} \tag{9}$$

$$\sum_{k=1}^{K} y_{k,t}^{\mathrm{g}} \leq K Y_{\mathrm{G}} \tag{10}$$

2.2.4. Constraints on Water Level of Power Station Reservoir and Its Fluctuation

The real-time water level of the power station reservoir is between certain upper and lower limits, and it must be ensured that the water level at the beginning and end of a dispatch cycle is consistent. By setting the water-to-electricity conversion coefficients η_{p} and η_{g} under the pumping and generating conditions, respectively, the water level of the reservoir and the power of the power station are linked:

$$E_{\mathrm{h,min}} \leq E_{\mathrm{h},t} \leq E_{\mathrm{h,max}} \tag{11}$$

$$E_{\mathrm{h},t+1} = E_{\mathrm{h},t} + \left(\eta_{\mathrm{p}} \sum_{k=1}^{K} P_{k,t}^{\mathrm{hp}} - \eta_{\mathrm{g}} \sum_{k=1}^{K} P_{k,t}^{\mathrm{hg}} \right) \Delta t \tag{12}$$

$$E_{\mathrm{h},t_0} = E_{\mathrm{h},t_{\mathrm{end}}} \tag{13}$$

where $E_{\mathrm{h,t}}$ is the real-time water level of the upper reservoir of the power station in the t-th period; $E_{\mathrm{h,max}}$ and $E_{\mathrm{h,min}}$ are, respectively, the upper and lower limits of the water level of the upper reservoir of the power station; t_0 and t_{end} are, respectively, the beginning and end of the dispatching cycle. The storage capacity constraint on the upper reservoir also represents the lower reservoir constraint.

2.2.5. Constraint on the Maximum Number of Start-Stop Switching Times of the Units

The maximum number of start-stop switching times are used to limit the start and stop of the unit to avoid frequent switching of the unit, thereby reducing the start-up and shut-down cost of PSH, which can be expressed as follows:

$$\sum_{t=2}^{T} \left| y_{k,t}^{\mathrm{p}} - y_{k,t-1}^{\mathrm{p}} \right| \leq n_{\mathrm{p}} \tag{14}$$

$$\sum_{t=2}^{T} \left| y_{k,t}^{\mathrm{g}} - y_{k,t-1}^{\mathrm{g}} \right| \leq n_{\mathrm{g}} \tag{15}$$

where n_{p} and n_{g} are, respectively, the number of start-stop switching times of the unit under the pumping and generating conditions in the dispatching cycle.

3. Operation and Configuration Optimization Model of EES

3.1. Power Curtailment after EES Optimization

On the basis of the operation optimization of the PSH station, EES can be configured to realize further accommodation of renewable energy, which is embodied in the charging and storing of electric energy during the power curtailment period, and discharging and releasing of electric energy during the rest period. It should be noted that if there is already power curtailment during this period, the discharge of EES will also exacerbate the power curtailment, so it is necessary to enforce the power curtailment. Therefore, the total amount of curtailed power after EES optimization is as follows:

$$W_{\mathrm{C,C}}{}' = W_{\mathrm{C,B}}{}' - \sum_{t=1}^{T} (P_{\mathrm{ch},t} - Y_{\mathrm{E}} P_{\mathrm{dis},t}) \Delta t \tag{16}$$

where $W_{\mathrm{C,C}}{}'$ is the preliminary total amount of curtailed power after EES optimization; $P_{\mathrm{ch},t}$ and $P_{\mathrm{dis},t}$ are, respectively, the charging and discharging power of EES in the t-th period; Y_{E} is a Boolean variable indicating whether there is power curtailment in this period, taking 1 to indicate yes, and taking 0 to indicate no.

Since the curtailed power after optimization will not be less than 0, the part of the curtailed power that is less than 0 after the EES compensation is set to 0, and the final total amount of curtailed power after EES optimization is:

$$W_{\mathrm{C,C}} = \max\left(W_{\mathrm{C,C}}{}', 0\right) \tag{17}$$

where $W_{\mathrm{C,C}}$ is the final total amount of curtailed power after EES optimization.

3.2. Objective Function of EES Operation Optimization

After the operation of PSH is optimized, based on the curtailed power after PSH optimization, the selection and parameters of EES are known, and the operation of EES is also reasonably arranged with the goal of minimizing the curtailed power. The synergy between PSH and EES can further improve the optimization effect of the curtailment of wind and PV power. The objective function of EES operation optimization is as follows:

$$\min F_2 = W_{C,C} \tag{18}$$

where F_2 is the total amount of curtailed power after PSH and EES optimization.

3.3. Objective Function of EES Configuration Optimization

Similarly, after the operation of PSH is optimized, the power curtailment and the parameters of EES are known, the rated power and rated capacity of EES are reasonably configured, and the charging and discharging power of EES is optimized. By reducing the total amount of power curtailment after optimization as much as possible, it can reach the expected power curtailment rate set in advance. With the goal of achieving the minimum rated power of EES required to achieve the expected power curtailment rate, the EES capacity configuration at this time is solved. Since there is a fixed multiple relationship between the rated capacity and rated power of EES in this paper, the rated capacity can be obtained when the rated power is solved. The objective function of EES configuration optimization is as follows:

$$\min F_3 = P_{ESS} \tag{19}$$

where F_3 is the rated power of EES.

3.4. Constraints of EES

3.4.1. Power Constraints of EES

Considering that excessive charging and discharging current will damage the performance of EES, the real-time operating power of EES cannot exceed the rated power:

$$0 \le P_{ch,t} \le y_{ch,t} P_{ESS} \tag{20}$$

$$0 \le P_{dis,t} \le y_{dis,t} P_{ESS} \tag{21}$$

where $y_{ch,t}$ and $y_{dis,t}$ are Boolean variables that characterize the charge and discharge states of EES, respectively, and satisfy $y_{ch,t} + y_{dis,t} \le 1$ at any time.

3.4.2. SOC Constraints of EES

The ratio of the remaining capacity of EES to its full charging capacity is the state of charge (SOC). The remaining capacity of EES must meet certain constraints during operation, that is, the SOC of EES must be between certain upper and lower limits. In addition, the consistent SOC at the beginning and end of each scheduling cycle of EES can ensure the periodicity of its continuous operation:

$$SOC(t) = SOC(t-1) + \frac{[\eta_{ch} P_{ch,t} - P_{dis,t}/\eta_{dis}]\Delta t}{E_{ESS}} \tag{22}$$

$$SOC(t_0) = SOC(t_{end}) \tag{23}$$

$$SOC_{min} \le SOC(t) \le SOC_{max} \tag{24}$$

where $SOC(t)$ is the SOC of EES in the t-th period; η_{ch} and η_{dis} are the energy conversion efficiencies of EES during charging and discharging, respectively; SOC_{max} and SOC_{min} are the upper and lower limits of the SOC of EES, respectively.

3.4.3. Constraint of Expected Power Curtailment Rate

The power curtailment rate after optimization should not be greater than the expected power curtailment rate, as shown in the following:

$$\sigma \le \sigma_0 \tag{25}$$

where σ_0 is the expected power curtailment rate.

4. Operation and Configuration Optimization Algorithms of PSH and EES

The optimization in this paper uses the YALMIP toolbox to call the CPLEX solver to solve. The CPLEX solver has the advantages of fast convergence speed and good robustness to linear problems, and it is suitable for solving the optimization model in this paper [31].

The overall idea of the optimization algorithm is shown in Figure 1, and the detailed steps are as follows:

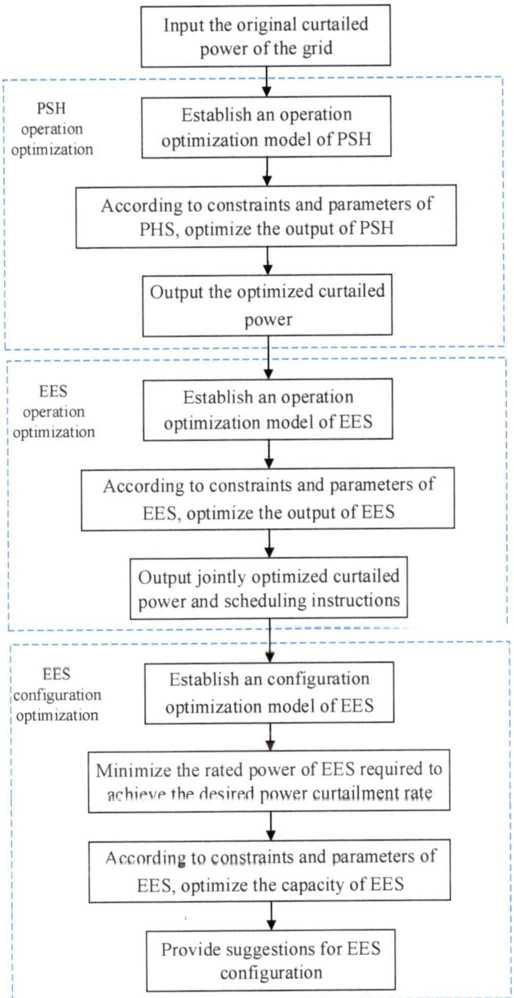

Figure 1. Flowchart of the optimization algorithm.

Step 1: Select an appropriate simulation period and input the original curtailed power of the power grid.

Step 2: Establish an operation optimization model of PSH and set constraints with the goal of minimizing the power curtailment.

Step 3: Based on the original curtailed power and parameters of PSH, use the CPLEX solver to solve the output of PSH. The optimized curtailed power is the output to the EES operation optimization model.

Step 4: Establish an operation optimization model of EES, with the goal of minimizing the curtailed power, set constraints, and solve the output of EES.

Step 5: Establish the configuration optimization model of EES, and then set the constraints to achieve the minimum rated power of EES required to achieve the expected power curtailment rate.

Step 6: Based on the curtailed power after the PSH optimization and parameters of EES, set the index of the expected power curtailment rate, and use the CPLEX solver to solve the configuration of EES.

Step 7: Compare and analyze the curtailed power before and after optimization, and then provide suggestions on the scheduling of PSH and EES, as well as the configuration of EES.

5. Case Study

5.1. Case Parameters

In order to verify the rationality of the above models and algorithms, this paper selects the parameters of PSH and the data of the curtailed power in an actual power grid for the case study.

5.1.1. Parameters of PSH

There are four PSH power stations in the power grid of the selected case, and the number of units and the upper and lower limits of the power of each power station are shown in Table 1. The difference between the fixed-speed unit and the variable-speed unit is mainly whether the power under the pumping condition is adjustable, and the power of the fixed-speed unit under the pumping condition is the full power state. This paper first discusses the case where all units are fixed-speed units, and the power range of PSH power generation is 50–100%. The total installed capacity of the four power stations is 6070 MW.

Table 1. Parameters of the units of PSH.

Serial Number of the Power Station	Number of Units	Capacity of a Single Unit (MW)	Total Installed Capacity (MW)
1	3	90	270
2	4	200	800
3	12	300	3600
4	4	350	1400

In this case, the four PSH power stations reach the upper limit of the water level after 6 h of pumping at full power, and the water levels at the beginning and end of the day are the same. The water-to-electricity conversion coefficients under the pumping condition and generating condition are set to be 251.62 and 318.24, respectively, and the unit is $m^3/(MW \cdot h)$ [27]. The start-stop switching times of the pumping and generating conditions are set to be three, that is, to ensure that each PSH power station does not pump and generate more than three times a day.

5.1.2. Parameters of Power Curtailment

The data of power curtailment in the case is selected from the actual power grid in an area. The data of power curtailment in a typical week is selected, one hour is a sampling point, and there are 168 sampling points in 7 days. The data of power curtailment is shown in Figure 2.

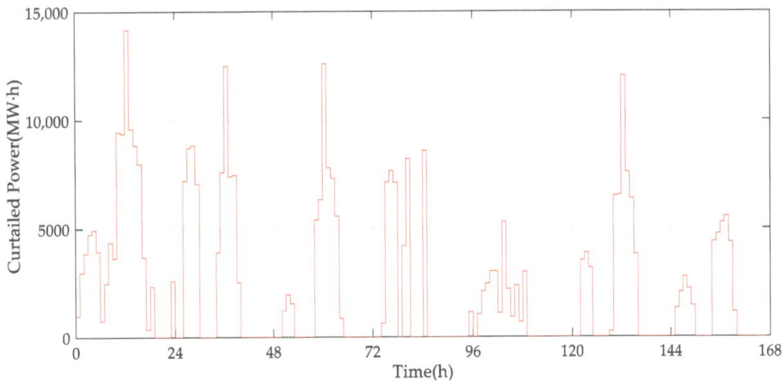

Figure 2. Original power curtailment data of the power grid.

5.1.3. Parameters of EES

In terms of the selection of EES, this paper directly selects the widely used lithium iron phosphate battery as an example. Its basic parameters are as follows: the ratio of the rated capacity and power of the battery is 2, and the charge-discharge efficiency is 90%. The upper limit of the SOC of the battery is 0.8, and the lower limit is 0.2. The purchase cost of the unit power is 600 yuan/kW, and the investment unit price of the unit capacity is 1600 yuan/(kW·h) [32].

5.2. Operation Optimization of PSH and EES

In order to verify the rationality of the above models and algorithms, this paper selects the parameters of PSH and the data of curtailed power in an actual power grid for case study.

For the operation optimization of the PSH power station, according to the objective function of PSH operation optimization and various constraints, the hourly output of PSH in a typical week is calculated, and the operation curve of a typical week is drawn, as shown in Figure 3.

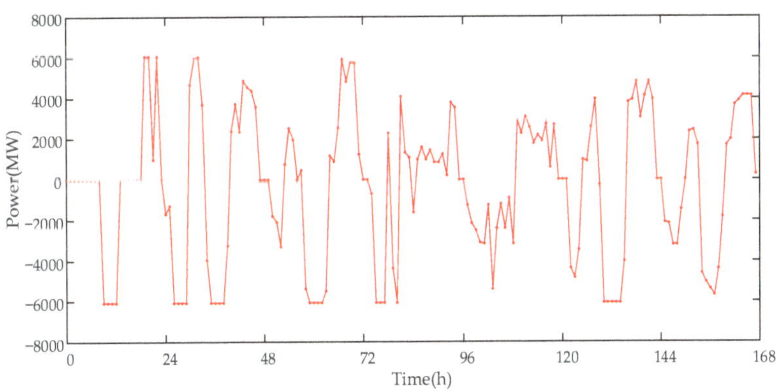

Figure 3. Operation curve of PSH in a typical week.

For the operation optimization of EES, according to the objective function of operation optimization and various constraints, the output of EES in a typical week is calculated. Since the configuration optimization of EES has not yet been carried out, the rated power of 2000 MW is directly selected as the assumption for the study of operation optimization. The operation curve of EES in a typical week is shown in Figure 4.

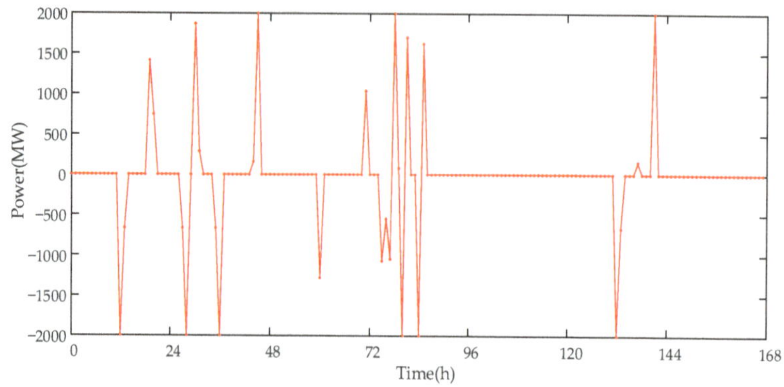

Figure 4. Operation curve of EES in a typical week.

The operation optimization results of PSH and EES are jointly analyzed, and the seven days of a typical week are further divided into three representative scenarios for specific analysis according to the pumping and generating (charging and discharging) and accommodation conditions.

5.2.1. Scenario 1

The first representative scenario is the simulation result of the second day, and the operation curve of PSH is shown in Figure 5. Among them, the output is negative when pumping water, and positive when generating electricity.

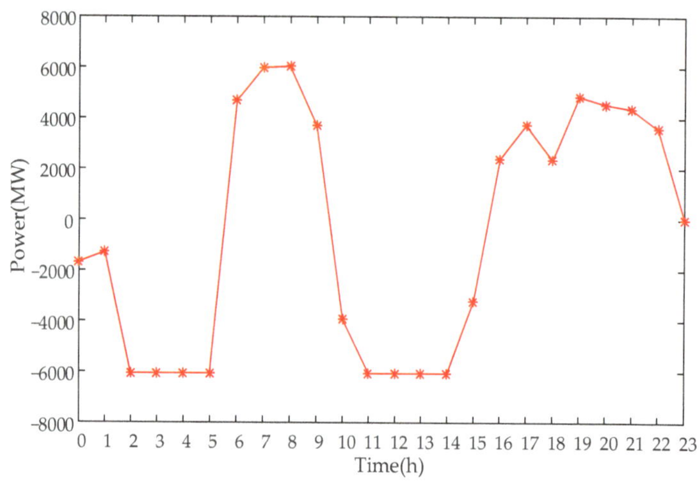

Figure 5. Operation curve of PSH (scenario 1).

The operation curve of EES is shown in Figure 6, where the output is negative when charging, and positive when discharging.

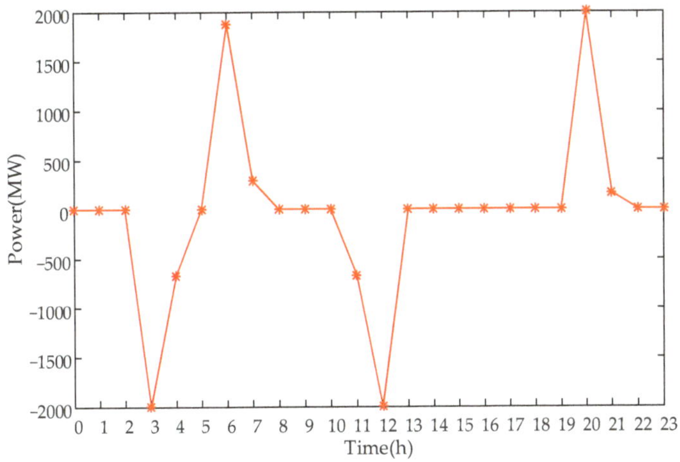

Figure 6. Operation curve of EES (scenario 1).

It can be seen from Figures 5 and 6 that the PSH and EES operations in this scenario are pumping twice and generating twice (charging twice and discharging twice).

In this case, the change in the SOC of EES is shown in Figure 7.

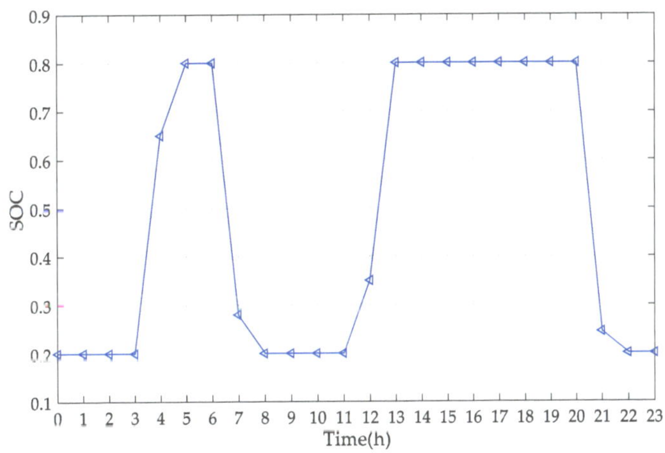

Figure 7. SOC curve of EES (scenario 1).

As can be seen from the above figure, the SOC of EES is within a limited range, meeting the requirements of the upper and lower limits of SOC, and the state of the beginning and end of the cycle is consistent.

The curtailed power before and after optimization are compared on a graph, and the scenario is analyzed in detail in relation to the changes in the curtailed power. The comparison of power curtailment is shown in Figure 8.

As can be seen from Figure 8, there is a large amount of power curtailment during certain periods of the day. The power curtailment is significantly reduced through the accommodation effect of PSH. However, due to the limited capacity of the PSH unit, the reduction may be limited. If we want to further reduce the power curtailment, we need to configure the EES.

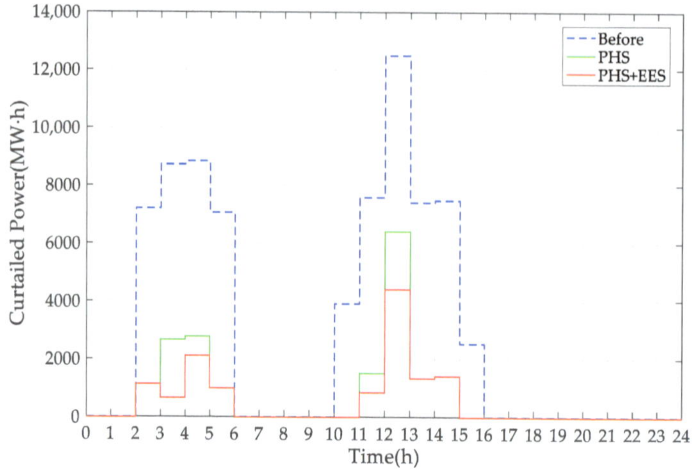

Figure 8. Comparison of power curtailment before and after optimization (scenario 1).

After adding the EES, the remaining part of the renewable energy is accommodated by the charging of the EES, which further reduces the power curtailment. Comparing the curtailed power before and after optimization, compared with the accommodation of PSH alone, the curtailed power has a further significant decrease after adding the EES. Therefore, the operation optimization of the PSH and EES is reasonable, which verifies the effectiveness of the operation strategy.

From the analysis of data changes, there is power curtailment from 2:00 to 5:00 and from 10:00 to 17:00 in this typical day. At this time, through the cooperative work of PSH and EES, the power curtailment has dropped significantly. The curtailed power reached the maximum value of 12,581.66 MW·h from 12:00 to 13:00. Due to the pumping of the PSH unit, the curtailed power dropped to 6511.66 MW·h. On this basis, the EES was charged for accommodation, and the curtailed power further decreased to 5232.31 MW·h. The optimization effect after adding the EES is related to the pre-selected EES capacity. If we want to get better optimization effect, we need to configure larger capacity EES. This verifies the feasibility of PSH and EES synergistically accommodating renewable energy.

5.2.2. Scenario 2

The second representative scenario is the simulation result of the 4th day, and the operation curve of PSH is shown in Figure 9.

The operation curve of EES is shown in Figure 10.

It can be seen from Figures 9 and 10 that the PSH and EES operations in this scenario are pumping three times and generating three times (charging three times and discharging three times).

In this case, the change in the SOC of EES is shown in Figure 11.

As can be seen from the above figure, the SOC of EES is within a limited range, meeting the requirements of the upper and lower limits of SOC, and the state of the beginning and end of the cycle is consistent.

The curtailed power before and after optimization are compared on a graph, and the scenario is analyzed in detail in relation to the changes in the curtailed power. The comparison of power curtailment is shown in Figure 12.

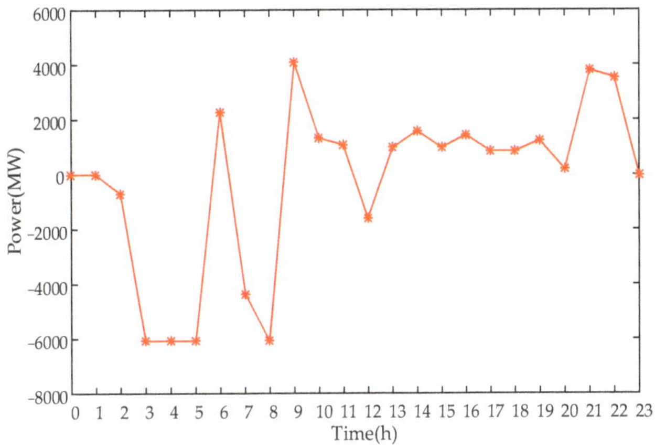

Figure 9. Operation curve of PSH (scenario 2).

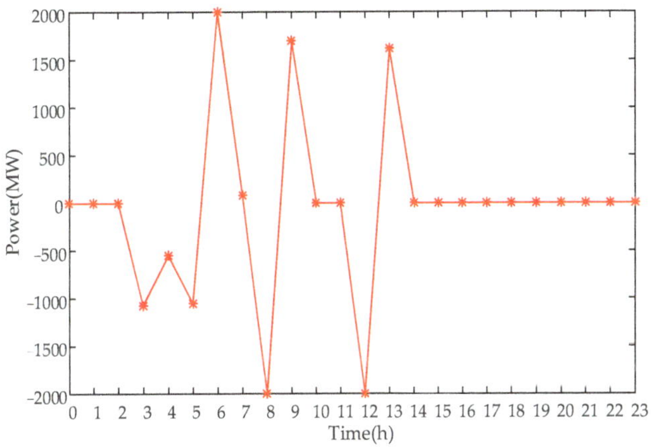

Figure 10. Operation curve of EES (scenario 2).

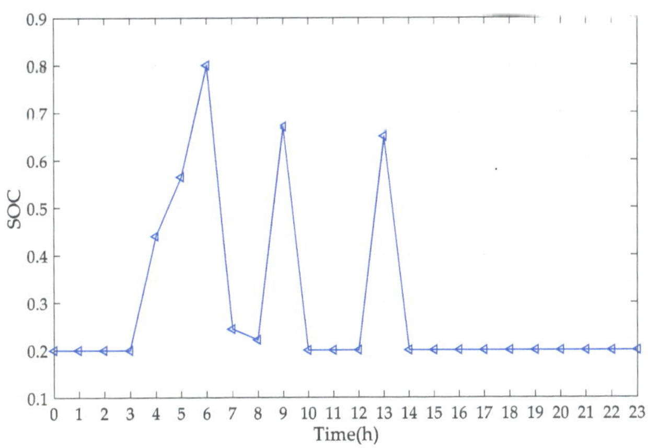

Figure 11. SOC curve of EES (scenario 2).

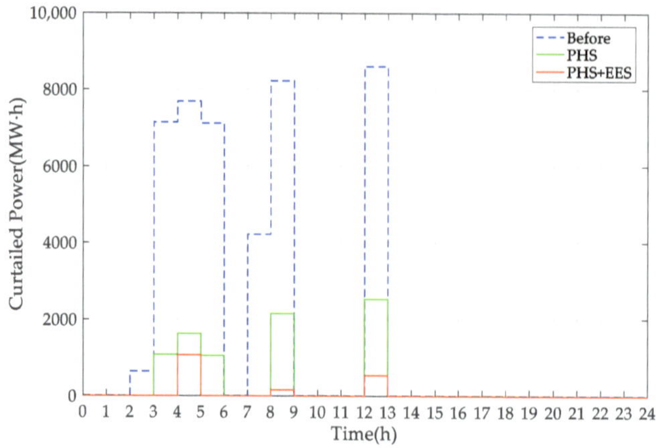

Figure 12. Comparison of power curtailment before and after optimization (scenario 2).

From the analysis of data changes, there is power curtailment from 2:00 to 6:00, 7:00 to 9:00, and 12:00 to 13:00 in this typical day. At this time, through the cooperative work of PSH and EES, the power curtailment has dropped significantly. The curtailed power reached the maximum value of 8602.84 MW·h from 12:00 to 13:00. Due to the pumping of the PSH unit, the curtailed power dropped to 2532.84 MW·h. On this basis, the EES was charged for accommodation, and the curtailed power further decreased to 532.84 MW·h.

5.2.3. Scenario 3

The third representative scenario is the simulation result of the 6th day, and the operation curve of the PSH is shown in Figure 13.

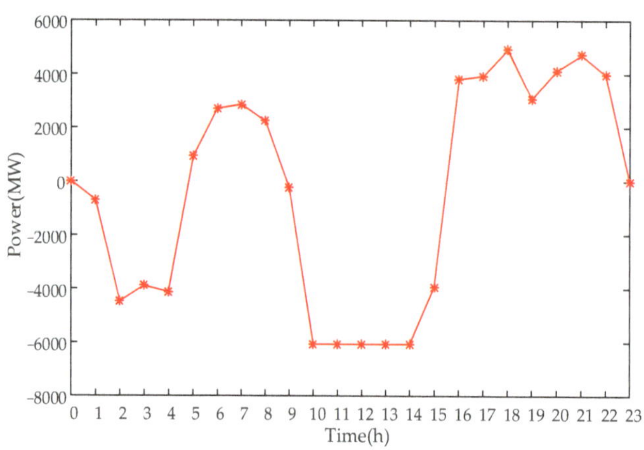

Figure 13. Operation curve of PSH (scenario 3).

The operation curve of the EES is shown in Figure 14.

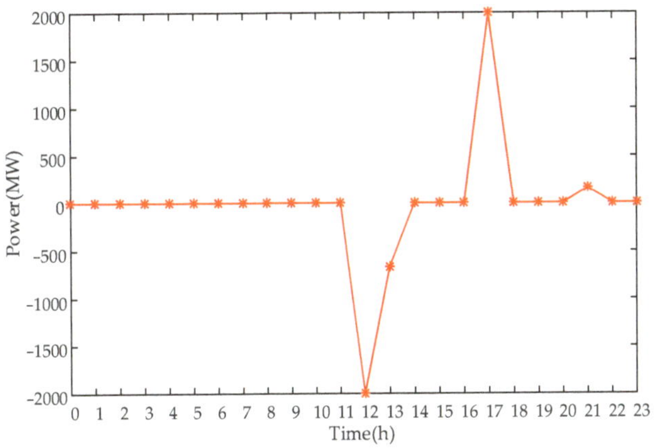

Figure 14. Operation curve of EES (scenario 3).

It can be seen from Figures 13 and 14 that the PSH and EES operations in this scenario are pumping twice and generating twice (charging once and discharging once).

In this case, the change in the SOC of EES is shown in Figure 15.

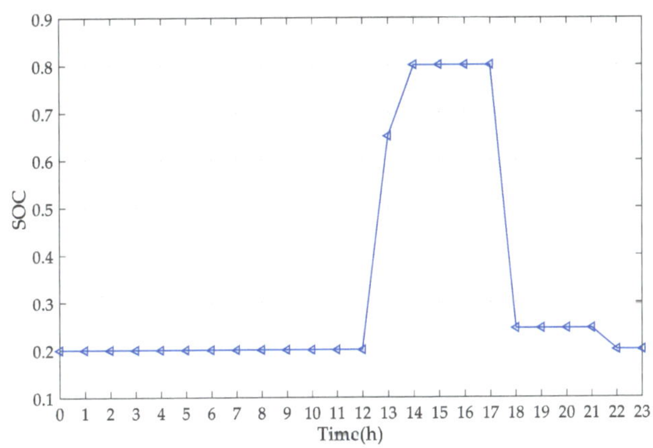

Figure 15. SOC curve of EES (scenario 3).

As can be seen from the above figure, the SOC of EES is within a limited range, meeting the requirements of the upper and lower limits of SOC, and the state of the beginning and end of the cycle is consistent.

The curtailed power before and after optimization are compared on a graph, and the scenario is analyzed in detail in relation to the changes in the curtailed power. The comparison of power curtailment is shown in Figure 16.

It can be seen from Figure 16 that from 2:00 to 5:00 on this typical day, since the curtailed power after the PSH operation has been completely accommodated, the EES only needs to be charged once and discharged once. The curtailed power reached the maximum value of 12,031.05 MW·h from 12:00 to 13:00. Due to the pumping of the PSH unit, the curtailed power dropped to 5961.05 MW·h. On this basis, the EES was charged for accommodation, and the curtailed power further decreased to 3961.05 MW·h.

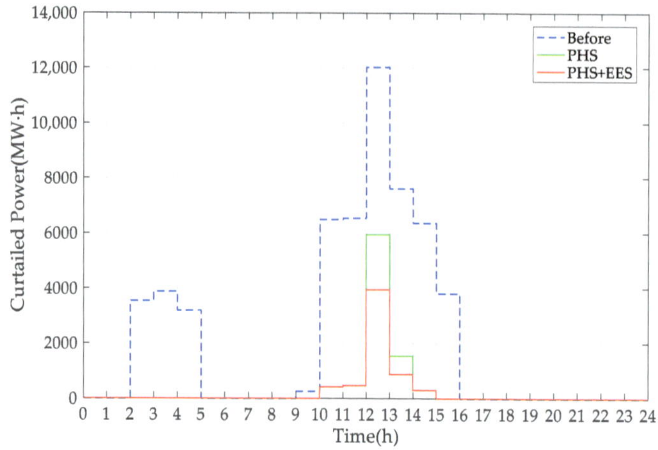

Figure 16. Comparison of power curtailment before and after optimization (scenario 3).

5.2.4. Configuration Optimization of EES

During the simulation cycle of this typical week, the operation of the EES is not considered, and the operation optimization of the PSH is carried out independently. The total amount of curtailed power before optimization is calculated from the original curtailed power, the total amount of curtailed power after adding PSH optimization is calculated from the optimization result, and the power curtailment rate is calculated from the renewable energy power generation, which is convenient for the next step of energy storage configuration optimization. The data are shown in Table 2.

Table 2. Optimization results of power curtailment and its rate.

Term	Date
Total curtailed power before optimization (MW·h)	385,760
Total curtailed power after PSH optimization (MW·h)	128,483
Percentage of drop in curtailed power	66.69%
Total renewable energy generation (MW·h)	2,301,608
Power curtailment rate	5.58%

After reducing power curtailment through PSH in the power grid, it may still fail to meet the requirements of the power curtailment rate. It is obviously more troublesome to expand the reservoir capacity of the PSH, but configuring EES can save costs and provide additional functions such as frequency regulation. Therefore, in order to achieve a further reduction in power curtailment and achieve the given power curtailment rate index, a certain capacity of the EES should be configured. In this scenario, it is necessary for PSH and EES to jointly accommodate renewable energy.

For the configuration optimization of EES, according to the objective function of the EES configuration and its constraints, the configuration of EES is solved based on the data of typical weeks. Among them, the expected power curtailment rate is set based on the data of the power curtailment rate in Table 2, and the rated power of EES that needs to be configured under different power curtailment rates is optimized. The configuration results of EES are shown in Table 3.

Table 3 shows the minimum cost of investment required to achieve the power curtailment rate target in various situations, which provides the power grid with recommendations for EES configuration. The relationship between the cost of investing in EES and the benefit of reducing power curtailment can be weighed according to needs. The above data are plotted to explore the relationship between the power curtailment rate and the investment cost of EES, as shown in Figure 17.

Table 3. Optimization results for the configuration of EES.

Power Curtailment Rate	Configured Power of EES (MW)	Configured Capacity of EES (MW·h)	Investment Cost of EES (Million Yuan)
5.58%	0	0	0
5.3%	673	1346	2557.4
5.1%	1149	2298	4366.2
4.9%	1665	3330	6327.0
4.7%	2201	4402	8363.8
4.5%	2955	5910	11,229.0
4.3%	4426	8852	16,818.8

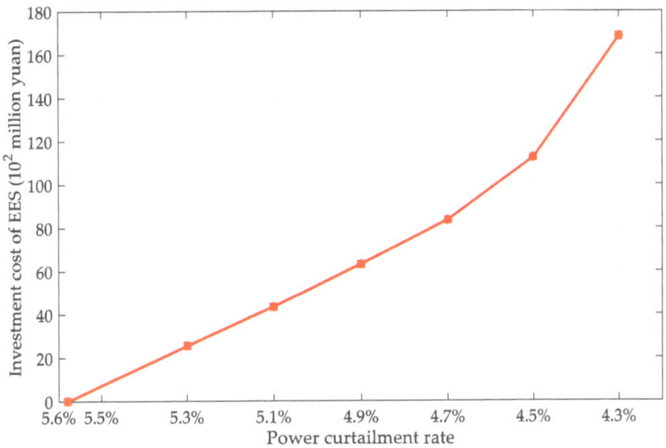

Figure 17. Investment cost of EES under different power curtailment rates.

It can be seen from Figure 17 that with the reduction of the power curtailment rate, it is necessary to configure larger capacity EES, which increases the investment cost. In addition, when the power curtailment rate is reduced to around 4.5%, the curve appears nonlinear, which means that when the power curtailment rate drops to a certain extent, to further improve the accommodation effect, it is necessary to configure incrementally larger capacity for the EES.

5.2.5. Involvement of Variable Speed Units

If all the units are fixed speed units, in the pumping condition, the units pumping water are in full power state, which results in more or less energy waste in certain periods. After replacing some fixed speed units with variable speed units, the power adjustment under pumping conditions is more flexible, and unbiased power compensation can be achieved in most time periods.

In this case, the two fixed speed units in PSH power station 3 are replaced with variable speed units of the same capacity, and the output range under pumping conditions is 70–100%. The simulation on the 7th day is used for comparison, and the comparison curves of the PSH operation after optimization on that day is shown in Figure 18.

It can be seen from the figure that after adding the variable speed units, the power under the pumping condition is significantly reduced. Specifically analyzing the optimization results, from 13:00 to 14:00, the original curtailed power during this period was 5560.15 MW·h. In the case of conventional units, the total pumping power of the PSH power station is 5690 MW; after adding the variable-speed unit, the total pumping power of the PSH power station is 5560.15 MW, which realizes the compensation of pumping power without deviation, which saves 129.85 MW of pumping power consumption compared with the conventional unit. Considering the total pumping power of the whole day, without adding variable speed units, the total pumping power of the PSH power station on that day is 39,039.24 MW; after adding variable speed units, the total pumping power is 36,958.32 MW. A total of 2080.92 MW of pumping power was saved on that day. Considering the long-period power station operation, the variable speed unit will bring considerable energy savings, which verifies the necessity of adding the variable speed unit.

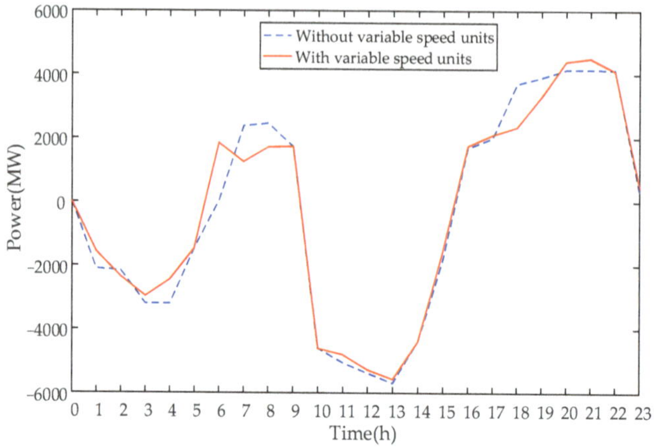

Figure 18. Comparison of PSH operation before and after adding variable speed units.

Comparing the total amount of curtailed power after optimization before and after adding variable speed units, it can be found that there is no difference in the reduction of the total amount of curtailed power on that day. This is because overcompensation without adding variable speed units will not affect the total amount of curtailed power after optimization. Therefore, it will not affect the configuration of EES. All in all, the variable speed unit can improve the flexibility of the unit combination, make the pumping compensation more accurate, and reduce the total power consumption of the PSH power station without changing the optimization effect. The comparative analysis in this part provides a basis for the appropriate use of variable speed units in the future construction of PSH power stations.

6. Conclusions

In this paper, aiming at improving the accommodation of renewable energy, a joint operation model of PSH and EES is proposed, and on this basis, the configuration of EES is studied.

In this paper, the proposed combined operation model of PSH and EES is used to reasonably arrange the operation of PSH and EES, so that the two cooperate to reduce the amount of curtailed power in the power grid and improve the accommodation of renewable energy. On this basis, using the proposed EES configuration optimization model, combined with the expected power curtailment rate, the corresponding EES configuration is solved, and the investment cost is calculated, which satisfies the actual situation for different needs for power curtailment rates. The rationality and effectiveness of the proposed optimization method and optimization model are verified by actual cases.

In the operation optimization of PSH, the influence of the participation of variable speed units on PSH operation is compared. Studies have shown that the addition of variable speed units will improve the flexibility of the unit's power output, make the pumping compensation more accurate, reduce the overall power consumption of PSH, and reduce energy waste. Therefore, the appropriate input of variable speed units should be considered in the construction of the actual PSH power station.

In future work, factors such as the hydraulic head will be considered to enrich the PSH model. The site selection optimization of EES will be studied considering the scenario of the actual power grid. Additionally, the complementarity of PSH and EES in terms of timescales, and start-stop costs, will be discussed more fully.

Author Contributions: Conceptualization, L.S. and F.Y.; methodology, L.S. and F.Y.; software, L.S., F.Y. and Y.L.; validation, L.S., F.Y. and T.Z.; formal analysis, L.S. and F.Y.; investigation, Y.L. and K.Y.L.; resources, F.W.; data curation, L.S. and F.Y.; writing—original draft preparation, L.S. and F.Y.; writing—review and editing, L.S., Y.L. and K.Y.L.; visualization, F.Y. and Y.L.; supervision, T.Z. and K.Y.L.; project administration, L.S. and F.W.; funding acquisition, T.Z. All authors have read and agreed to the published version of the manuscript.

Funding: This work was supported in part by State Key Laboratory of Smart Grid Protection and Control under Grant SGNR0000KJJS2200297 and in part by National Natural Science Foundation of China under Grant 52061635102.

Institutional Review Board Statement: Not applicable.

Informed Consent Statement: Not applicable.

Data Availability Statement: Not applicable.

Acknowledgments: We would like to extend special thanks to the editors and reviewers for their insightful advice and comments on the manuscript.

Conflicts of Interest: The authors declare no conflict of interest.

References

1. Santika, W.G.; Anisuzzaman, M.; Bahri, P.A.; Shafiullah, G.M.; Rupf, G.V.; Urmee, T. From goals to joules: A quantitative approach of interlinkages between energy and the sustainable development goals. *Energy Res. Soc. Sci.* **2019**, *50*, 201–214. [CrossRef]
2. Zhang, Y.; Zang, W.; Zheng, J.; Cappietti, L.; Zhang, J.; Zheng, Y.; Fernandez-Rodriguez, E. The influence of waves propagating with the current on the wake of a tidal stream turbine. *Appl. Energy* **2021**, *290*, 116729. [CrossRef]
3. Sun, C.; Negro, E.; Vezzù, K.; Pagot, G.; Cavinato, G.; Nale, A.; Bang, Y.H.; Noto, V.D. Hybrid inorganic-organic proton-conducting membranes based on SPEEK doped with WO3 nanoparticles for application in vanadium redox flow batteries. *Electrochim. Acta* **2019**, *309*, 311–325. [CrossRef]
4. Liu, Z.; Wang, H.; Zhou, B.; Yang, D.; Li, G.; Yang, B.; Xi, C.; Hu, B. Optimal operation strategy for wind–hydrogen–water power grids facing offshore wind power accommodation. *Sustainability* **2022**, *14*, 6871. [CrossRef]
5. Che, L.; Liu, X.; Zhu, X.; Cui, M.; Li, Z. Intra-interval security assessment in power systems with high wind penetration. *IEEE Trans. Sustain. Energy* **2019**, *10*, 1890–1903. [CrossRef]
6. Karavas, C.; Arvanitis, K.G.; Papadakis, G. Optimal technical and economic configuration of photovoltaic powered reverse osmosis desalination systems operating in autonomous mode. *Desalination* **2019**, *466*, 97–106. [CrossRef]
7. Ma, W.; Wang, W.; Wu, X.; Hu, R.; Tang, F.; Zhang, W. Control strategy of a hybrid energy storage system to smooth photovoltaic power fluctuations considering photovoltaic output power curtailment. *Sustainability* **2019**, *11*, 1324. [CrossRef]
8. Denholm, P.; Mai, T. Timescales of energy storage needed for reducing renewable energy curtailment. *Renew. Energy* **2019**, *130*, 388–399. [CrossRef]
9. Solomon, A.A.; Bogdanov, D.; Breyer, C. Curtailment-storage-penetration nexus in the energy transition. *Appl. Energy* **2019**, *235*, 1351–1368. [CrossRef]
10. Li, N.; Hedman, K.W. Enhanced pumped hydro storage utilization using policy functions. *IEEE Trans. Power Syst.* **2017**, *32*, 1089–1102. [CrossRef]
11. Wang, W.; Li, C.; Liao, X.; Qin, H. Study on unit commitment problem considering pumped storage and renewable energy via a novel binary artificial sheep algorithm. *Appl. Energy* **2017**, *187*, 612–626. [CrossRef]
12. Alharbi, H.; Bhattacharya, K. Participation of pumped hydro storage in energy and performance-based regulation markets. *IEEE Trans. Power Syst.* **2020**, *35*, 4307–4323. [CrossRef]
13. Pérez-Díaz, J.I.; Jiménez, J. Contribution of a pumped-storage hydropower plant to reduce the scheduling costs of an isolated power system with high wind power penetration. *Energy* **2016**, *109*, 92–104. [CrossRef]
14. Wang, W.; Tai, G.; Pei, J.; Pavesi, G.; Yuan, S. Numerical investigation of the effect of the closure law of wicket gates on the transient characteristics of pump-turbine in pump mode. *Renew. Energy* **2022**, *194*, 719–733. [CrossRef]
15. Zhang, Y.; Zhang, Z.; Zheng, J.; Zhang, J.; Zheng, Y.; Zang, W.; Lin, X.; Fernandez-Rodriguez, E. Experimental investigation into effects of boundary proximity and blockage on horizontal-axis tidal turbine wake. *Ocean Eng.* **2021**, *225*, 108829. [CrossRef]
16. Asef, P.; Milan, M.; Lapthorn, A.; Padmanaban, S. Future trends and aging analysis of battery energy storage systems for electric vehicles. *Sustainability* **2021**, *13*, 13779. [CrossRef]
17. Byers, C.; Botterud, A. Additional capacity value from synergy of variable renewable energy and energy storage. *IEEE Trans. Sustain. Energy* **2020**, *11*, 1106–1109. [CrossRef]

18. Dawood, F.; Shafiullah, G.; Anda, M. Stand-alone microgrid with 100% renewable energy: A case study with hybrid solar PV-battery-hydrogen. *Sustainability* **2020**, *12*, 2047. [CrossRef]
19. Li, X.; Wang, S. Energy management and operational control methods for grid battery energy storage systems. *CSEE J. Power Energy Syst.* **2019**, *7*, 1026–1040. [CrossRef]
20. Li, J.; Niu, M.; Wang, S.; Zhou, J.; Yuan, X. Operation and control analysis of 100 MW class battery energy storage station on grid side in Jiangsu power grid of China. *Autom. Electr. Power Syst.* **2020**, *44*, 28–35. [CrossRef]
21. Pan, H.; Hu, Y.; Chen, L. Room-temperature stationary sodium-ion batteries for large-scale electric energy storage. *Energy Environ. Sci.* **2013**, *6*, 2338–2360. [CrossRef]
22. Zhang, J.; Liu, T.; Cheng, X.; Xia, M.; Zheng, R.; Peng, N.; Yu, H.; Shui, M.; Shu, J. Development status and future prospect of non-aqueous potassium ion batteries for large scale energy storage. *Nano Energy* **2019**, *60*, 340–361. [CrossRef]
23. Zhang, H.; Sun, C. Cost-effective iron-based aqueous redox flow batteries for large-scale energy storage application: A review. *J. Power Sources* **2021**, *493*, 229445. [CrossRef]
24. Capron, O.; Samba, A.; Omar, N.; Coosemans, T.; Bossche, P.V.d.; Van Mierlo, J. Lithium-ion batteries: Thermal behaviour investigation of unbalanced modules. *Sustainability* **2015**, *7*, 8374–8398. [CrossRef]
25. Ma, T.; Yang, H.; Lu, L.; Peng, J. Technical feasibility study on a standalone hybrid solar-wind system with pumped hydro storage for a remote island in Hong Kong. *Renew. Energy* **2014**, *69*, 7–15. [CrossRef]
26. Xia, P.; Deng, C.; Chen, Y.; Yao, W. MILP based robust short-term scheduling for wind–thermal–hydro power system with pumped hydro energy storage. *IEEE Access* **2019**, *7*, 30261–30275. [CrossRef]
27. Lin, L.; Yue, X.; Xu, B.; Sun, Y.; Wei, M. Sequence and strategy of pumped storage-thermal combined peak shaving considering benefits of pumped storage and deep regulation of thermal power. *Power System Technol.* **2021**, *45*, 20–32. [CrossRef]
28. Dui, X.; Zhu, G.; Yao, L. Two-stage optimization of battery energy storage capacity to decrease wind power curtailment in grid-connected wind farms. *IEEE Trans. Power Syst.* **2018**, *33*, 3296–3305. [CrossRef]
29. Zhang, Z.; Mishra, Y.; Dou, C.; Yue, D.; Zhang, B.; Tian, Y. Steady-state voltage regulation with reduced photovoltaic power curtailment. *IEEE J. Photovolt.* **2020**, *10*, 1853–1863. [CrossRef]
30. Ye, Z.; Li, X.; Jiang, F.; Chen, L.; Wang, Y.; Dai, S. Hierarchical optimization economic dispatching of combined wind-PV-thermal-energy storage system considering the optimal energy abandonment rate. *Power System Technol.* **2021**, *45*, 241–251. [CrossRef]
31. Yang, B.; Wang, J.; Chen, Y.; Li, D.; Sun, L. Optimal sizing and placement of energy storage system in power grids: A state-of-the-art one-stop handbook. *J. Energy Storage* **2020**, *32*, 1–30. [CrossRef]
32. Jiang, K.; Li, H.; Li, W.; Cheng, S. On several battery technologies for power grids. *Autom. Electr. Power Syst.* **2013**, *37*, 47–53. [CrossRef]

sustainability

MDPI

Article

Research on Flow Field Characteristics in Water Jet Nozzle and Surface Damage Caused by Target Impact

Qinghong Zhang, Zhouhao Shi, Weidong Shi *, Zhanshan Xie, Linwei Tan and Yongfei Yang

School of Mechanical Engineering, Nantong University, Nantong 226019, China;
2009310029@stmail.ntu.edu.cn (Q.Z.); 1610011046@stmail.ntu.edu.cn (Z.S.); xiezhanshanntu@outlook.com (Z.X.); tanlinweintu@outlook.com (L.T.); yangyongfeintu@outlook.com (Y.Y.)
* Correspondence: wdshi@ujs.edu.cn

Abstract: As a new processing method, water jet processing technology has risen rapidly due to its wide range of applications, no pollution, and zero discharge. In this paper, the flow characteristics and failure characteristics of ultra-high-pressure gas-liquid jet in the range of 300 MPa are analyzed by numerical calculation. The research conclusion shows that the jet atomization diffusion is caused by the friction between the liquid medium and the surrounding gas, the mixed flow of broken water droplets and cavitation. The jet diffusion process is essentially the energy exchange process between the jet in the core area and the turbulent flow in the atomization area. The distribution of the turbulent kinetic energy in the atomization area can determine the degree of jet diffusion and the rate of energy decay. The water jet impacted the surface of the target to form a crater-like annular erosion pit. With the increase of the impact pressure, the deformation showed an overall increasing trend, and the increasing trend increased significantly. The central depression of the erosion area is caused by the damage of the material by the stagnation pressure in the core area. The flow characteristics of gas-liquid flow in the process of formation and diffusion in the high-pressure water jet nozzle are explored from the microscopic point of view, and it also provides a theoretical basis for equipment optimization in engineering.

Keywords: high-pressure water jet; flow field characteristics; impact simulation; matrix damage

Citation: Zhang, Q.; Shi, Z.; Shi, W.; Xie, Z.; Tan, L.; Yang, Y. Research on Flow Field Characteristics in Water Jet Nozzle and Surface Damage Caused by Target Impact. *Sustainability* **2022**, *14*, 9074. https://doi.org/10.3390/su14159074

Academic Editors: Bin Huang, Yuquan Zhang, Jianhua Zhang and Ling Zhou

Received: 4 July 2022
Accepted: 20 July 2022
Published: 25 July 2022

1. Introduction

Water jet is a high-pressure or ultra-high-pressure pump that pressurizes water or a mixed fluid with solid medium, and then ejects it at supersonic or subsonic speed through a nozzle to form one or more high-speed water jets to achieve cleaning and cutting, rock-breaking mining and other purposes [1–4]. In the process of promoting industrial green development, ultra-high-pressure water rust removal is one of the most representative process innovation measures currently being practiced.

T. Mabrouki et al. [5] presented a numerical study to explore the impact of water jet on the impact erosion of non-coated and coated metal sheets, and they developed a finite element model to study and analyze in detail the impact of water jet on the material. Mabrouki and Raissi [6] performed numerical calculations using the Dyna 3D program to simulate the interaction of the jet with the target. Chillman et al. [7] used ultra-high-pressure water jet technology to remove the coating of titanium alloy without damaging the substrate. The jet removal effect of titanium alloys was evaluated by the test method and surface micro-morphology analysis. Guha et al. [8] studied the effects of different target distances on pressure attenuation and pressure distribution through experimental and numerical studies, and concluded that there is a linear relationship between the pressure attenuation on the jet axis and the target distance. In the field of theoretical analysis of jet impingement and coating removal mechanism, Kunaporn et al. [9] established a mathematical model to analyze the pressure distribution of water jet impinging on the aluminum alloy surface under different target distances, and used it to predict the contact pressure range.

Glover et al. [10] established a mathematical model for calculating the width of the impurity layer removed by fixed and moving jets, and used high-speed photography to measure the jet cleaning width under different exposure times, and verified the calculation results. Azhari et al. [11] analyzed the three-dimensional surface micro-morphology of 304 stainless steel treated with different pressure jets, and found that repeated cleaning with different pressure jets can improve the surface finish and hardness of the workpiece and prolong the fatigue life of the workpiece. Nayak et al. [12] verified the influence of parameters such as target distance, pressure, abrasive grain size and abrasive flow on the material removal rate and surface in water jet machining, and obtained the law of the change of material surface machining quality with water jet parameters. Rihem et al. [13] developed and verified the transient finite element three-dimensional simulation of water jet, and proved that the results of multi group droplet simulation are closer to the experimental surface performance than that of single group droplet simulation. Barsukov et al. [14] simulated and analyzed the abrasive water jet, and studied the effects of abrasive particles, contact angle and other factors on water jet cutting and abrasive spalling. Hamid et al. [15] provided experimental and mathematical data, found that the cross-sectional velocity distribution of the free jet emitted from the conical nozzle is similar. Aganin et al. [16] studied the efficiency and effect of water jet. At the same time, they proposed theoretical calculation formulas for flow, velocity and contraction coefficient. Mieszala et al. [17] analyzed the erosion mechanism during abrasive water jet machining and found that different crystal structures and surface microstructures have an impact on material removal.

However, for ultra-high-pressure water jets, the linear smooth solution of the Navier-Stokes equations has not been a breakthrough. Many of these phenomena have not been fully understood, especially the understanding of ultra-high-pressure jet flow characteristics is quite superficial, and the research conclusions at the mechanism level are not perfect [18–21]. In this paper, through numerical calculation, the flow characteristics and damage characteristics of ultra-high- pressure jets in the range of 300 MPa have been studied in many aspects. Under the existing equipment and technical conditions, the water jet rust removal efficiency and quality can be improved, and the water jet rust removal cost can be reduced. It is of great significance to realize the large-scale promotion and application of high-pressure water jet rust removal on the surface of ships.

2. Theoretical Model and Calculation Method

2.1. Geometry and Meshing

In the research of water jet, the jet flow field is usually divided into an inner flow field and an outer flow field with the jet nozzle as the boundary. The inner flow field is mainly the inner flow field of the nozzle and the inlet section. The inner flow field is mainly used to study the formation process of the jet. The outer flow field is mainly composed of a complete jet beam, external air and workpiece. The outer flow field is mainly used to explore the processing performance of the water jet process and the divergence of the jet outside the nozzle. Therefore, the overall internal and external flow domains are simplified and established as shown in Figure 1. The main fluid domain is divided into four regions: the inlet section, the nozzle section, the target area, and the outlet section.

Since the hemispherical water body in the inlet section is not suitable for structured grid division, the overall grid adopts a hybrid grid. The Pointwise 18.3R1 software (Pointwise, Fort Worth, TX, USA) is used for mesh generation. The CFX 2020 software (ANSYS, Canonsburg, PA, USA) is used for numerical calculation. The hemisphere of the inlet section is meshed with unstructured meshes, and the rest is meshed with structured meshes. The wall boundary of the nozzle area is locally densified. The grid independence test is evaluated based on the maximum pressure of the impacted surface of the target. The results are shown in Table 1.

Table 1. Grid independence test.

Test Group	Number of Grid Cells	Maximum Pressure of the Impacted Surface of the Target (MPa)
A1	324,651	59.439
A2	632,218	71.965
A3	975,376	79.731
A4	1,262,945	82.686
A5	1,794,662	83.127
A6	2,036,387	83.238
A7	2,560,398	83.367

Under the principle of ensuring that the grid error is less than 1% and reducing the amount of calculation as much as possible, the number of grids is finally determined to be between 1 million and 2 million.

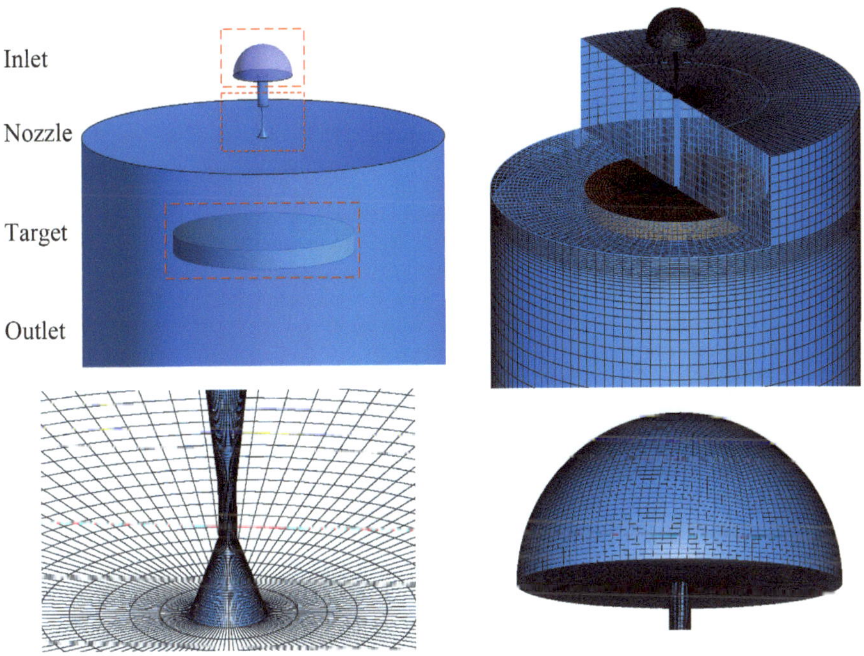

Figure 1. Geometry model and meshing.

2.2. Boundary Conditions and Parameter Settings

For practical conditions, the inlet section should use a pressure inlet. However, if the pressure inlet is directly used, an overflow error will occur. Therefore, most experts and scholars convert the pressure inlet to the velocity inlet through the Bernoulli equation through Equation (1):

$$Z + \frac{p}{\rho g} + \frac{u^2}{2g} = C \tag{1}$$

In the formula: Z is the position of the fluid, p is the pressure of the fluid, ρ is the density of the fluid, g is the acceleration of gravity, u is the velocity of the fluid, and C is a constant.

However, there is a problem with this approach. The Bernoulli equation comes from the conservation of energy at its root. The Bernoulli equation applies to incompressible ideal fluids [22]. And for ultra-high-pressure water jets, since water becomes compressible at a high enough pressure, the calculations will suffer from compressibility. Under the pressure of 250 MPa, the compressibility of the ultra-high-pressure water jet reaches 0.08715 [16,23], so the following three equations are used to calculate the jet velocity with water as the compressible liquid:

$$\rho_l = \rho_0 \left(1 + \frac{p}{nk_0}\right)^n \quad (2)$$

$$U_{bern} = \sqrt{\frac{2p}{\rho}} \quad (3)$$

where ρ_0 is the density of water at 25 degrees Celsius under standard atmospheric pressure, n is the density index, k_0 is the reference bulk modulus to describe the compressibility of water, which depends on temperature and pressure. U_{bern} is the jet velocity of the incompressible water body derived from Bernoulli's equation. U_{th} is the theoretical jet velocity considering compressibility. U_{th} can be expressed as:

$$U_{th} = \sqrt{\frac{2nk_0}{\rho_0(1-n)}\left[\left(1 + \frac{P}{nk_0}\right)^{1-n} - 1\right]} \quad (4)$$

After consulting the literature [24–26], we determined that the shape of the inlet is hemispherical. On the one hand, it can effectively prevent the occurrence of overflow errors, and on the other hand, the velocity of the internal flow field is closer to the real situation. In terms of inlet velocity, the pressure curve is programmed so that the inlet pressure of the hemisphere changes gradually over time. Under this equation, the inlet pressure changes with a linear uniform acceleration. After the pressure change process ends, the high-pressure water jet will continue to impact the target with an inlet pressure of 300 MPa (as shown in Figure 2b). The outlet section adopts an open boundary, and the set pressure value is one atmosphere.

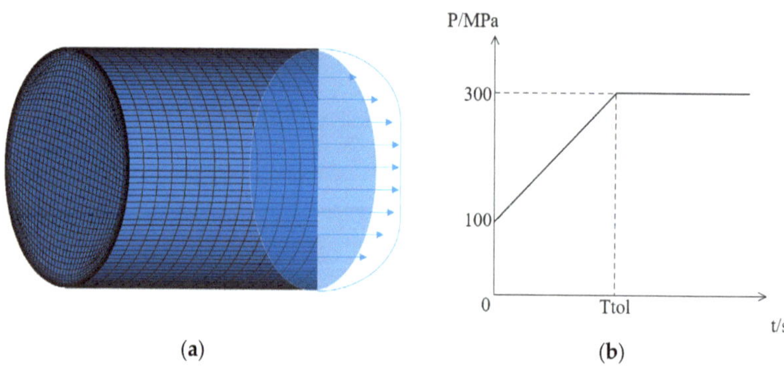

Figure 2. Setting of inlet boundary conditions. (**a**) Velocity field of the flow field in the nozzle. (**b**) Inlet pressure curve.

From the point of view of the jetting method, the jet belongs to the free intercepting flow of one medium injected into another medium. The development of this kind of flow is influenced by the interaction of corresponding fluid theory such as nozzle pressure fluctuation, liquid surface tension, and gas-liquid phase shear. Among them, the solution of various equations for jet breaking (hereinafter referred to as the breaking mechanism) is more complicated and cumbersome. For the relatively mature jet breaking calculation, experts and scholars classify and calculate the jet according to the breaking mechanism through the jet characteristics such as Reynolds number, Weber number, and Ohnesorge number. The classification basis is shown in Table 2.

Table 2. Classic crushing mode.

Broken Mode	Jet Characteristics
Rayleigh broken	liquid Weber number $w_{eL} > 8$ and the air Weber number $w_{eg} < 0.4$
Atomization and crushing	Air Weber number $w_{eg} > 40.3$

When high-pressure water flows through the microfluidic channel of the nozzle, the water pressure energy is converted into kinetic energy, forming a high-speed water jet. Due to the dynamic viscosity of the fluid and the friction of the air, the velocity of the micro-water jet at the front end will gradually decay. The microflow at the edge of the jet is slowed down due to air friction and entrainment. The microfluidic velocity in the flow field is reduced to zero, while the jet beam is broken into droplets. Of course, jet breakage and atomization will greatly affect the rust removal effect of water jet. However, since the detailed mechanism of spray atomization is still unclear, most of the existing atomization models are empirical or semi-empirical models, and it is difficult to comprehensively predict the spray characteristics under a wide range of conditions. In addition, there is another difficulty that the parameter calibration is relatively complicated when using the atomization model under various injection conditions. Therefore, the crushing model and the atomization model used in this paper are compared and verified to prove the predictive performance of these models.

3. Results and Analysis of Numerical Calculation

3.1. Analysis of Flow Field Calculation Results

In the numerical simulation, the crushing model is divided into primary crushing and secondary crushing. The turbulent primary crushing theory adopts the classical Kelvin-Helmholtz (KH) crushing model. It can predict with high fidelity the morphological transition of liquid column to liquid droplet under different injection pressure and ambient pressure conditions. For the secondary crushing of the droplets, the KH-RT atomization model derived by Kelvin-Helmholtz (KH) and Rayleigh-Taylor (RT) was used. The crushing model and the atomization model are calculated in low-speed and high-speed water jets, respectively. The liquid column using the KH atomization model is broken once at a low speed, and the trajectory of the liquid droplet and the breaking point are clearly visible, as shown in Figure 3a. In the high-speed state, the small droplets appear atomization after the secondary crushing, as shown in Figure 3b.

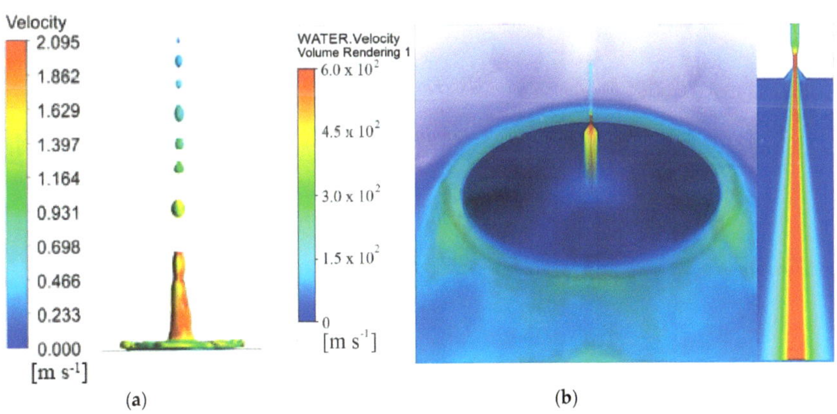

Figure 3. Calculation results of the crushing model and the atomization model. (**a**) Numerical simulation of KH crushing model. (**b**) Numerical simulation of KH-RT atomization model.

A partial enlarged view of the jet velocity vector of the flow field in the nozzle is shown in Figure 4. Six planes are intercepted at equal distances from the cylindrical section of the nozzle to the axis of the diffuser section. We can see that when the fluid flows through the nozzle, there is no vector arrow of the vapor velocity from the constricted section of the nozzle to the cylindrical section. The position where the vapor flow rate first begins to be generated is at the edge of the junction of the cylindrical section and the diffuser section, which generally presents a ring shape (as shown in Figure 4a NO.1–NO.2). This indicates that cavitation do not form inside the nozzle under the boundary condition of the high pressure inlet. When the fluid passes through the nozzle cylindrical section to the nozzle diffusion section, the liquid medium at the edge of the high-pressure water column produces friction with the surrounding gas medium. This causes the liquid at the edge of the high-pressure water column to gradually break up into discrete droplets. The liquid at the edge mixes with the surrounding gas medium to form a gas-liquid mixed flow. In addition, the pressure release of the nozzle diffuser produces a large velocity gradient, which reduces the pressure to below the saturated vapor pressure. The liquid forms cavitation based on the vapor.

In order to prove the existence of cavitation bubbles, the author also conducts a separate submerged jet calculation for the nozzle section. Even if the flow field is submerged in water to ensure that there is no gas medium in the nozzle, it can be clearly found in the figure that the gas-liquid mixed fluid still exists, and the result is shown in the Figure 4b shown. At the same time, a red part of the liquid in the shape of a copper coin is returned on the lower end face of the diffusion section, and the backflow position is between NO.5–NO.6 in Figure 4a. This indicates that the liquid in the diffusion section is between the positions of NO.2 and NO.4, and the discontinuity of the water jet is mainly guided by the mixing of cavitation bubbles and high-pressure water jets. At the position of NO.5–NO.6, the reflux gas and the broken small liquid droplets start to mix, resulting in the turbulent atomization of the water jet. That is, whether it is a free water jet in the air or a submerged water jet, the generation of cavitation bubbles will cause the discontinuity of the water jet and dominate the flow field inside the nozzle [9,27–29].

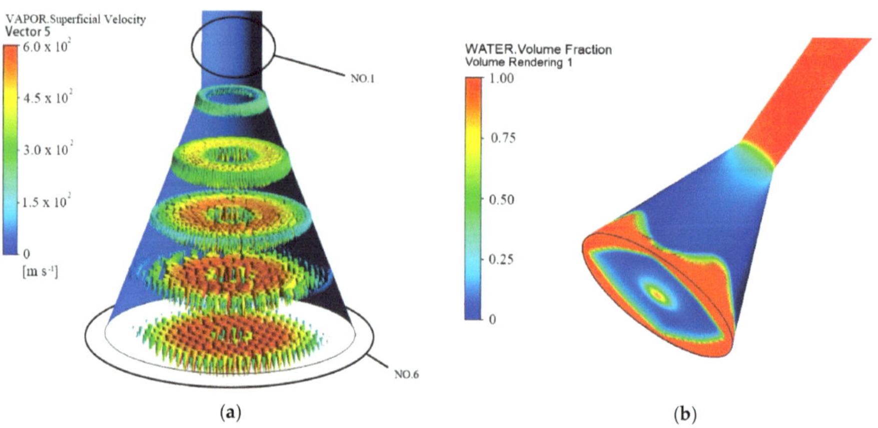

Figure 4. Cloud diagram of the flow field in the nozzle. (**a**) The velocity vector cloud diagram of the flow field in the nozzle. (**b**) Cloud diagram of the volume fraction of the flow field in the nozzle.

Figure 5 is a cloud diagram of the spatial distribution of the velocity field of the ultra-high-pressure water jet impacting the target. In order to completely analyze the causes of jet energy attenuation and the water beam diffusion mechanism, the obtained results were shielded in the low-velocity domain, and the water jet impact was studied in blocks. The internal and external flow fields are divided into three parts: cavitation jet (I), free jet (II), stagnation, and high-speed diffusion (III) in the nozzle.

It can be clearly seen from Figure 5 that after the jet enters the external flow field from the nozzle, the jet gradually begins to diverge, the velocity in the central region gradually decreases, and the area of the turbulent region increases rapidly. This is due to the violent friction between the edge of the high-speed jet and the air, which reduces the continuity of the jet, and the liquid column is broken into liquid droplets and mixed with the air. This discrete liquid phase and gas phase also have great speed and energy. It is mixed in the outer layer of the jet and has a great disturbance in the core region. As the distance of the jet increases, the liquid inside the jet is disturbed and gradually diffuses into the turbulent region until the jet is completely atomized. It can be seen that the gas-liquid turbulence in the outer layer of the jet is the main characterization of the jet diffusion. A detailed analysis of the turbulent kinetic energy characteristics of the inner and outer layers in the diameter direction of the jet and the turbulent kinetic energy characteristics of the flow along the axis direction.

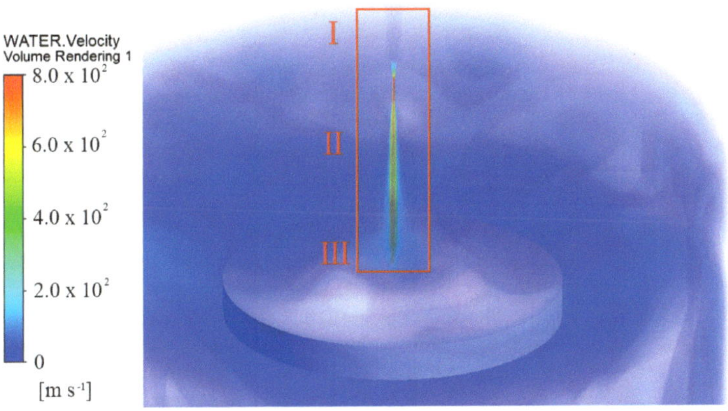

Figure 5. The spatial distribution cloud diagram of the velocity field of the ultra-high-pressure water jet impacting the target body.

The free jet in the outer flow field can be divided into a cluster section, a diffuser section, and a diverging section along the nozzle axis direction. The direction perpendicular to the nozzle axis can be divided into core area and atomization area.

In order to better analyze the free jet flow law in the external flow field, the spatial distribution of the turbulent kinetic energy of the water jet was calculated, and then the turbulent kinetic energy cloud map in the direction of the nozzle axis was intercepted, as shown in Figure 6. It can be clearly found from Figure 6 that there is a clear boundary between the core area and the atomization area. There is a vigorous energy exchange at the boundary, and the closer to the interface, the more intense the energy interaction. Therefore, it can be approximated that the peak region of turbulent kinetic energy can be used as the separation surface of the inner and outer jets, and the valley region of turbulent kinetic energy can be regarded as the separation surface of the outer jet and the air. At the same time, it is found that the disturbance in the core area of the jet at the nozzle outlet exists but the disturbance is small, indicating that the extremely high velocity in the nozzle is the dominant effect in the cluster section. In the jet diffusion section, the disturbance of the jet core area increases, the radius of the core area decreases, and the velocity decreases, and finally merges with the turbulent area completely.

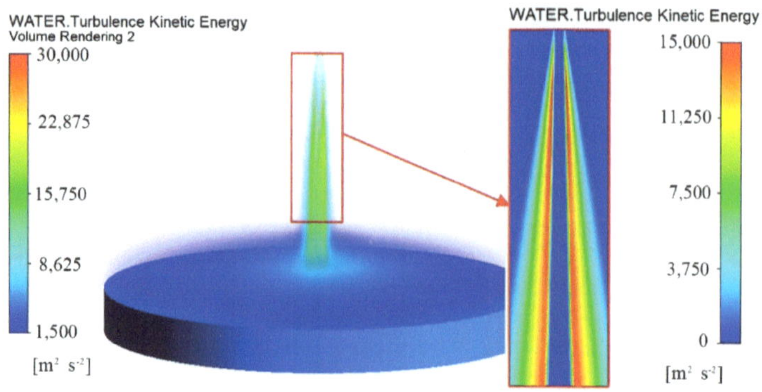

Figure 6. Turbulent kinetic energy cloud diagram in the axial direction of the water jet nozzle.

3.2. Numerical Simulation Analysis of Matrix Damage

The stagnation and high-speed diffusion zone is a special fluid domain formed by the target blocking the high-pressure water jet, which is subdivided into two parts. As the pressure of water increases to more than 300 MPa, it is compressed and accelerated by the cylindrical section of the nozzle to form a continuous jet with a velocity of 800 m/s or even greater. After the flow direction of the original water jet is blocked, part of the fluid is sputtered around, and there is a huge energy to form a circular high-speed diffusion zone. The high-pressure water jet in the center of the core area forms a high-pressure stagnation area due to the wrapping and stagnation of the peripheral flow field, and the extremely high pressure generated by it causes the material to fail and deform.

Fluid-structure interaction can only deform through surface pressure conduction, and cannot simulate the damage of stagnation pressure to materials and the accumulated damage of materials. Therefore, the pressure data obtained in the flow field simulation is fitted through the data, and the pressure surface pressure of 200 MPa–300 MPa is transmitted to Abaqus through the cylindrical coordinate system for solid field simulation. The ultra-high-pressure water jet impinges on the target, extracts the pressure value on the impacted surface, and obtains the jet intensity diagram, as shown in Figure 7a. The pressure curve of the high-pressure water jet impact surface is fitted along the radial direction, and the fitting curve adopts a quartic polynomial, as shown in Figure 7b, to ensure that the correlation coefficient R^2 reaches more than 0.99. The overall time is 0.1 s to minimize the effect of accumulated damage on the surface of the material.

The plastic deformation of the target in the depth direction at 0.1 s was collected under different pressures, and the deformation curve was drawn as shown in Figure 8. It can be clearly seen from Figure 8 that with the increase of the impact pressure, the deformation generally presents an increasing trend. And the increasing trend is obviously rising, which shows that the ultra-high pressure water jet pressure has a great influence on the impact damage of the substrate surface. Since the surface of the substrate is not constrained, the impacted surface is a free boundary, the formation of the central jet causes the material close to the central jet to expand to the side away from the center, and micro-bulges appear at the edge of the impact center area. In terms of three-dimensional space, the impacted surface presents a ring shape.

In terms of action time, the maximum deformation amount in the depth direction in the process of 0–0.1 s under different pressures was collected, and the graph was drawn as shown in Figure 9. The deformation increment is analyzed and compared. When the pressure just touches the target at 0.01 s, the deformation has not yet reached the level visible to the naked eye. From 0.01 to 0.08 s, the pressure acts on the surface of the target, and the deformation increases rapidly, reaching the level of 10^{-3} mm and producing tiny pits. The material in the depth direction pushes and accumulates in the breadth direction, forming a crater-shaped impact corrosion surface.

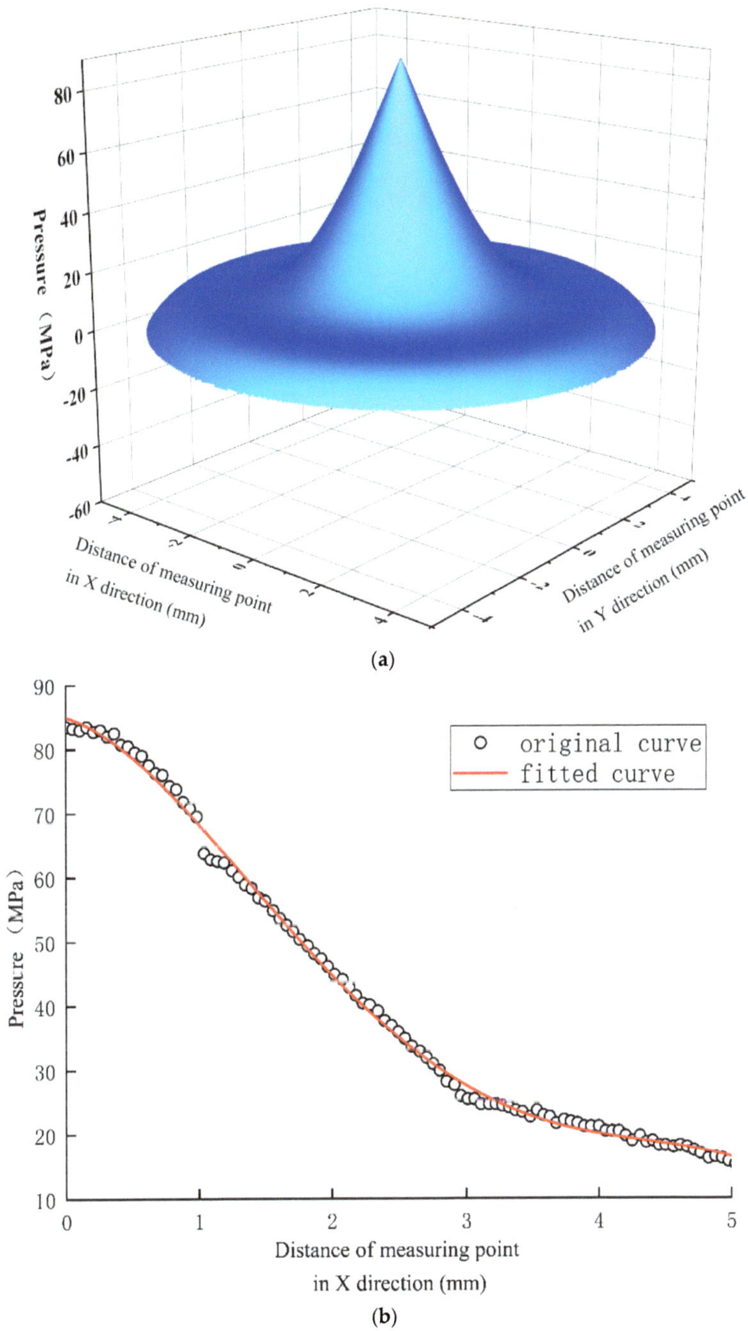

Figure 7. Jet intensity and fitting curve diagram. (a) Jet intensity diagram. (b) Fitting curve.

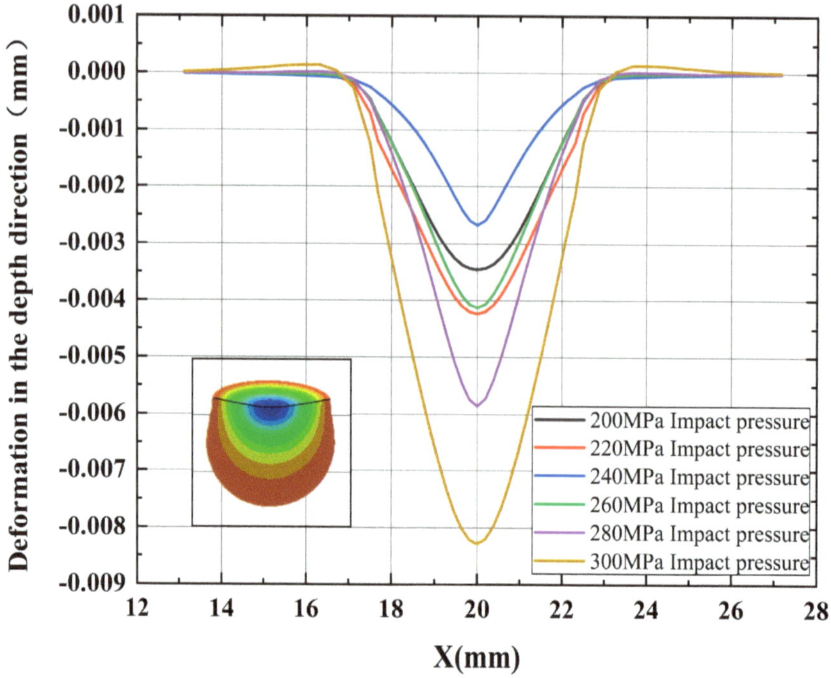

Figure 8. The plastic deformation curve of the target under different pressures.

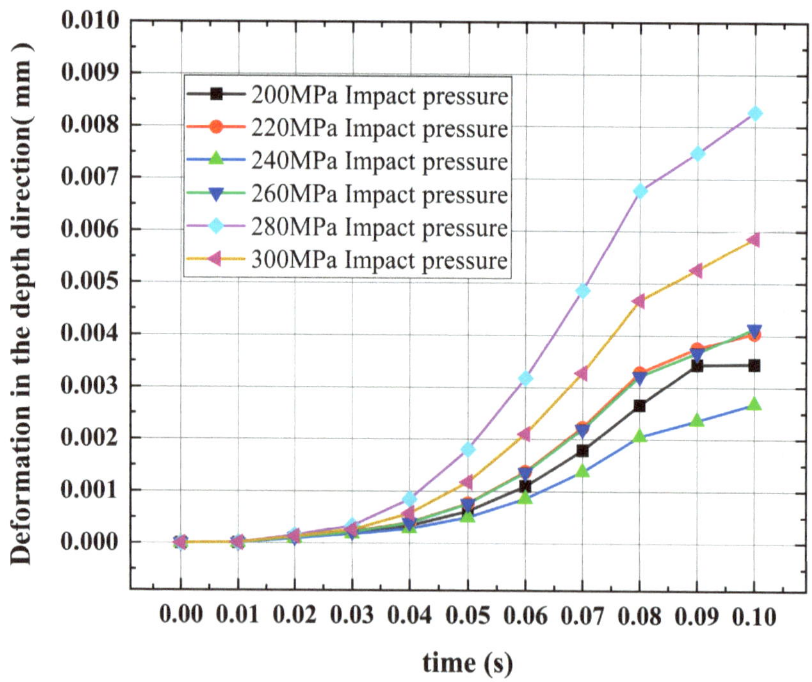

Figure 9. Plastic deformation diagram under different action time.

4. Experimental Analysis

4.1. Experimental Design

The ultra-high pressure water jet experiment was carried out on the PJ2000/15ES experimental platform, and the experimental platform is shown in Figure 10. The experimental platform is located in the Institute of Fluid Machinery and Marine Engineering Equipment of Nantong University (Nantong, China). The manufacturer of the experimental platform is Jiangsu Water Energy Metal Technology Company (Changzhou, China). The experimental system is mainly composed of high-pressure pump, high-pressure pipeline, control system, protection device, water inlet system and other main components. The high pressure pump and motor are connected by elastic coupling. High pressure water is delivered to the nozzle through high pressure pipes. The rated pressure of the experimental platform is 200 MPa, and the rated flow rate is 15 L/min. The sample material is structural steel.

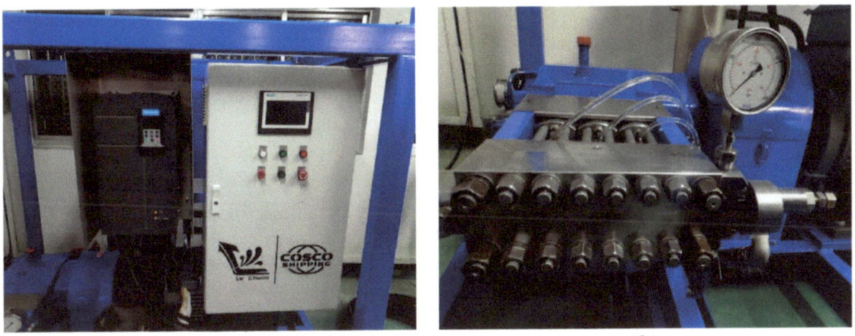

(a) (b)

Figure 10. Water jet experimental platform. (**a**) Control system. (**b**) ultra-high pressure pump.

4.2. Analysis of Experimental Results

In the continuous jet condition, Figure 11 shows the surface morphology of the sample after the experiment. Due to the impact of high-speed water jets, the pressure in the depth direction extends to the width direction. The pressure accumulates below the surface layer, producing plastic strain. The matrix forms a crater like surface morphology. This is consistent with the surface morphology produced by the experiment, as shown in Figure 11. In terms of the shape of surface pits, the experimental results are well correlated with the finite element simulation results. This shows the validity of the finite element model of ultra-high-pressure water jet.

(a) (b) (c)

Figure 11. SEM images of the material surface after water jet erosion at different pressures. (**a**) 200 MPa pressure. (**b**) 240 MPa pressure. (**c**) 300 MPa pressure.

It can be clearly seen from Figure 11 that under the impact pressure of 200 MPa, the surface of the sample is damaged, and the damage range is small. Under the impact pressure of 240 MPa, the damage range of the sample surface increases. Under the impact pressure of 300 MPa, the damage range of the sample surface increases significantly. With the increase of the impact pressure, the overall deformation showed an increasing trend, and the increasing trend increased significantly. This shows that the ultra-high-pressure water jet pressure has a great influence on the impact damage of the substrate surface. When the water jet impinges on the workpiece surface, the fluid velocity drops sharply. Fluid flow is blocked in the stagnation zone. The ultra-high pressure in the stagnation zone leads to the destruction of the matrix.

5. Conclusions

(1) The water jet is divided into three regions along the axial direction: the cavitation jet in the nozzle, the free jet, and the stagnation zone. It is divided into a core area and an atomization area along the radial direction. The distribution of turbulent kinetic energy in the atomization area can be used as the basis for judging the degree of jet diffusion and the rate of energy decay. The peak point of turbulent kinetic energy can be used as the boundary between the inner and outer jets.

(2) After the water jet impactes the surface of the target, a crater-like crater is formed. With the increase of impact pressure, the overall deformation showes an increasing trend. And the increasing trend is obviously rising. Without considering the accumulated damage, the jet pressure is the most important factor.

(3) The depression of the erosion area is caused by the destruction of the material by the stagnation pressure. Accumulated damage has a serious impact on the surface erosion damage of the target.

Author Contributions: Conceptualization, Z.S., W.S., Z.X., L.T. and Y.Y.; Data curation, Q.Z., Z.S. and W.S.; Investigation, Z.S., W.S., Z.X., L.T. and Y.Y.; Methodology, W.S., Z.X., L.T. and Y.Y.; Project administration, W.S. and L.T.; Resources, Z.S. and W.S.; Software, Q.Z. and Z.S.; Supervision, W.S. and Y.Y.; Writing—original draft, Q.Z.; Writing—review & editing, Q.Z. All authors have read and agreed to the published version of the manuscript.

Funding: This work was supported by the National Key R&D Program of China (Grant No. 2019YFB2005300), National High-tech Ship Scientific Research Project of China (Grant No. MC-202031-Z07), National Natural Science Foundation of China (Grant Nos. 51979138, 52109106).

Conflicts of Interest: The authors declare no conflict of interest.

References

1. Duplessis, M.P.; Hashish, M. High Energy Water Jet Cutting Equations for Wood. *J. Eng. Ind.* **1978**, *100*, 452. [CrossRef]
2. Leu, M.C.; Meng, P.; Geskin, S.; Tismeneskiy, L. Mathematical Modeling and Experimental Verification of Stationary Waterjet Cleaning Process. *J. Manuf. Sci. Eng.* **1998**, *120*, 571–579. [CrossRef]
3. Meng, P.; Geskin, E.S.; Leu, M.C.; Li, F.; Tismeneskiy, L. An Analytical and Experimental Study of Cleaning with Moving Waterjets. *J. Manuf. Sci. Eng.* **1998**, *120*, 580–589. [CrossRef]
4. Guha, A.; Barron, R.M.; Balachandar, R. Numerical simulation of high-speed turbulent water jets in air. *J. Hydraul. Res.* **2010**, *48*, 119–124. [CrossRef]
5. Mabrouki, T.; Raissi, K.; Cornier, A. Numerical simulation and experimental study of the interaction between a pure high-velocity waterjet and targets: Contribution to investigate the decoating process. *Wear* **2000**, *239*, 260–273. [CrossRef]
6. Mabrouki, T.; Raissi, K. Stripping process modelling: Interaction between a moving waterjet and coated target. *Int. J. Mach. Tools Manuf.* **2002**, *42*, 1247–1258. [CrossRef]
7. Ramulu, M.; Hashish, M.; Cantrell, A. High Pressure Waterjets—An Innovative Means of Alpha Case Removal for Superplastically Formed Titanium Alloys. *Key Eng. Mater.* **2010**, *433*, 103–111.
8. Guha, A.; Barron, R.M.; Balachandar, R. An experimental and numerical study of water jet cleaning process. *J. Mater. Process. Technol.* **2011**, *211*, 610–618. [CrossRef]
9. Kunaporn, S.; Ramulu, M.; Hashish, M. Mathematical Modeling of Ultra-High-Pressure Waterjet Peening. *J. Eng. Mater. Technol.* **2005**, *127*, 186–191. [CrossRef]
10. Glover, H.; Brass, T.; Bhagat, R.; Davidson, J.; Pratt, L.; Wilson, D. Cleaning of complex soil layers on vertical walls by fixed and moving impinging liquid jets. *J. Food Eng.* **2016**, *178*, 95–109. [CrossRef]

11. Azhari, A.; Schindler, C.; Hilbert, K.; Godard, C.; Kerscher, E. Influence of waterjet peening and smoothing on the material surface and properties of stainless steel 304. *Surf. Coat. Technol.* **2014**, *258*, 1176–1182. [CrossRef]
12. Nayak, B.B.; Abhishek, K.; Mahapatra, S.S.; Das, D. Application of WPCA based taguchi method for multi-response optimization of abrasive jet machining process. *Mater. Today Proc.* **2018**, *5*, 5138–5144. [CrossRef]
13. Amri, R.; Laamouri, A.; Manchoul, S.; Fathallah, R. Transient dynamic finite element simulation for prediction of surface integrity induced by waterjet peening. *Proc. Inst. Mech. Eng. Part L J. Mater. Des. Appl.* **2019**, *234*, 105–119. [CrossRef]
14. Barsukov, G.V.; Zhuravleva, T.A.; Kozhus, O.G. The Research and Simulation (Modelling) of the Destruction of a Single Abrasive Particle during Waterjet Cutting. *Mater. Sci. Forum* **2020**, *989*, 235–241. [CrossRef]
15. Hamid, A.; Atan, R. Spray characteristics of jet–swirl nozzles for thrust chamber injector. *Aerosp. Sci. Technol.* **2009**, *13*, 192–196. [CrossRef]
16. Aganin, A.; Guseva, T. The Influence of the End Shape of the Liquid Jet on Its Impact onto a Dry Wall. *Uchenye Zap. Kazan. Univ. Ser. Fiz.-Mat. Nauki* **2019**, *161*, 39–52. [CrossRef]
17. Mieszala, M.; Torrubia, P.L.; Axinte, D.; Schwiedrzik, J.; Guo, Y.; Mischler, S.; Michler, J.; Philippe, L. Erosion mechanisms during abrasive waterjet machining: Model microstructures and single particle experiments. *J. Mater. Process. Technol.* **2017**, *247*, 92–102. [CrossRef]
18. Thakur, R.K.; Singh, K.K.; Ramkumar, J. Impact of nanoclay filler reinforcement on CFRP composite performance during abrasive water jet machining. *Mater. Manuf. Process.* **2021**, *36*, 1264–1273. [CrossRef]
19. Lohar, S.R.; Kubade, P.R. Investigation of Effect of Abrasive Water Jet Machining (AWJM) Process Parameters on Performance Characteristics of High Carbon High Chromium Steel (AISI D3). *IARJSET* **2017**, *4*, 152–158. [CrossRef]
20. Liu, J.; Chen, Y.; Zhu, Y. Breaking Mechanism and Damage Evolution Rule of Ultra-High-Pressure Water Jet Impacting Steel Fiber Reinforced Concrete. *Adv. Civ. Eng.* **2021**, *2021*, 6697409. [CrossRef]
21. Zhang, Y.; Zang, W.; Zheng, J.; Cappietti, L.; Zhang, J.; Zheng, Y.; Fernandez-Rodriguez, E. The influence of waves propagating with the current on the wake of a tidal stream turbine. *Appl. Energy* **2021**, *290*, 116729. [CrossRef]
22. Sembrat, J.; Miller, P.; Skavdahl, J.; Frenzel, L. Conservation of Historic Metals by Waterjetting Techniques. *J. Archit. Conserv.* **2005**, *11*, 121–146. [CrossRef]
23. Han, Y.; Zhou, L.; Bai, L.; Shi, W.; Agarwal, R. Comparison and validation of various turbulence models for U-bend flow with a magnetic resonance velocimetry experiment. *Phys. Fluids* **2021**, *33*, 125117. [CrossRef]
24. Chillman, A.; Ramulu, M.; Hashish, M. Waterjet and Water-Air Jet Surface Processing of a Titanium Alloy: A Parametric Evaluation. *J. Manuf. Sci. Eng.* **2010**, *132*, 011012. [CrossRef]
25. Ayed, Y.; Robert, C.; Germain, G.; Ammar, A. Development of a numerical model for the understanding of the chip formation in high-pressure water-jet assisted machining. *Finite Elem. Anal. Des.* **2016**, *108*, 1–8. [CrossRef]
26. Yildizeli, A.; Cadirci, S. Multi-objective optimization of multiple impinging jet system through genetic algorithm. *Int. J. Heat Mass Transf.* **2020**, *158*, 119978. [CrossRef]
27. Znamenskaya, I.A.; Nersesyan, D.A.; Sysoev, N.N.; Koroteeva, E.Y.; Shirshov, Y.N. An optical study of high-pressure water-jet dynamics. *Mosc. Univ. Phys. Bull.* **2016**, *71*, 405–412. [CrossRef]
28. Henning, A.; Liu, H.T.; Olsen, C. Economic and Technical Efficiency of High Performance Abrasive Waterjet Cutting. *J. Press. Vessel Technol.* **2012**, *134*, 21405. [CrossRef]
29. Chillman, A.; Hashish, M.; Ramulu, M. Energy Based Modeling of Ultra High-Pressure Waterjet Surface Preparation Processes. *J. Press. Vessel Technol.* **2011**, *133*, 061205. [CrossRef]

MDPI AG
Grosspeteranlage 5
4052 Basel
Switzerland
Tel.: +41 61 683 77 34

Sustainability Editorial Office
E-mail: sustainability@mdpi.com
www.mdpi.com/journal/sustainability